Nursing Laboratory and Diagnostic Tests Demystified

Notice

Nursing Laboratory and Diagnostic Tests Demystified

Jim Keogh, RN

New York Chicago San Francisco Lisbon London
Madrid Mexico City Milan New Delhi San Juan
Seoul Singapore Sydney Toronto

The McGraw-Hill Companies

Nursing Laboratory and Diagnostic Tests Demystified

1 2 3 4 5 6 7 8 9 0 DOC/DOC 14 13 12 11 10 9

ISBN 978-0-07-162380-3
MHID 0-07-162380-9

This book was set in Times Roman by Glyph International.
The editors were Joseph Morita and Karen Davis.
The production supervisor was Phil Galea.
Production management was provided by Somya Rustagi, Glyph International.
The index was prepared by Arc Films, Inc.
RR Donnelley was printer and binder.

This book is printed on acid-free paper.

Library of Congress Cataloging-in-Publication Data

Keogh, James Edward, 1948-
Nursing laboratory and diagnostic tests demystified / Jim Keogh.
p. ; cm.
Includes index.
ISBN-13: 978-0-07-162380-3 (pbk.)
ISBN-10: 0-07-162380-9 (pbk.)
1. Diagnosis, Laboratory. 2. Nursing. I. Title.
[DNLM: 1. Laboratory Techniques and Procedures. 2. Nursing Diagnosis—methods. 3. Nursing Care—methods. WY 100.4 K37n 2010]
RB37.K46 2010
616.07'5—dc22

2009035167

This book is dedicated to Anne, Sandy, Joanne, Amber-Leigh Christine, Shawn, and Eric, without whose help and support this book could not have been written.

CONTENTS

INTRODUCTION

The patient hears the words medical test and cringes because these words bring back memories of being prodded, stuck, and poked at, which was not a pleasant experience. And there is the dreaded wait for test results. In medical shows on television, test results are known almost immediately, but in real life it can take days—days of anticipation and worry.

Medical tests are scary because of the unknown. The patient tells the healthcare provider about their aches, pains, and annoying discomforts. The patient answers seemingly unrelated questions while watching the healthcare provider piece together the puzzle in her mind. And then just when the patient expects a prescription for medication to make it all better, the healthcare provider says. "I'd like to run a few tests."

With a stare reminisce of deer in headlights, the patient freezes thinking "My God, I have cancer." That silence implies the patient has no further question, giving the cue for the healthcare provider to leave. This is also the cue for the nurse to step in to help put the patient at ease by anticipating the questions that the patient probably wants answered.

In this book you will learn about common medical tests. You will learn

- What they are
- Why they are ordered
- How they are performed
- How the patient should prepare for those tests
- How long it takes to receive test results
- How to understand those test results

Consider this book as a book of answers to questions that patients want answered even though they might not be able to ask those questions at the moment.

A LOOK INSIDE

Medical tests can be challenging to learn unless you follow the step-by-step approach that is used in Medical Testing Demystified. Topics are presented in an order in which many students like to learn them—starting with basic components and then gradually moving on to those that are a little more complex to understand.

Each chapter follows a time-tested formula that explains the topic in an easy-to-read style. You can then compare your knowledge with what you are expected to know by taking chapter tests and the final exam. There is little room for you to go adrift.

Chapter 1 Hematology Tests

Hematology is the study of blood, blood diseases, and organs that form blood. Hematology clinical laboratory tests are used to examine blood and blood components to determine if they are within normal limits. Values outside the normal limits might be signs of a disease. Hematology tests count the number of white and red blood cells and platelets. In addition, these tests measure the time necessary for blood to clot and the capability of blood to carry oxygen throughout the body. These tests also determine inflammation and infection in the patient and the type of infection. This chapter covers hematology tests.

Chapter 2 Electrolytes

Electrolytes are salts that are electrically charged ions used to maintain voltage across cell membranes and carry electrical impulses within the body. The concentration of electrolytes within the body is on constant change. However, the kidneys make adjustments to keep electrolytes in balance. Electrolyte tests are referred to as an electrolyte panel, basic metabolic panel, or comprehensive metabolic panel. An electrolyte panel measures only electrolytes in a sample of blood. Basic metabolic panel and comprehensive metabolic panel measure electrolytes and other components. You will learn about electrolytes and electrolyte tests in this chapter.

Chapter 3 Arterial Blood Gases Tests

Arterial blood gases tests indicate how well the patient's lungs transfer oxygen and carbon dioxide to and from blood. Blood taken from an artery is analyzed to identify the concentration of oxygen, carbon dioxide, and bicarbonate in arterial blood. In addition, this test determines the pH value of the blood sample. The concentration of blood is altered through the body, depending on the patient's condition. The body is able to compensate for these changes by adjusting the levels of acid and bicarbonate in the blood. Arterial blood gases tests are discussed in this chapter.

Chapter 4 Liver Tests

The liver is the largest gland in the body that makes and secretes substances. It synthesizes albumin, which maintains blood volume and clotting factors. The liver also synthesizes, stores, and metabolizes fatty acids and cholesterol. Fatty acids are used for energy by the body. The liver stores and metabolizes carbohydrates. Carbohydrates are converted into glucose for energy. It also forms and secretes bile. Bile contains acids that help the intestines absorb fats and vitamins A, D, E, and K, which are fat-soluble vitamins. In addition, the liver clears the body of medication and harmful chemicals such as bilirubin, which result from the metabolism of aged red blood cells, and ammonia, which is the result of metabolism of proteins. The liver transforms these chemicals into components that are easily excreted by the body in urine or stool. These tests are discussed in this chapter.

Chapter 5 Cardiac Enzymes and Markers Tests

Cardiac muscle contains enzymes. In a myocardial infarction, cardiac muscle is damaged causing the release of cardiac enzymes into the blood stream. When a patient is suspected of having a myocardial infarction, the healthcare provider will order the cardiac enzymes and cardiac markers tests to determine if cardiac muscle enzymes appear in the patient's blood. There are several cardiac enzymes and cardiac markers tests commonly used by healthcare providers to confirm a myocardial infarction. You will learn about these tests in this chapter.

Chapter 6 Serologic Tests

The presence of foreign protein in the body from a microorganism or from mismatched donated blood causes a reaction of the body's immune system. This reaction produces antibodies that destroy the foreign protein by metabolizing it into components that can be excreted safely by the body. Serologic tests examine the patient's blood serum for antibodies. Healthcare providers order serologic tests for a number of purposes. These include diagnosing an infection, to determine if the patient has developed immunity to specific antigens, to determine a patient's blood type, and to determine if the patient has an autoimmune disorder. In this chapter you will learn about common serologic tests.

Chapter 7 Endocrine Tests

The endocrine system sends hormones via blood vessels to regulate bodily functions, including metabolism, growth, mood, and tissue function. Hormones are created, stored, and released by glands and act as messengers, signaling other glands and organs to react in a specific manner. Hormones are released based on existing hormone levels in the blood in order to keep hormonal levels in balance. Diseases can dysregulate the release of hormones, resulting in under- or overproduction of one or more hormones. Endocrine tests are administered to assess if the patient is experiencing an endocrine disease. These tests are discussed in this chapter.

Chapter 8 Glucose Tests

There are two pancreatic endocrine hormones secreted by the islet cells in the pancreas. These are insulin and glucagon. Both are secreted based on blood glucose levels. When blood glucose is elevated, the pancreas secretes insulin, which causes glucose to cross the cell membrane allowing it to be used for energy, resulting in a decrease in blood glucose. Blood glucose must be maintained within a narrow range, which occurs naturally with the secretion of insulin and glucagon. Healthcare providers order glucose tests to monitor the blood glucose level. Based on the results of these tests, the healthcare provider may administer insulin or glucose to the patient. You will learn about glucose tests in this chapter.

Chapter 9 Tumor Markers

A tumor is uncontrollable growth of cells that may be malignant (cancerous) or benign (noncancerous). Blood tests are performed to detect the presence of tumor markers. A tumor marker is a substance, usually a protein, which is produced either by tumor cells or by other cells in response to the presence of the tumor. The presence of a tumor marker does not mean that the patient has cancer. Conditions other than cancer can also generate the tumor marker. Likewise, the absence of a tumor marker does not mean that the patient is cancer free because many times early stages of cancer do not produce a tumor marker. In this chapter you will learn about these tests.

Chapter 10 Pregnancy and Genetic Tests

Healthcare providers have an arsenal of tests that provide clues to the underlying cause of infertility and risk for genetic disorders that can affect the fetus or newborn. These tests are ordered to screen patients for disorders when there are no telltale signs or symptoms, and other tests are ordered when the healthcare provider is looking to confirm signs or symptoms that are outside the normal parameters. In this chapter you are introduced to these tests and learn what they are, how they are performed, how to prepare the patient for the test, and how to interpret the test result.

Chapter 11 Tests for Infection

Signs and symptoms of an infection typically indicate that a microorganism has invaded the patient's body. There has been a tendency for healthcare providers to prescribe an antibiotic at the first sign of an infection. In doing so the healthcare provider assumes that the patient is experiencing a bacterial infection and that the antibiotic will eliminate the bacteria. A sample of the patient's blood or infected tissue is sent to the laboratory where the microorganism is encouraged to replicate in a culture dish. Laboratory specialists then conduct tests to identify the microorganism. Once identified, the microorganism is exposed to medication known to kill it. The laboratory specialist determines the best medication and the minimum dose to administer to the patient that will kill the microorganism. In this way, the proper dose of the right medication can be prescribed, reducing the risk that the microorganism will become resistant to the medication. In this chapter you learn about tests used to identify microorganisms and tests to identify medications to kill the microorganism.

Chapter 12 Renal Function Tests

Metabolic waste is carried by the blood to the kidneys. The glomerulus in the kidneys acts as a filter to remove waste from the blood, which is collected in a tubule as urine. Metabolic waste such as sodium, potassium, and phosphorus can be reused by the body and are returned to the blood by the kidneys. The remaining waste is excreted as urine. Renal function is measured in percentages. A person with two healthy kidneys has 100% renal function. Likewise, a person with one healthy kidney and one kidney in total renal failure is said to have 50% renal function. A person will experience health problems if he/she has 25% or less renal function. Dialysis is typically ordered for a patient with less than 15% renal function. There are two tests used to determine renal function. These are blood urea nitrogen (BUN) and creatine. You will learn about these tests in this chapter.

Chapter 13 Pancreatic and Lipid Metabolism Tests

The pancreas produces insulin and glucagon along with digestive enzymes that are used by the small intestine to break down carbohydrates, protein, and fat. Healthcare providers administer pancreatic tests to determine pancreatic function.

Lipids are compounds used to store energy, to develop cell membranes, and are elements of vitamins and hormones. Lipids combine with protein to form lipoprotein. Common lipoproteins in the body are cholesterol and triglycerides. Healthcare providers order lipid metabolism tests to determine the level of lipids in the patient's blood stream. You will learn about these pancreatic and lipid tests in this chapter.

Chapter 14 Diagnostic Radiology Tests

Radiologic tests using X-rays enable the healthcare provider to view inside the body without opening the skin. Although an X-ray provides a primitive view when compared with CT, CAT scans, and MRI, it remains a cost-effective way to identify many common disorders. X-rays are based on the principle that the X-ray is absorbed by dense objects and will pass through lesser dense objects. Dense objects such as bone appear white on the X-ray file and lesser dense objects, such as air, appear black or a lighter shade of gray such as fluid and fat. There are many kinds of X-ray tests that a healthcare provider can order, each focusing on a particular area of the body. You will learn about each of these tests in this chapter.

Chapter 15 Computed Tomography Scan

A computed tomography (CT) scan makes detailed images of structures within the body using a doughnut-shaped X-ray machine. While the patient lies within the scanner, an X-ray beam rotates around the patient creating an image that represents a thin slice of the patient. Each rotation takes less than a second. All slices are stored on a computer. The computer is used to reassemble slices of the patient, enabling the healthcare provider to identify any abnormalities. Typically, the healthcare provider will print the image of any slices that indicate an abnormality, which is then saved with the patient's chart. You will learn about different types of CT scans in this chapter.

Chapter 16 Ultrasound Scan

An ultrasound scan creates an image of organs and structures inside the body, using sound waves similar in concept to how ship crews are able to identify underwater objects while on the surface of the water. An ultrasound scan is commonly ordered instead of a CT scan or MRI because it is less expensive and in many situations provides the healthcare provider with sufficient information to assist in reaching a diagnosis. In this chapter you will learn about different kinds of ultrasound scans.

Chapter 17 Magnetic Resonance Imaging

Magnetic resonance imaging (MRI) uses pulsating radio waves in a magnetic field to produce an image of inside the patient's body. The patient lies on his/her back on a table. A coil is placed around the area of the patient that is being scanned and a belt is placed around the patient to detect breathing. The table moves into the magnetic field and the belt triggers the MRI scan so that breathing does not interfere with capturing the image. The MRI produces digital images that are displayed on a computer screen

and can be stored for further review by the patient's healthcare team. The MRI creates images that are more detailed than images produced by a CT scan, X-ray, or ultrasound. In this chapter you will learn about different kinds of MRI tests.

Chapter 18 Positron Emission Tomography Scan

A positron emission tomography (PET) scan is a nuclear medicine test that creates a roadmap of blood flow in the patient's body, enabling the healthcare provider to visualize abnormal blood flow to the patient's tissues and organs. A radioactive chemical called a tracer and a special camera that detects the tracer inside the patient's body are the keys to a PET scan. These images show the tracer containing blood as the blood makes its way into organs and tissues giving the healthcare provider a clear picture of blood flow within the body. In this chapter you will learn about the PET scan.

Chapter 19 Cardiovascular Tests and Procedures

Cardiovascular tests are performed to assess the patient's heart and vascular system to determine if the blood is adequately pumped and flowing throughout the patient's body. These tests measure cardiac contraction, the risk for coronary artery disease, and are used to identify blockage to coronary arteries and blood vessels to the extremities. When a blockage is identified, the healthcare provider can perform one of several procedures to restore blood flow. The blockage might be surgically removed or pressed against the wall of the blood vessel and held in place by a stent. Alternatively, the healthcare provider may surgically bypass the blocked blood vessels using a vein from the patient's leg or by using an artificial blood vessel. In this chapter you will learn about these tests and procedures.

Chapter 20 Lung Tests and Procedures

The lungs exchange carbon dioxide and oxygen on the hemoglobin in red blood cells. In order to do so effectively, the lungs must be able to expand and retract and blood must freely flow to the lungs. When the patient experiences signs and symptoms of lung disorder and disease, the healthcare provider tests the lungs and orders procedures to evaluate the respiratory system. The healthcare provider can examine the respiratory tract using a bronchoscopy and remove samples of suspicious tissue for microscopic examination. The capacity and function of the lungs are measured using several pulmonary function tests. Blood flow to the lungs is monitored by a lung scan and by a performing pulmonary angiogram to identify restriction or blockage of blood flow to the lungs. This is also performed using CT imaging. In this chapter you will learn about these tests and procedures.

Chapter 21 Tests and Procedures for Females

Female patients routinely undergo breast and cervical examinations for signs of cysts, growths, abnormal tissue, structural abnormalities, and infection. Many of these tests enable the healthcare provider to take a tissue sample or perform a biopsy on abnormal tissue. If the tissue sample is identified to be cancerous, the cancerous organ is removed. In this chapter you will learn about tests and procedures that are performed to test for disorders and repair those disorders.

Chapter 22 Maternity Tests

There are several tests that are performed during pregnancy and shortly after childbirth to assess the health of the fetus and newborn. In a high-risk pregnancy, the healthcare provider might perform a chorionic villus sampling or amniocentesis early on in the pregnancy to determine if the fetus has a genetic disorder or other health issues. Later in the pregnancy, the healthcare provider performs a biophysical profile of the fetus to determine the overall health of the fetus. This is where an assessment is made of the fetal heart rate, breathing and body movements, muscle tone, and the volume of amniotic fluid. You will learn about all these tests in this chapter.

Chapter 23 Chest, Abdominal, Urinary Tract Tests and Procedures

When there are suspected disorders of the upper gastrointestinal tract, thyroid gland, liver, gallbladder, kidneys, spleen, urinary tract, and other organs in the upper part of the body, the healthcare provider is likely to order a number of tests to uncover the underlying problem. Some tests enable the healthcare provider to look down the esophagus to examine the stomach, duodenum, and the bile and pancreatic ducts and to take a biopsy or, in some cases, to remove an obstruction. Other tests enable the healthcare provider to scan the liver, spleen, gallbladder, and kidneys by using contrast material to highlight the structure of the organ. Images of the organ are captured with a camera and studied to uncover diseases and disorders. You will learn about these and other tests and procedures in this chapter.

Chapter 24 Bone and Muscle Tests

Aching bones and muscles might be from a cause other than overexercising. It could be a sign of an underlying disorder that needs immediate medical attention. Healthcare providers are able to assess the reason for the patient's discomfort by testing his/her bones and muscles. Healthcare providers have an assortment of tests and procedure that are used to investigate signs of a disorder. An image of the bone can be taken by using a bone scan, where a radioactive tracer highlights the structure of the bone or a myelogram that uses contrast material to bring out the structure of the spine into view. An arthrogram is used to create an image of soft tissues and structures of a joint. A bone mineral density test is ordered to determine the thickness of bone, looking for the first sign of osteoporosis. You will learn about these tests and procedures in this chapter.

Chapter 25 Tests for Males

There are a number of medical test and procedures that are specifically designed to diagnose and treat disorders that affect men. There are a group of tests and procedures focused on fertility. Men are susceptible to developing an enlarged prostate gland, which could be caused by prostate cancer. Prostatic cancer cells are in part fueled by testosterone, which is produced by the testicles. The healthcare provider might perform an orchiectomy, which is the surgical removal of one or both testicles. This reduces the level of testosterone in the patient's body. In this chapter you will learn about these tests and procedure.

Chapter 26 Skin Tests

Skin is the largest and the most visible organ in the body and is susceptible to wrinkles, blemishes, growths including both nonmelanoma and melanoma, and infection. Healthcare providers perform an assortment of tests and procedures to diagnose and treat skin conditions. Lesions, warts, and blemishes that make skin unsightly and unhealthy can be removed by performing one of a number of procedures. Infected skin can be treated once the microorganism that causes the infection is identified. The healthcare provider is able to identify the microorganism by performing a wound culture, where a sample of the infected tissue is placed in an environment that is favorable for the growth of microorganism (culture) and then identified. Once the laboratory determines which medication will fight the microorganism, the medication is administered to the microorganism to determine which medication kills the microorganism. You will learn about these tests and procedures in this chapter.

Chapter 27 Sinus, Ears, Nose, Throat Tests and Procedures

Snoring, headaches, frequent infections, and turning the volume on the TV high could be signs of an underlying problem. The healthcare provider can perform a number of tests to assess the sinus, ears, nose, and throat. You will learn about these tests in this chapter.

Chapter 28 Vision Tests and Procedures

Light rays pass through the cornea, the pupil, and lenses which focus the ray of light on to the retina located at the back of the eye. When light rays are not properly focused on the retina, the patient is unable to see clearly. In this chapter you will learn about tests that are used to diagnose problems with sight and disorders that can lead to loss of vision. You will also learn about procedures that can be performed to treat vision disorders.

CHAPTER 1

Hematology Tests

Hematology is the study of blood, blood diseases, and organs that form blood. Hematology clinical laboratory tests are used to examine blood and blood components to determine if they are within normal limits. Values outside the normal limits might be signs of a disease.

Hematology tests count the number of white and red blood cells and platelets. In addition, these tests measure the time necessary for blood to clot and the capability of blood to carry oxygen throughout the body. Hematology tests also determine inflammation and infection in the patient and the type of infection.

In this chapter you'll learn how to collect a blood specimen and learn about commonly performed hematology tests.

Learning Objectives

1. How to Collect Blood Specimen from a Vein
2. How to Collect Blood Specimen from a Heel Stick
3. How to Collect Blood Specimen from a Finger Stick
4. Blood Type Test
5. Partial Thromboplastin Time (PTT)
6. Total Serum Protein
7. Blood Alcohol
8. Lead
9. Serum Osmolality
10. Uric Acid in Blood
11. C-Reactive Protein (CRP)
12. Complete Blood Count (CBC)
13. Chemistry Screen
14. Vitamin B_{12}
15. Cold Agglutinins
16. Toxicology Tests (Tox Screen)
17. Folic Acid
18. Gastrin
19. Ferritin
20. Lactic Acid
21. Prothrombin Time
22. Reticulocyte Count
23. Schilling Test
24. Sedimentation Rate (SR)
25. Iron (Fe)
26. Serum Protein Electrophoresis (SPE)

Key Words

ABO
Agglutinins
Albumin
Antidiuretic hormone (ADH)
Antigens
Blood group antigens
Coumadin
C-reactive protein
Erythrocyte indices
Ferritin
Fibrinogen
Gastrin
Globulin
Heparin
Intrinsic factor
Lactic acid
Osmolality
Parietal cells
Partial thromboplastin time (PTT)
Phlebitis
Prothrombin time (PT)
Purine
Reticulocyte
Rh antigen
Rh factor
Rouleaux formation
Thromboplastin
Toxin
Transferrin protein

How to Collect Blood Specimen from a Vein 1

Healthcare facilities provide training for collection of blood specimen. Here are the basic steps that are necessary to collect a blood sample.

1. Wrap a tourniquet around the patient's upper arm to stop blood flow, making veins easier to identify.
2. Clean the puncture site with alcohol.
3. Insert the needle into the vein with the bevel up.
4. Attach the appropriate test tube to the needle. Allow the blood to fill the test tube.
5. Remove the tourniquet to restore blood flow.
6. Place a gauze pad over the site while withdrawing the needle.
7. Apply firm pressure to the site until bleeding has stopped.

TEACH THE PATIENT

- Explain that
  - The tourniquet may feel tight.
  - The patient may feel a pinch or nothing at all when the needle is inserted into the vein.
  - There might be a small bruise at the site. Keeping pressure on the site reduces the chance of bruising.
  - Taking anticoagulants (aspirin, Coumadin) may require keeping pressure on the site for more than 10 minutes to stop the bleeding.
  - The vein may become swollen after the test (phlebitis). The patient should call their healthcare provider and apply a warm compress to reduce the swelling.

How to Collect Blood Specimen from a Heel Stick 2

Several drops of blood are collected from the heel of a baby.

1. Clean the heel with alcohol.
2. Puncture the heel with a small sterile lancer.
3. Collect several drops of blood in a small test tube.
4. Place a gauze pad over the site.
5. Maintain pressure until bleeding stops.
6. Apply a small bandage.

TEACH THE PARENT

- Explain
  - That the patient may feel a pinch or nothing at all when the lancer punctures the skin.
  - There will be a bandage on the site for a short-time period.
  - That a small bruise might appear at the site.

How to Collect Blood Specimen from a Finger Stick 3

Several drops of blood are collected from the finger.

1. Clean the finger with alcohol.
2. Puncture the finger with a small sterile lancer.
3. Collect several drops in a small test tube.
4. Place a gauze pad over the site.
5. Maintain pressure until bleeding stops.
6. Apply a small bandage.

TEACH THE PATIENT

- Explain
  - That the patient may feel a pinch or nothing at all when the lancer punctures the skin.
  - There will be a bandage on the site for a short-time period.
  - A small bruise might appear at the site.

Blood Type Test 4

Blood is identified by an antigen on the surface of red blood cells. Major types of these antigens are blood group antigens and Rh antigen. There are four types of blood group antigens that are determined by performing the ABO test.

- Type A: Has the A antigen and antibodies in plasma against B antigen.
- Type B: Has the B antigen and antibodies in plasma against A antigen.
- Type O: Has neither the A antigen nor the B antigen and antibodies in plasma against A antigen and B antigen.
- Type AB: Has both the A antigen and the B antigen and no antibodies in plasma against A antigen and B antigen.

Red blood cells may have the Rh antigen attached to them, sometimes called the Rh factor, and is determined by the Rh test.

- Rh positive (+): The Rh antigen is presented on the red blood cells.
- Rh negative (–): The Rh antigen is not presented on the red blood cells.

A patient's blood type is described as a combination of blood group antigen and Rh antigen by using the blood type letter(s) followed by a plus (+) or minus (–) sign, indicating if the Rh antigen is present. For example, type A means that the patient has the A antigen but doesn't have the Rh antigen attached to the red blood cells.

Hint

The test will also examine minor antigens that, attached to red blood cells, can cause an adverse blood transfusion reaction.

WHAT IS BEING MEASURED?

- Blood group antigens on red blood cells
- Rh antigens on red blood cells

HOW IS THE TEST PERFORMED?

See How to Collect Blood Specimen from a Vein.

RATIONALE FOR THE TEST

- Assess
  - Blood compatibility for a blood transfusion and organ transplant
  - If a pregnant woman is Rh positive or negative

NURSING IMPLICATIONS

- Assess the patient for conditions that might affect the test results.
  - The patient has taken methyldopa, levodopa, or cephalexin. These medications can cause a false Rh positive test result.
  - Recent X-ray with contrast.
  - Bone marrow transplant.
  - A blood transfusion in the previous 3 months.
  - Has or has had cancer or leukemia.

UNDERSTANDING THE RESULTS

- The result is available in an hour.
- The following chart lists blood type compatibility between a recipient (the patient) and a donor.

Blood Transfusion Compatibility Chart	
Recipient	**Donor**
A–	A– O–
A+	A– A+ O– O+
B–	B– O–
B+	B– B+ O– O+
AB–	AB– O–
AB+	AB– AB+ A– A+ B– B+ O– O+
O–	O–
O+	O– O+

Hint

It was once thought that type O negative is the universal blood donor, meaning that a patient can receive O negative blood regardless of the patient's blood type. Type AB positive is the universal recipient. It is now understood that these blood types contain antibodies that can cause a transfusion reaction. Before any transfusion, a sample of the recipient's blood is cross-matched (mixed) with the donor's blood in the laboratory to determine compatibility.

TEACH THE PATIENT

- No special preparation is needed for the test.
- Explain how the test is performed.
- Explain the conditions that can negatively affect the test results.

Partial Thromboplastin Time 5

When bleeding occurs, 12 blood clotting factors cause the blood to coagulate to stop the bleeding. Coagulation of blood is affected by blood clotting factors, which can be absent and of decreased or increased levels, or by changes in the way blood clotting factors function. In addition, clotting inhibitors can reduce the effectiveness of clotting factors. The partial thromboplastin time (PTT) test measures the clotting time of blood.

HINT

Another blood clotting test is prothrombin time (PT). The heparin neutralization assay is performed to determine if substances other than heparin cause an increase in PTT.

WHAT IS BEING MEASURED?

- The time necessary for blood to clot

HOW IS THE TEST PERFORMED?

- The test may be performed frequently for patients who experience a clotting or bleeding problem. The test is also performed before invasive procedures and before surgery.
- The test is performed frequently for patients taking heparin until the healthcare provider determines the proper dosage for the patient.
- See How to Collect Blood Specimen from a Vein.

RATIONALE FOR THE TEST

- Assess
  - The blood's ability to clot prior to any invasive procedure
  - For hemophilia
  - For lupus anticoagulant syndrome or antiphospholipid antibody syndrome caused when the antibodies attack blood clotting factors
  - The effectiveness of the dose of heparin administered to the patient (PPT test)

NURSING IMPLICATIONS

- Assess
  - Prescription and nonprescription medications taken by the patient. Some medications such as aspirin and antihistamines affect the results of the test.
  - Herbal and natural remedies taken by the patient because these remedies can affect the test results.

- A patient with a high PTT value is at risk for bleeding and bruising. Pressure must be applied to the injection site longer than for a patient with a normal PTT value.

UNDERSTANDING THE RESULTS

- The result is available quickly. The laboratory determines normal values based on calibration of testing equipment with a control test. Test results are reported as high, normal, or low based on the laboratory's control test.
- Generally normal range is
  - PTT: 30 to 40 seconds.
  - Proper heparin dose: The PTT is 1.5 to 2.5 seconds greater than the normal value.
- Longer times may indicate
  - Hemophilia.
  - von Willebrand disease.
  - A blood clotting factor is low or absent.
  - Nephrotic syndrome.
  - Cirrhosis.
  - Antiphospholipid antibody syndrome.
  - Lupus anticoagulant syndrome.
  - Factor XII deficiency.
  - Disseminated intravascular coagulation (DIC).
  - The patient has been administered heparin.
  - Hypofibrinogenemia.

HINT

Patients who have inherited bleeding disorders may have normal PTT values.

TEACH THE PATIENT

- Explain
  - Why blood sample is taken.

- Why samples may be taken frequently until the healthcare provider determines the therapeutic dose of heparin for the patient.
- That taking aspirin, antihistamines, and herbal and natural remedies may affect the test result.

Total Serum Protein 6

The total serum protein test assesses the levels of albumin, globulin, and total protein in a blood sample. The result compares the ratio of albumin to globulin. Protein is not stored. It is continuously metabolized into amino acids, which are used to make enzymes, hormones, and new proteins.

- Albumin is a protein produced by the liver that keeps blood from leaking from blood vessels. Albumin is also important for tissue growth and healing because it carries medicine to tissues.
- Globulin is a group of proteins made by the liver and the immune system that binds with hemoglobin and transports iron and metals in the blood to help fight infection. Globulin is composed of three different proteins: alpha, beta, and gamma.

A test for total serum protein reports separate values for total protein, albumin, and globulin. The amounts of albumin and globulin are also compared (albumin/globulin ratio).

WHAT IS BEING MEASURED?

- Total protein in the blood sample particularly albumin and globulin

HOW IS THE TEST PERFORMED?

- See How to Collect Blood Specimen from a Vein.

RATIONALE FOR THE TEST

- Screen for
  - Liver and kidney functions (albumin)
  - Malnutrition (albumin)
  - Cause of edema, ascites, and pulmonary edema (albumin)

- Risk of infection (globulin)
- Multiple myeloma (globulin)
- Macroglobulinemia (globulin)

NURSING IMPLICATIONS

- Assess if the patient
  - Is taking androgens (male sex hormones), estrogen, corticosteroids, insulin, and growth hormone, which can affect the test results
  - Has a chronic illness that interferes with nutrition
  - Has any recent injuries or infections
  - Is pregnant
  - Has been on extended bed rest

UNDERSTANDING THE RESULTS

- Test results are available in 12 hours. The laboratory determines normal values based on calibration of testing equipment with a control test. Test results are reported as high, normal, or low based on the laboratory's control test.
- The healthcare provider will likely order a serum protein electrophoresis if there are abnormal globulin levels.
- Generally the normal range is
  - Total protein: 5.5 to 9.0 g/dL
  - Albumin: 3.5 to 5.5 g/dL
  - Globulin: 2.0 to 3.5 g/dL
  - Albumin/globulin ratio: Greater than 1.0
- High albumin levels may indicate
  - Severe dehydration
- High globulin levels may indicate
  - Liver disease
  - Kidney disease
  - Hodgkin lymphoma
  - Leukemia

  - Hemolytic anemia
  - Macroglobulinemia
  - Rheumatoid arthritis
  - Lupus
  - Sarcoidosis
  - Autoimmune hepatitis
  - Tuberculosis
- Low albumin levels may indicate
  - Malnutrition
  - Kidney disease
  - Liver disease
  - Uncontrolled diabetes
  - Hodgkin lymphoma
  - Severe burns
  - Systemic lupus erythematosus (SLE)
  - Rheumatoid arthritis
  - Heart failure
  - Hyperthyroidism
  - Crohn disease
  - Celiac disease
- A ratio less than 1 indicates abnormal function in the body.

HINT

Liver damage reduces the protein level in blood, but this level may not be reduced for 2 weeks after the damage because protein stays in blood for up to 18 days.

TEACH THE PATIENT

- Explain
  - Why the blood sample is taken.
  - The function of albumin and globulin and the meaning of abnormal values.

Blood Alcohol 7

Alcohol depresses the central nervous system (CNS) when large amounts of alcohol enter the blood. Alcohol is absorbed within a few minutes and peaks within an hour. Food decreases absorption of alcohol. Alcohol is mostly metabolized by the liver and excreted in urine and by breathing. The blood alcohol test measures the level of alcohol in the blood.

HINT

A taximeter that measures alcohol levels in the patient's breath is another test to determine the level of alcohol consumed by the patient.

CAUTION

The patient's signed consent may be required before the test is administered because the result of the blood alcohol test can have legal repercussions for the patient. Consult the healthcare facility's policies regarding administering the blood alcohol test.

WHAT IS BEING MEASURED?

- Alcohol level in the blood

HOW IS THE TEST PERFORMED?

- See How to Collect Blood Specimen from a Vein.

RATIONALE FOR THE TEST

- Screen for
  - Intoxication
  - The reason for altered mental status (AMS)
  - Ingested alcohol

NURSING IMPLICATIONS

- The result of the blood alcohol test represents the level of alcohol at the time the test was administered and cannot predict the amount of alcohol consumed by the patient because alcohol continues to be metabolized by the liver.

- Assess if the patient
  - Has taken sedatives, kava, ginseng, or antihistamines, since these can enhance the sedative effect of alcohol.
  - Has diabetic ketoacidosis.
  - Ingested methanol or isopropyl alcohol.
  - Has vital signs. Alcohol ingestion can depress respiration and the central nervous system.
- Assess if the insertion site was cleaned with alcohol, since this can affect the test results.
- Take safety precautions, since the patient may have impaired reaction time and decreased vision.
- The speed in which alcohol is metabolized is influenced by
  - The proof of each drink. The higher the proof, the higher the blood alcohol level.
  - The patient's weight. Blood alcohol level decreases as weight increases, since additional weight represents increased water that dilutes the alcohol.
  - The number of drinks the patient ingested each hour. It takes approximately an hour to metabolize a drink. The greater the number of drinks per hour, the higher the blood alcohol level.
  - The patient's age. The older patient will have a higher blood alcohol level than a young patient who consumed the same quantity of alcohol because the older patient's metabolism is slower.
  - The amount of food ingested by the patient. The higher the ingestion of food, the lower the blood alcohol level, since the food can absorb some alcohol.
  - Women have a higher blood alcohol level than men because they have less water than men in their body.
  - The use of mixers such as club soda increases the absorption of alcohol, resulting in a higher blood alcohol level.
- Quick assessment
  - 0.03 to 0.059 blood alcohol level
    - Talkativeness
    - Decrease in inhibition
    - Decrease in alertness, coordination, and concentration

- 0.06 to 0.1 blood alcohol level
  - Decrease in reflexes, peripheral vision, depth perception, and reasoning
  - Blunted feelings
  - Impaired sexual pleasure
- 0.11 to 0.20 blood alcohol level
  - Staggering
  - Slurred speech
  - Decreased reaction time
  - Emotional swings
  - Over expression
- 0.21 to 0.29 blood alcohol level
  - Blackout of memory
  - Stupor
  - Loss of consciousness
  - Impaired sensations
- 0.30 to 0.39 blood alcohol level
  - Decrease in breathing
  - Decrease in heart rate
  - Severe depression
  - Unconsciousness
  - Risk of death
- Greater than 4 blood alcohol level
  - Death

UNDERSTANDING THE RESULTS

- Test results are available quickly. The laboratory determines normal values based on calibration of testing equipment with a control test. Test results are reported as high, normal, or low based on the laboratory's control test.
- Normal range: 0.0
- Abnormal level: Greater than 0.0
- Intoxication (National Highway Traffic Safety Administration): 0.08 g/dL or 80 mg/dL or 17 mmol/L

TEACH THE PATIENT

- Explain
  - Why blood sample is taken.
  - That the patient consent is needed to administer the test unless ordered by legal authorities. Check your healthcare facility's policy.
  - The test result.

Lead 8

Lead affects growth and development if lead-tainted water, paint chip, food, or dust is ingested or if lead is in contact with the skin. Pregnant women can pass lead to the fetus or to the newborn through breast milk. Children who are in early development are at risk for permanent growth impairment if they ingest lead. The lead blood test measures the level of lead in the blood sample.

HINT

The healthcare provider may order the urine aminolevulinic acid (ALA) test to determine the extent of lead poisoning (not for children). The lead mobilization urine test is performed during chelation therapy to assess if the therapy is removing lead in urine.

WHAT IS BEING MEASURED?

- Lead levels in the blood

HOW IS THE TEST PERFORMED?

- See How to Collect Blood Specimen from a Vein.
- See How to Collect Blood Specimen from a Heel Stick.
- See How to Collect Blood Specimen from a Finger Stick.

RATIONALE FOR THE TEST

- Assess
  - For lead poisoning
  - The treatment of lead poisoning

NURSING IMPLICATIONS

- Assess if the patient is taking herbal medicine.
- Test results of 10 mcg/dL or above from a finger or heel stick require blood to be drawn from a vein to confirm the results, since the blood sample might be contaminated from lead at the site of the stick.
- Consult your healthcare institution's policy to determine if a positive test result must be reported to the local health department. Lead levels above 10 mcg/dL on two occasions typically must be reported, so the source of the lead can be found. Tests are repeated within a few days or in a month following the initial test.
- Quick assessment using the United States Centers for Disease Control and Prevention (CDC) classes
    - Learning problems: Class 1 or 2
    - Slow growth: Class 2
    - Hearing problems: Class 2
    - Headache: Class 3
    - Neurologic problems: Class 3
    - Weight loss: Class 3
    - Anemia: Class 4
    - Seizures: Class 4
    - Brain damage: Class 5

UNDERSTANDING THE RESULTS

- Test results are available in 1 week. The laboratory determines normal values based on calibration of testing equipment with a control test. Test results are reported as high, normal, or low based on the laboratory's control test.
- Normal range: 0 mcg/dL
- Abnormal Class 1: 1 to 9 mcg/dL
- Abnormal Class 2A: 10 to 14 mcg/dL
- Abnormal Class 2B: 15 to 19 mcg/dL
- Abnormal Class 3: 20 to 44 mcg/dL
- Abnormal Class 4: 45 to 69 mcg/dL
- Abnormal Class 5: Greater than 69 mcg/dL

Caution

Businesses whose employees work with lead must test lead levels in their employees and provide the employee with documentation if lead level in their blood is 40 mcg/dL or greater as required by the United States Occupational Safety and Health Administration (OSHA) regulation. Employees with a level higher than 45 mcg/dL should be seen by their healthcare providers.

TEACH THE PATIENT

- Explain.
  - Why blood sample is taken.
  - That multiple samples may be taken.
  - Why the healthcare provider must report certain results to the local health department.

Serum Osmolality 9

Serum osmolality is the number of particles of substances that are dissolved in the serum (liquid). These substances include glucose, chloride, sodium, proteins, and bicarbonate. Serum osmolality should be within a normal range—a balance between fluid and particles of substances.

Serum osmolality is controlled adjusting the fluid output of the kidneys, using the antidiuretic hormone (ADH) produced by the pituitary gland. ADH is a vasopressin that reduces fluid output from the kidneys. Therefore, when ADH is released into the blood stream, it increases fluid in the blood. A decrease in ADH production increases fluid output by the kidneys and decreases fluid in the blood.

A decrease in fluid results in an increase in serum osmolality—there is less fluid in the blood. This signals the pituitary gland to release ADH, which stimulates the kidneys to retain fluid, thereby increasing fluid in the blood and decreasing serum osmolality—fluid level in the blood is restored.

An increase in fluid results in a decrease in serum osmolality—there is more fluid in the blood. This signals the pituitary gland to stop releasing ADH, which causes the kidneys to increase the output of fluid, thereby decreasing fluid in the blood and increasing serum osmolality—fluid level in the blood is restored.

Hint

The healthcare provider may test urine osmolality. The result of the urine test is compared with the serum osmolality to estimate kidney function.

WHAT IS BEING MEASURED?

- Serum osmolality of blood

HOW IS THE TEST PERFORMED?

- See How to Collect Blood Specimen from a Vein.

RATIONALE FOR THE TEST

- Screen
  - For dehydration.
  - For overhydration.
  - The quantity of poison ingested by the patient.
  - For the underlying cause of a seizure.
  - For the underlying cause of a coma.
  - The function of the hypothalamus. The hypothalamus signals the pituitary gland to release ADH.
  - The presence of syndrome of inappropriate antidiuretic hormone (SIADH) secretion.

NURSING IMPLICATIONS

- Assess if the patient
  - Has ingested alcohol, since this can affect the test results
  - Recently received a blood transfusion

UNDERSTANDING THE RESULTS

- Test results are available in 4 hours.
- Normal range: 280 to 300 mOsm/kg of water.
- High values may indicate
  - Buildup of urea in the blood
  - Poisoning
  - High level of salt in the blood
  - Diabetes insipidus (DI)

    - High level of glucose in the blood
    - Dehydration
- Low values may indicate
    - Lung cancer
    - Overproduction of ADH
    - Overhydration
    - Low salt level in the blood

TEACH THE PATIENT

- Explain
    - Why blood sample is taken.
    - Not to ingest alcohol 12 hours before the test is administered.

Uric Acid in Blood 10

Some foods contain purine. Uric acid is produced when purine is metabolized. Uric acid enters the blood and is then excreted by the kidneys through the urine and a small amount in stool. The uric acid test measures the level of uric acid in the blood.

Hint

The healthcare provider may order a 24-hour uric acid urine test.

WHAT IS BEING MEASURED?

- Uric acid level in blood

HOW IS THE TEST PERFORMED?

- See How to Collect Blood Specimen from a Vein.

RATIONALE FOR THE TEST

- Screen for
    - Uric acid kidney stones

- Gout
- Adverse reaction of radiation therapy and chemotherapy
- Assess treatment of hyperuricemia.

NURSING IMPLICATIONS

- Assess
  - The patient for starvation, strenuous exercise, and a high-protein diet, since these can raise the level of uric acid
  - If the patient is taking aspirin, theophylline, diuretics, niacin, caffeine, vitamin C, ascorbic acid, epinephrine, levodopa, warfarin, diazoxide, cisplatin, cyclosporine nicotinic acid, phenothiazines, tacrolimus, methyldopa, or ethambutol, which can affect the test results
  - If the patient is pregnant, since uric acid level can increase during pregnancy that assists the healthcare provider in diagnosing preeclampsia
  - If the patient has eaten liver, red meats, game meat, herring, sardines, scallops, or beer
- Uric acid levels are higher in the morning and lower in the evening, therefore note the time when the test is administered.

UNDERSTANDING THE RESULTS

- Test results are available in 2 days. The laboratory determines normal values based on calibration of testing equipment with a control test. Test results are reported as high, normal, or low based on the laboratory's control test.
- Normal uric acid range is
  - Men: 3.4 to 7.0 mg/dL
  - Women: 2.4 to 6.0 mg/dL
  - Children: 2.5 to 5.5 mg/dL
- High uric acid values may indicate
  - Kidney disease
  - Multiple myeloma
  - Lymphoma
  - Hemolytic anemia

  - Heart failure
  - Leukemia
  - Sickle cell anemia
  - Preeclampsia
  - Cirrhosis
  - Alcohol abuse
  - Hypoparathyroidism
  - Hypothyroidism
  - Starvation
  - Lead poisoning
  - Malnutrition
  - Psoriasis
  - Lesch-Nyhan syndrome
  - Obesity
  - High-purine diet
- Low uric acid values may indicate
  - Low-protein diet
  - SIADH
  - Wilson disease
  - Cancer
  - Liver disease
  - Taking Aloprim, Zyloprim, Benemid, and Probalan, sulfinpyrazone, 1,500 mg or more aspirin each day

TEACH THE PATIENT

- Explain
  - Why blood sample is taken.
  - That the patient should avoid eating liver, red meat, game meat, herring, sardines, scallops, and beer for 24 hours before the test is administered.
  - That the healthcare provider may ask the patient to stop taking the following for 2 weeks: aspirin, theophylline, diuretics, niacin, caffeine,

vitamin C, ascorbic acid, epinephrine, levodopa, warfarin, diazoxide, cisplatin, cyclosporine, nicotinic acid, phenothiazines, tacrolimus, methyldopa, or ethambutol.

HINT

Uric acid crystals can form in joints leading to gout even when uric acid levels are normal. A high level of uric acid does not mean that the patient has gout. Gout is diagnosed by testing fluid from the affected join for uric acid crystals.

C-Reactive Protein 11

C-reactive protein (CRP) is produced as part of the inflammatory process and attached to the invading microorganism or damaged cells enhancing phagocytosis in the destruction of the microorganism or damaged cell. The CRP test measures the CRP level in the blood. A high level indicates inflammation, but not the source of the inflammation. The healthcare provider will order other tests to identify the source of the inflammation.

HINT

The healthcare provider may order a high-sensitivity CRP test (hs-CRP) to determine if inflammation has damaged the inner lining of arteries increasing the risk of a myocardial infarction. The healthcare provider may order the total cholesterol test and high-density lipoprotein (HDL) cholesterol test. These test results combined with the result from the CRP test help to determine the risk of cardiac problems.

WHAT IS BEING MEASURED?

- The CRP level in blood

HOW IS THE TEST PERFORMED?

- See How to Collect Blood Specimen from a Vein.

RATIONALE FOR THE TEST

- Screen for inflammation, arthritis, autoimmunity, irritable bowel syndrome (IBS), and smoking

- Assess
    - The treatment of inflammation
    - The treatment of diseases that cause inflammation
    - The response to cancer treatment

Hint

The CRP level peaks 6 hours following surgery and decreases by the third day. Infection as a result of the surgery is suspected if the CRP level remains high after the third day. Elevated CRP level prior to surgery increases the risk of infection of the following surgery.

NURSING IMPLICATIONS

- Assess if the patient
    - Has exercised immediately prior to administering the test
    - Has taken nonsteroidal anti-inflammatory drugs (NSAIDs), birth control pills, aspirin, hormone replacement therapy, pravastatin, or corticosteroids
    - Has an intrauterine device (IUD)
    - Is pregnant
    - Is obese

UNDERSTANDING THE RESULTS

- Test results are available in 1 day. The laboratory determines normal values based on calibration of testing equipment with a control test. Test results are reported as high, normal, or low based on the laboratory's control test.
- Normal CRP range: 0 to 1 mg/mL
- High CRP values may indicate
    - Inflammation
    - Infection
    - Risk for a myocardial infarction

Hint

The C-peptide level quickly raises, then returns to normal if treatment of infection or cancer is successful.

TEACH THE PATIENT

- Explain
  - Why blood sample is taken.
  - That the patient should not exercise before the test.
  - That the healthcare provider may ask the patient to stop taking NSAIDs, birth control pills, aspirin, hormone replacement therapy, pravastatin, or corticosteroids for 2 weeks.

Complete Blood Count 12

The complete blood count (CBC) test measures several components of blood to assess the patient for various disorders, it is part of a routine blood screening. Components measured are

- Leukocyte count (WBC): Leukocytes normally increase when infection is present, but can also increase in the absence of infection if the patient has leukemia.
- Leukocyte cell type (WBC differential): There are five major types of leukocyte cells, each having a role in the immune process. These are neutrophils, lymphocytes, monocytes, eosinophils, and basophils. The quantity of each leukocyte cell type provides important information in diagnosing the patient's condition.
- Erythrocyte count (RBC): Erythrocyte cells carry oxygen and carbon dioxide.
- Erythrocyte indices
  - Mean corpuscular volume (MCV): This is the size of erythrocytes.
  - Mean corpuscular hemoglobin (MCH): This is the amount of hemoglobin in an erythrocyte cell.
  - Mean corpuscular hemoglobin concentration (MCHC): This is the concentration of hemoglobin in an erythrocyte cell.
  - Red cell distribution width (RDW): This shows the different sizes of erythrocyte cells.
- Hematocrit (HCT, packed cell volume): The hematocrit test measures the volume in percentage taken up by erythrocytes in the patient's blood.

- Hemoglobin (Hgb): The hemoglobin test measures the amount of hemoglobin in blood. Hemoglobin is the part of an erythrocyte that carries oxygen.
- Thrombocyte count (platelet): Platelets form blood clots.

HINT

The healthcare provider may also order the erythrocyte sedimentation rate (ESR) test to detect inflammation and the reticulocyte count to identify the number of immature leukocytes.

WHAT IS BEING MEASURED?

- Leukocyte, erythrocyte, and thrombocyte levels in the blood

HOW IS THE TEST PERFORMED?

- See How to Collect Blood Specimen from Vein.
- See How to Collect Blood Specimen from a Heel Stick.

RATIONALE FOR THE TEST

- Screen for
  - Anemia
  - Infection
  - Leukemia
  - Risk of bleeding
  - Asthma
  - Allergies
  - The cause of bruising
  - The cause of fatigue
  - Polycythemia
  - Blood loss

NURSING IMPLICATIONS

- Assess if the patient
  - Is taking antibiotics, steroids, Equanil, thiazide diuretics, Miltown, quinidine, Meprospan, or chemotherapy

- Has high levels of triglycerides, since these can affect the hemoglobin test
- Has an enlarged spleen, since this can reduce the platelet count
- Is pregnant
- Smokes
- Is stressed
- Has exercised before the test is administered
- Note
  - How long the tourniquet was tied when acquiring the blood sample.
  - If the blood sample clumps in the test tube, indicating that the sample is contaminated.

UNDERSTANDING THE RESULTS

- Test results are available quickly. The laboratory determines normal values based on calibration of testing equipment with a control test. Test results are reported as high, normal, or low based on the laboratory's control test.
- Normal WBC count range
  - Men: 5,000 to 10,000 mcL^3
  - Nonpregnant women: 4,500 to 11,000 mcL^3
  - Pregnant women
    - First trimester: 6,600 to 14,100 mcL^3
    - Second trimester: 6,900 to 17,100 mcL^3
    - Third trimester: 5,900 to 14,700 mcL^3
    - Postpartum: 9,700 to 25,700 mcL^3
- Normal leukocyte cell type range
  - Neutrophils: 55% to 70%
  - Band neutrophils: 0% to 3%
  - Lymphocytes: 20% to 40%
  - Monocytes: 2% to 8%
  - Eosinophils: 1% to 4%
  - Basophils: 0.5% to 1%
- Normal RBC count range
  - Men: 4.7 to 6.1 million mcL

  - Women: 4.2 to 5.4 million mcL
  - Children: 4.0 to 5.5 million mcL
  - Newborn: 4.8 to 7.1 million mcL
- Normal hematocrit range
  - Men: 42% to 52%
  - Women: 37% to 47%
  - Pregnant women
    - First trimester: 35% to 46%
    - Second trimester: 30% to 42%
    - Third trimester: 34% to 44%
    - Postpartum: 30% to 44%
  - Children: 32% to 44%
  - Newborn: 44% to 64%
- Normal hemoglobin range
  - Men: 14 to 18 g/dL
  - Women: 12 to 16 g/dL
  - Pregnant women
    - First trimester: 11.4 to 15 g/dL
    - Second trimester: 10 to 14.3 g/dL
    - Third trimester: 10.2 to 14.4 g/dL
    - Postpartum: 10.4 to 18 g/dL
  - Children: 9.5 to 15.5 g/dL
  - Newborn: 14 to 24 g/dL
- Normal erythrocyte indices range
  - MCV: 80 to 95 fL
  - MCH: 27 to 31 pg
  - MCHC: 32 to 36 g/dL
- Normal red cell distribution width range: 11% to 14.5%
- Normal thrombocyte count range
  - Adults and children: 150,000 to 400,000 mm^3
  - Babies: 200,000 to 475,000 mm^3
  - Newborn: 150,000 to 300,000 mm^3

- High WBC count might indicate
  - Inflammation
  - Infection
  - Leukemia
  - Tissue damage
  - Stress
  - Malnutrition
  - Burns
  - Lupus
  - Kidney failure
  - Rheumatoid arthritis
  - Tuberculosis
  - Thyroid gland problems
- High RBC count might indicate
  - Lung disease
  - Alcoholism
  - Polycythemia vera
  - Smoking
  - Kidney disease
  - Dehydration
  - Burns
  - Sweating
  - Vomiting
  - Diarrhea
  - Spurious polycythemia
  - Carbon monoxide (CO) exposure
  - That the patient is using diuretics
- High thrombocyte count might indicate
  - That the patient is bleeding
  - Cancer
  - Bone marrow problems
  - Iron deficiency

- High MCV might indicate
  - Folate deficiency
  - Vitamin B_{12} deficiency
- Low WBC count might indicate
  - Alcoholism
  - Lupus
  - Cushing syndrome
  - AIDS
  - Aplastic anemia
  - Enlarged spleen
  - Viral infection
  - Malaria
  - That the patient is undergoing chemotherapy
- Low RBC count might indicate
  - Anemia
  - Addison disease
  - Sickle cell disease
  - Inflammatory bowel disease
  - Cancer
  - Peptic ulcer
  - Lead poisoning
  - Removal of the spleen
  - Pernicious anemia
  - Heavy menstrual bleeding
- Low thrombocyte count might indicate
  - Risk for bleeding
  - Idiopathic thrombocytopenic purpura (ITP)
  - Pregnancy
  - Enlarged spleen
- Low mean corpuscular volume might indicate
  - Iron deficiency
  - Thalassemia deficiency

HINT

Hematocrit and hemoglobin results indicate the same disorders as high and low erythrocyte (RBC) count. The mean corpuscular hemoglobin result indicates the same disorders as high and low mean corpuscular volume.

TEACH THE PATIENT

- Explain
  - Why the blood sample is taken.
  - That the patient should not exercise before the test is administered.
  - That the patient should tell the healthcare provider if he/she smokes.
  - That the healthcare provider may request that the patient stop taking antibiotics, steroids, Equanil, thiazide diuretics, Miltown, quinidine, or Meprospan for 2 weeks prior to the test.

Chemistry Screen 13

A chemistry screen test examines various components of the blood and is used to assess the patient's overall health. The various types of chemistry screens are called chem-20, chem-12, chem-7, and so on, where the number represents the number of blood components examined in the test.

HINT

A chemistry screen test is also called sequential multichannel analysis (SMA) or sequential multichannel analysis with computer (SMAC), followed by the number of components being examined. Other common names are comprehensive metabolic panel (CMP) or basic metabolic panel (BMP).

The following tests are included in a chemistry screen. Refer to each test for details.

- Albumin
- Alkaline phosphatase (ALP)
- Alanine aminotransferase (ALT)
- Aspartate aminotransferase (AST)
- Bilirubin
- Blood glucose

- Blood urea nitrogen (BUN)
- Calcium (Ca) in blood
- Carbon dioxide (CO_2)
- Chloride (Cl)
- Cholesterol and triglyceride tests
- Creatinine and creatinine clearance
- Lactic acid
- Phosphate in blood
- Potassium (K) in blood
- Sodium (Na) in blood
- Total serum protein
- Uric acid in blood

Vitamin B_{12} 14

Vitamin B_{12} is required for cell growth and metabolism and is stored in the liver. The vitamin B_{12} test measures the level of vitamin B_{12} in a blood sample.

HINT

The healthcare provider may order the folic acid test along with the vitamin B_{12} test. The healthcare provider may order a Schilling test to assess the patient's ability to absorb vitamin B_{12}.

WHAT IS BEING MEASURED?

- Vitamin B_{12} level in the blood

HOW IS THE TEST PERFORMED?

- See How to Collect Blood Specimen from a Vein.

RATIONALE FOR THE TEST

- Screen for
  - The underlying cause of anemia, peripheral neuropathy, and dementia
  - Vitamin B_{12} deficiency

NURSING IMPLICATIONS

- Assess if the patient
  - Has eaten or drunk 12 hours before the test is administered
  - Has taken vitamin C
  - Is pregnant
  - Is breast-feeding
  - Has been diagnosed with pernicious anemia
  - Has undergone dye testing less than a week prior to the test
  - Has taken birth control pills, Prilosec, neomycin, Dilantin, Protonix, Glucophage, Prevacid, colchicines, para-aminosalicylic acid, rabeprazole, triamterene, or methotrexate

UNDERSTANDING THE RESULTS

- Test results are available quickly. The laboratory determines normal values based on calibration of testing equipment with a control test. Test results are reported as high, normal, or low based on the laboratory's control test.
- Normal vitamin B_{12} range: 160 to 950 pg/mL
- High values of vitamin B_{12} may indicate
  - Hepatitis
  - Cirrhosis
  - Leukemia
- Low values of vitamin B_{12} may indicate
  - Folic acid deficiency anemia
  - Pernicious anemia
  - Hyperthyroidism
  - Fish tapeworm infection
  - Multiple myeloma

TEACH THE PATIENT

- Explain
  - Why the blood sample is taken.
  - Not to eat or drink 12 hours before the test.

- That the healthcare provider may request that the patient stop taking birth control pills, Prilosec, neomycin, Dilantin, Protonix, Glucophage, Prevacid, colchicines, para-aminosalicylic acid, rabeprazole, triamterene, vitamin C, or methotrexate for 2 weeks prior to the test.

Cold Agglutinins 15

Agglutinins are antibodies that cause red blood cells to aggregate forming a clump called Rouleaux formation at low temperatures as an immune reaction to an infection. High levels of agglutinins can impede blood flow to the extremities when exposed to cold, resulting in tissue damage unless the extremity is warmed. High levels of agglutinins can cause hemolytic anemia. The cold agglutinins test measures the level of agglutinins in a blood sample.

WHAT IS BEING MEASURED?

- Agglutinins level in the blood

HOW IS THE TEST PERFORMED?

- See How to Collect Blood Specimen from a Vein.

RATIONALE FOR THE TEST

- Screen for
  - Hemolytic anemia
  - The underlying cause for pneumonia

NURSING IMPLICATIONS

- Assess if the patient has
  - Taken penicillin or cephalosporins
  - Measles, malaria, congenital syphilis, pneumonia, chickenpox, anemia, infectious mononucleosis, cirrhosis, or multiple myeloma, since these diseases can produce a false positive test result

UNDERSTANDING THE RESULTS

- Test results are available quickly. The laboratory determines normal values based on calibration of testing equipment with a control test.

Test results are reported as high, normal, or low based on the laboratory's control test.

- Results are reported as a titer. A titer specifies how much of the sample of blood is diluted with saline before antibodies are no longer detected. The titer is reported as a ratio of parts of the blood sample and saline. The higher the second number of the ratio, the greater the number of antibodies in the blood sample.
- Normal cold agglutinins titer range: Less than 1 to 40 (1:40).
- High cold agglutinins titer may indicate
  - Pneumonia
  - Hepatitis C
  - Cirrhosis
  - Infectious mononucleosis
  - Rheumatoid arthritis
  - Malaria
  - Hemolytic anemia
  - Multiple myeloma
  - Lymphoma
  - Cytomegalovirus (CMV)
  - Scleroderma
  - Risk for thrombosis

TEACH THE PATIENT

- Explain
  - Why blood sample is taken.
  - That the healthcare provider may request that the patient stop taking penicillin or cephalosporins for 2 weeks prior to the test.

Toxicology Tests (Tox Screen) 16

A toxin is a substance that disrupts the body's function and includes prescription medication, nonprescription medication, and illegal medication. Toxicology tests measure the levels of one or a series of toxins in a blood sample.

Hint

The healthcare provider may order a urine toxicology test because traces of toxins can remain in urine longer than in blood. The healthcare provider may order a saliva toxicology test.

WHAT IS BEING MEASURED?

- Toxin level in the blood

HOW IS THE TEST PERFORMED?

- See How to Collect Blood Specimen from a Vein.

RATIONALE FOR THE TEST

- Screen for
  - The underlying cause of the patient's unusual behavior
  - The underlying cause why the patient is unconscious
  - Use of medication (toxin)

NURSING IMPLICATIONS

- Ask the patient to sign a consent form before administering the test unless the patient is incapacitated and unable to provide written consent.
- Ask the patient to list medications, vitamins, and herbal supplements that have been taken in the previous 4 days prior to the test.
- Assess if the patient has
  - Taken cough medicine in the previous 4 days
  - Eaten poppy seeds in the previous 4 days
- The chain of custody must be adhered to if the patient is suspected of drug abuse. Each person who handles the blood sample maintains the chain of custody by signing the chain of custody document. The chain of custody document must be attached to the test result.
- Note on the specimen the time when the blood sample is taken.

UNDERSTANDING THE RESULTS

- Test results are available quickly.
- Normal toxin range: 0 except for therapeutic levels of medication taken by the patient.

- Abnormal toxin range: Greater than 0 or higher than the therapeutic levels of medication taken by the patient.

HINT

The healthcare provider may order a quantitative toxicology test to determine the level of toxin in the blood sample if the toxicology test is positive.

CAUTION

A positive test result must be confirmed by two or more testing methods to rule out a false positive result.

TEACH THE PATIENT

- Explain
  - Why blood sample is taken.
  - That the patient must provide written consent otherwise the toxicology test cannot be administered unless the patient is incapacitated and unable to provide written consent.
  - That patient should identify any medication, vitamins, herbal supplements, cough medicine, and poppy seeds that have been taken in the previous 4 days.

Folic Acid 17

Folic acid is a type of vitamin B that is necessary for cell development and maintenance. Women who are planning to become pregnant should increase the intake of folic acid to reduce the risk of spina bifida, and cleft lip and palate. The folic acid test measures the level of folic acid in blood.

HINT

The healthcare provider may order a vitamin B_{12} blood test along with the folic acid blood test.

WHAT IS BEING MEASURED?

- Folic acid levels in the blood

HOW IS THE TEST PERFORMED?

- The patient must refrain from eating and drinking other than water for 8 hours before the test.
- See How to Collect Blood Specimen from a Vein.

RATIONALE FOR THE TEST

- Screen for
  - Malnutrition
  - Anemia
  - Malabsorption
  - Sufficient level of folic acid to reduce the risk of birth defects
  - Treatment of folic acid deficiency

NURSING IMPLICATIONS

- Assess if the patient has
  - Eaten or drunk except water for 8 hours before the test
  - Vitamin B_{12} anemia
  - Iron deficiency anemia
  - Taken Dyrenium, Proloprim, Dilantin, pentamidine, phenobarbital, Mysoline, or methotrexate, since these medications may affect the test result

UNDERSTANDING THE RESULTS

- Folic acid test results are available quickly. The laboratory determines normal values based on calibration of testing equipment with a control test. Test results are reported as high, normal, or low based on the laboratory's control test.
- Normal folate range
  - In plasma
    - Adult: 2 to 20 ng/mL
    - Children: 5 to 21 ng/mL
  - In red blood cells
    - Adult: 140 to 628 ng/mL
    - Children: Greater than 160 ng/mL

- High folic acid may indicate
  - Increased dietary folic acid or supplements
  - Vitamin B_{12} deficiency
- Low folic acid may indicate
  - Alcohol abuse
  - Anorexia nervosa
  - Crohn disease
  - Cirrhosis
  - Celiac disease
  - Hemolytic anemia
  - Kidney disease
  - Cancer

TEACH THE PATIENT

- Explain
  - Why blood sample is taken.
  - That the patient must not eat or drink except water for 8 hours before the test.
  - That the healthcare provider may ask the patient to stop taking Dyrenium, Proloprim, Dilantin, pentamidine, phenobarbital Mysoline, or methotrexate for 2 weeks prior to the test.

Gastrin 18

Gastrin is a hormone produced by the G cells of the stomach lining when food enters the stomach. Gastrin stimulates the parietal cells in the stomach to secrete hydrochloric acid (HCl) that is used in digestion. Gastrin also stimulates the product of pepsin, which is a digestive enzyme. The gastrin test measures the level of gastrin in the blood.

HINT

The healthcare provider may order an intravenous (IV) secretin test, where secretin, a digestive hormone, is injected into a vein and blood samples are taken immediately, then every 5 minutes for 15 minutes, and then at 30 minutes.

WHAT IS BEING MEASURED?

- Gastrin levels in blood

HOW IS THE TEST PERFORMED?

- Refrain from drinking alcohol 1 day before the test is administered.
- Refrain from eating and drinking except for water 12 hours before the test.
- No smoking 4 hours before the test.
- Refrain from drinking water 1 hour before the test.
- Refrain from taking Rolaids, Prilosec, Pepcid, Zantac, or Tums 12 hours before the test.
- Rest 30 minutes before the test is administered.
- See How to Collect Blood Specimen from a Vein.

RATIONALE FOR THE TEST

- Screen for
  - Zollinger-Ellison syndrome
  - Pernicious anemia
  - Pancreatic tumor
  - G-cell hyperplasia

NURSING IMPLICATIONS

- Assess if the patient has
  - Refrained from drinking alcohol 1 day before the test is administered
  - Refrained from eating and drinking except for water 12 hours before the test
  - Refrained from smoking 4 hours before the test
  - Refrained from drinking water 1 hour before the test
  - Refrained from resting 30 minutes before the test was administered
  - Taken calcium supplement or medication containing calcium
  - Taken tricyclic antidepressants, Urised, Lomotil, or Motofen
  - Taken Rolaids, Prilosec, Pepcid, Zantac, or Tums 12 hours before the test

- Hypoglycemia
- Cirrhosis, kidney failure, or rheumatoid arthritis
- Undergone small bowel resection or peptic ulcer surgery

UNDERSTANDING THE RESULTS

- Gastrin test results are available in 2 days. The laboratory determines normal values based on calibration of testing equipment with a control test. Test results are reported as high, normal, or low based on the laboratory's control test.
- Normal gastrin range
  - Adult: Less than 200 pg/mL
  - Children: Less than 125 pg/mL
- High gastrin level may indicate
  - Kidney failure
  - Pernicious anemia
  - Zollinger-Ellison syndrome
  - Hypercalcemia
  - Peptic ulcers
  - G-cell hyperplasia
  - Hyperparathyroidism
  - Stomach cancer
  - Sarcoidosis
  - Small bowel resection
- Low gastrin level may indicate
  - Hypothyroidism

TEACH THE PATIENT

- Explain
  - Why blood sample is taken.
  - That the healthcare provider may want the patient to stop taking tricyclic antidepressants, Urised, Lomotil, calcium supplements, or Motofen for 2 weeks before the test.

- Refrain from
  - Drinking alcohol 1 day before the test is administered
  - Eating and drinking except for water 12 hours before the test
  - Smoking 4 hours before the test
  - Drinking water 1 hour before the test
  - Taking Rolaids, Prilosec, Pepcid, Zantac, or Tums 12 hours before the test
- Rest 30 minutes before the test is administered.

Ferritin 19

Ferritin is a protein found in bone marrow, liver, skeletal muscles, and the spleen that binds to iron. The ferritin test measures the level of ferritin in blood to determine the amount of iron in the body.

HINT

Blood should contain a small amount of ferritin since most ferritin is bound to iron.

WHAT IS BEING MEASURED?

- Ferritin level in blood

HOW IS THE TEST PERFORMED?

- See How to Collect Blood Specimen from a Vein.

RATIONALE FOR THE TEST

- Screen for
  - Hemochromatosis (excess iron)
  - Iron deficiency anemia
  - Treatment of hemochromatosis and iron deficiency anemia
  - Inflammation

NURSING IMPLICATIONS

- Assess if the patient has
  - A diet high in red meat

- Had a radioactive scan 3 days prior to the test
- Taken birth control pills or antithyroid medication
- Had a blood transfusion 4 months prior to the test
- Inflammation, since inflammation increases ferritin levels
- Had changes in the menstrual cycle recently

- Note the age of the patient since ferritin level increases with age.

UNDERSTANDING THE RESULTS

- The ferritin test results are available quickly. The laboratory determines normal values based on calibration of testing equipment with a control test. Test results are reported as high, normal, or low based on the laboratory's control test.
- Normal ferritin test results
  - Men: 12 to 300 ng/mL
  - Women: 10 to 150 ng/mL
  - 6 months to 15 years: 7 to 142 ng/mL
  - 2 to 5 months: 50 to 200 ng/mL
  - 1 month: 200 to 600 ng/mL
  - Newborn: 25 to 200 ng/mL
- High ferritin level may indicate
  - Hemochromatosis
  - Thalassemia
  - Alcoholism
  - Iron deficiency anemia
  - Hepatitis
  - Cirrhosis
  - Hodgkin disease
  - Arthritis
  - Leukemia
  - Lupus
  - Excess dietary iron
- Low ferritin level may indicate
  - Iron deficiency

- Excessive menstrual bleeding
- Bleeding
- Colon cancer
- Ulcers
- Colon polyps
- Pregnancy
- Insufficient dietary iron

TEACH THE PATIENT

- Explain why the sample is taken.

Lactic Acid 20

Muscle cells convert glucose into lactic acid and use lactic acid for energy when oxygen levels are low during strain of heart failure, exercise, shock, and sepsis. Lactic acid is not used for energy when there is a normal oxygen level in the blood. Lactic acid is metabolized by the liver. Liver disorders can result in lactic acidosis because of a high level of lactic acid in the blood. The lactic acid test measures the level of lactic acid in blood.

HINT

The healthcare provider may order an arterial blood gas test to measure lactic acid in blood.

WHAT IS BEING MEASURED?

- Lactic acid level in blood

HOW IS THE TEST PERFORMED?

- Avoid eating or drinking 10 hours before the test.
- Avoid exercising 12 hours before the test.
- Avoid clenching the fist during the test.
- See How to Collect Blood Specimen from a Vein.

RATIONALE FOR THE TEST

- Screen for
  - Lactic acidosis
  - The underlying cause of acidosis
  - Tissue oxygenation

NURSING IMPLICATIONS

- Assess if the patient
  - Has exercised 12 hours prior to the test
  - Has eaten or drunk 10 hours before test
  - Had clenched fist when the blood sample was drawn since clenched fist may increase lactic acid level
  - Is taking alcohol, Glucophage, isoniazid, epinephrine, or Tylenol 12 hours prior to the test
  - Has an infection
- Note the time the tourniquet has been in place when the blood sample is drawn, since the longer the tourniquet is in place, there is a risk of increased lactic acid level.

UNDERSTANDING THE RESULTS

- The lactic acid test results are available in 1 day. The laboratory determines normal values based on calibration of testing equipment with a control test. Test results are reported as high, normal, or low based on the laboratory's control test.
- Normal lactic acid test results: 5 to 20 mg/dL
- High lactic acid level may indicate
  - Lactic acidosis
  - Heart failure
  - Pulmonary embolism
  - Anemia
  - Dehydration
  - Leukemia

- Cirrhosis
- Disruption of blood flow
- Alcohol poisoning
- The patient exercised prior to the test
- The patient had taken Glucophage, or isoniazid prior to the test
- Excess dietary iron

TEACH THE PATIENT

- Explain
  - Why the sample is taken.
  - That the healthcare provider may ask the patient to refrain from taking alcohol, Glucophage, isoniazid, epinephrine, or Tylenol 12 hours prior to the test.
- Avoid
  - Eating or drinking 10 hours before the test
  - Exercising 12 hours before the test
  - Clenching the fist during the test

Prothrombin Time 21

There are 12 factors that must be present to coagulate (clot) blood. Prothrombin is clotting factor II, synthesized by the liver with the assistance of vitamin K. When a blood vessel is injured, prothrombin is converted to thrombin, a protein, which forms a blood clot with other proteins to stop the bleeding. The prothrombin time (PT) test is the time necessary for plasma to clot. The healthcare provider orders the PT test to assess the patient's risk for bleeding and to assess the therapeutic effect of anticoagulate medication.

HINT

The healthcare provider is likely to order the International Normalized Ratio (INR) test, which was established by the World Health Organization (WHO) as a standard for measuring blood coagulation. Also the healthcare provider is likely to order the PTT test to measure blood coagulation and the CBC to measure the platelet count.

WHAT IS BEING MEASURED?

- The PT test is the time necessary for plasma to clot.

HOW IS THE TEST PERFORMED?

- The same test must be taken at the same time each day.
- See How to Collect Blood Specimen from a Vein.

RATIONALE FOR THE TEST

- Screen for
  - The risk for bleeding
  - The therapeutic level of Coumadin
  - Bleeding disorders
  - Vitamin K deficiency
  - Liver function

NURSING IMPLICATIONS

- Assess if the patient
  - Is vomiting or having diarrhea
  - Abuses alcohol
  - Has eaten pork liver, broccoli, beef liver, soybean, chickpeas, turnip greens, or kale
  - Is taking vitamin K, Tagamet, birth control pills, aspirin, antibiotics, hormone replacement therapy, or Coumadin
  - Has taken laxatives

UNDERSTANDING THE RESULTS

- The PT test results are available quickly. The laboratory determines normal values based on calibration of testing equipment with a control test. Test results are reported as high, normal, or low based on the laboratory's control test.
- Normal test results
  - PT: 10 to 13 seconds
  - INR: 1 to 1.4 seconds

    - Coumadin PT: 1.5 to 2.5 times normal PT
    - Coumadin INR: 2 to 3 times normal PT
- Longer PT may indicate
    - Cirrhosis
    - Risk for bleeding
    - Vitamin K deficiency
    - Disseminated intravascular coagulation (DIC)
    - Too much Coumadin or heparin
- Low Coumadin PT may indicate
    - Not enough Coumadin

TEACH THE PATIENT

- Explain
    - Why sample is taken.
    - That several samples may be taken at the same time each day to assess the therapeutic level of Coumadin, if the patient is taking this medication.

Reticulocyte Count 22

Reticulocyte is an immature red blood cell that is released by bone marrow and develops into a mature red blood cell in 2 days. The reticulocyte count test determines the amount of reticulocyte in a blood sample.

HINT

The healthcare provider may order the reticulocyte index (RI).

WHAT IS BEING MEASURED?

- The percentage of reticulocyte in blood

HOW IS THE TEST PERFORMED?

- See How to Collect Blood Specimen from a Vein.

RATIONALE FOR THE TEST

- Screen for anemia.
- Assess the treatment of anemia.

NURSING IMPLICATIONS

- Assess if the patient
  - Is undergoing radiation therapy
  - Is pregnant
  - Has had a blood transfusion in the previous week
  - Has undergone a prostate biopsy in the past 8 weeks
  - Is taking Bactrim, Septra, corticotrophin, Imuran, levodopa, Chloromycetin, methotrexate, or Cosmegen

UNDERSTANDING THE RESULTS

- The reticulocyte test results are available in 1 day. The laboratory determines normal values based on calibration of testing equipment with a control test. Test results are reported as high, normal, or low based on the laboratory's control test.
- Normal test results: 0.5% to 2.5%
- Higher levels may indicate
  - Hemolysis
  - Anemia
  - The patient is being treated for anemia
  - The patient is in a high altitude location
  - The patient has been hemorrhaging
- Low levels may indicate
  - Aplastic anemia
  - Iron deficiency anemia
  - Folic acid deficiency
  - Vitamin B_{12} deficiency
  - Radiation exposure
  - Bone marrow cancer

TEACH THE PATIENT

- Explain
  - Why the sample is taken.
  - That the healthcare provider may ask the patient to avoid taking Bactrim, Septra, corticotrophin, Imuran, levodopa, Chloromycetin, methotrexate, and Cosmegen before the test.

Schilling Test 23

Vitamin B_{12} is the key to cell metabolism and energy production. Vitamin B_{12} attaches to the intrinsic factor produced by the parietal cells in the stomach. The intrinsic factor protects vitamin B_{12} from intestinal bacteria and enables absorption of vitamin B_{12} by the intestines. The Schilling test measures the absorption of vitamin B_{12}. There are two parts to the Schilling test:

- Part 1: The patient ingests vitamin B_{12} that is radioactively tagged. A 24-hour urine sample is examined for the presence of vitamin B_{12}. Up to 25% of the ingested vitamin B_{12} will normally be detected in the 24-hour urine sample. Little or no vitamin B_{12} detected is a sign of vitamin B_{12} malabsorption.
- Part 2: If part 1 is abnormal, then the healthcare provider may order part 2. The patient ingests radioactively tagged vitamin B_{12} and the intrinsic factor. A 24-hour urine sample is examined for the presence of vitamin B_{12}. Detection of vitamin B_{12} is a sign that the patient has pernicious anemia. Absence of vitamin B_{12} is a sign of an intestinal absorption problem.

HINT

The healthcare provider may order a 48- or 72-hour urine sample if the patient has kidney disease. The healthcare provider may also order the methylmalonic acid (MMA) test and the homocysteine test.

CAUTION

The Schilling test is not performed on a pregnant woman or a woman who is breast-feeding.

WHAT IS BEING MEASURED?

- The presence of vitamin B_{12} in a 24-hour urine sample

HOW IS THE TEST PERFORMED?

- Part 1
  - Avoid eating or drinking except water for 12 hours before the test is administered.
  - Avoid taking vitamin B_{12} supplements for 3 days before the test is administered.
  - Avoid taking laxatives for 24 hours before the test is administered.
  - The patient swallows a small capsule containing radioactively tagged vitamin B_{12}.
  - The patient receives an intramuscular (IM) injection of nonradioactive vitamin B_{12} 2 hours after ingesting the capsule. This prevents the radioactively tagged vitamin B_{12} from binding with tissues.
- Part 2 (within 7 days of part 1)
  - Avoid eating or drinking except water for 12 hours before the test is administered.
  - Avoid taking vitamin B_{12} supplements for 3 days before the test is administered.
  - Avoid taking laxatives for 24 hours before the test is administered.
  - The patient swallows a small capsule containing radioactively tagged vitamin B_{12} and the intrinsic factor.
  - The patient receives an IM injection of nonradioactive vitamin B_{12} 2 hours after ingesting the capsule. This prevents the radioactively tagged vitamin B_{12} from binding with tissues.

HINT

The amount of radioactivity in the radioactive vitamin B_{12} capsule is very low and poses no risk.

CAUTION

The patient may have an allergic reaction to radioactive vitamin B_{12}, but this is rare.

RATIONALE FOR THE TEST

- Screen for
  - Absorption of vitamin B_{12}
  - Production of the intrinsic factor

NURSING IMPLICATIONS

- Assess if the patient
  - Is pregnant
  - Properly collected the 24-hour urine sample
  - Is taking Mycitracin, Dilantin, or colchicine
  - Has kidney disease
  - Used a laxative prior to administration of the test
  - Has had a radioactive scan 2 weeks prior to the test

UNDERSTANDING THE RESULTS

- The Schilling test results are available quickly. The laboratory determines normal values based on calibration of testing equipment with a control test. Test results are reported as high, normal, or low based on the laboratory's control test.
- Normal test results: 8% to 10% of vitamin B_{12} is found in urine.
- Lower levels may indicate
  - Part 1
    - Pernicious anemia
    - Celiac disease
    - Hypothyroidism
    - Cirrhosis
    - Hepatitis
  - Part 2
    - Celiac disease
    - Hypothyroidism
    - Cirrhosis
    - Hepatitis

TEACH THE PATIENT

- Explain
  - Why sample is taken.
  - That the patient will not feel any discomfort when ingesting the capsule.

- That the radioactivity in the radioactive vitamin B_{12} is low and not dangerous.
- And demonstrate how to take a 24-hour urine sample.
- That the healthcare provider may ask the patient to refrain from taking Mycitracin, Dilantin, or colchicines prior to the test.

Sedimentation Rate 24

An increase in fibrinogen in blood during the inflammatory process causes erythrocytes (RBC) to adhere to each other, forming a stack called rouleaux. The sedimentation rate (SR) test measures how many millimeters per hour erythrocytes settle to the bottom of a test tube. Rouleaux settles quicker than erythrocytes, therefore the increased sedimentation rate indicates that the patient has inflammation.

HINT

The healthcare provider might order other tests in addition to the SR test to diagnose inflammation. The healthcare provider might order the CRP test to determine if the patient has inflammation.

CAUTION

Not all inflammation increases the SR, therefore a normal SR does not rule out inflammation.

WHAT IS BEING MEASURED?

- The SR test measures how many millimeters per hour erythrocytes settle to the bottom of a test tube.

HOW IS THE TEST PERFORMED?

- See How to Collect Blood Specimen from a Vein.

RATIONALE FOR THE TEST

- Screen for inflammation.
- Assess treatment of inflammation.

NURSING IMPLICATIONS

- Assess if the patient
  - Is pregnant
  - Has her menstrual period
  - Has anemia
- Note the patient's age, since SR increases with age.

UNDERSTANDING THE RESULTS

- The SR test results are available quickly. The laboratory determines normal values based on calibration of testing equipment with a control test. Test results are reported as high, normal, or low based on the laboratory's control test.
- Normal test results
  - Men: 0 to 15 mm/h
  - Women: 0 to 20 mm/h
  - Children: 0 to 10 mm/h
  - Newborn: 0 to 2 mm/h
- Higher levels may indicate
  - Inflammation
  - Rheumatoid arthritis
  - Systemic lupus erythematosus
  - Appendicitis
  - Pneumonia
  - Chronic kidney disease
  - Lymphoma
  - Pelvic inflammatory disease
  - Graves disease
  - Giant cell arteritis
  - Polymyalgia rheumatica
  - Toxemia of pregnancy
- Lower levels may indicate
  - Sickle cell disease

- Hyperglycemia
- Polycythemia

TEACH THE PATIENT

- Explain
  - Why sample is taken.
  - That additional tests are necessary to diagnose inflammation.
  - That normal results do not rule out inflammation.

Iron 25

Iron (Fe) is a mineral in food that is needed for cell growth. Once metabolized, iron binds to the transferrin protein, which transports iron to bone marrow and other tissues. The iron test measures the amount of iron that is bound to transferrin. There are three iron tests:

- Total iron-binding capacity (TIBC) test: This test measures the capacity of the blood to carry iron by determining the amount of iron needed to bind to all the available transferring protein.
- Serum iron test: This test measures the amount of circulating iron in the blood.
- Transferrin saturation test: This is the percentage of serum iron of TIBC.

Hint

The healthcare provider may order the ferritin test, the siderocyte stain test, and a CBC along with the iron tests.

WHAT IS BEING MEASURED?

- Serum iron in blood

HOW IS THE TEST PERFORMED?

- Avoid taking iron supplements 12 hours before the test is administered.
- Avoid taking vitamin B_{12} supplements two days before the test.
- Draw the blood in the morning to ensure the highest level of iron.
- See How to Collect Blood Specimen from a Vein.

RATIONALE FOR THE TEST

- Screen for
    - Nutritional status of the patient
    - Iron deficiency anemia
    - Hemochromatosis
- Assess treatment of iron deficiency anemia.

NURSING IMPLICATIONS

- Assess if the patient
    - Has taken iron supplements 12 hours before the test is administered
    - Has taken vitamin B_{12} supplements 2 days before the test
    - Is sleep deprived
    - Is stressed
    - Has received a blood transfusion 4 months prior to the test
    - Is taking estrogen, birth control pills, Chloromycetin, aspirin, corticotrophin, and St. John's wort
- Note if the sample was taken in the morning.

UNDERSTANDING THE RESULTS

- The iron test results are available quickly. The laboratory determines normal values based on calibration of testing equipment with a control test. Test results are reported as high, normal, or low based on the laboratory's control test.
- Normal test results
    - Serum iron
        - Men: 80 to 180 mcg/dL
        - Women: 60 to 160 mcg/dL
        - Children: 50 to 120 mcg/dL
    - TIBC: 250 to 450 mcg/dL
    - Transferrin saturation
        - Men: 20% to 50%
        - Women: 15% to 50%

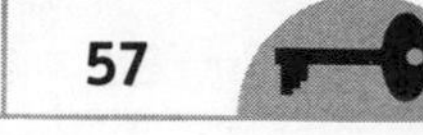

- High level may indicate
  - Hemochromatosis
  - Kidney disease
  - Cirrhosis
  - Lead poisoning
- Low level may indicate
  - Iron deficiency anemia
  - Iron deficient diet
  - Bleeding
  - Pregnancy
  - Rapid growth

TEACH THE PATIENT

- Explain
  - Why the sample is taken.
  - That the sample must be taken in the morning.
  - That the healthcare provider may ask the patient to refrain from taking iron supplements, vitamin B_{12} supplements, estrogen, birth control pills, Chloromycetin, aspirin, corticotrophin, and St. John's wort prior to the test.

Serum Protein Electrophoresis 26

Blood serum contains two groups of protein. These are albumin and globulin. Serum protein electrophoresis (SPE) separates albumin and globulin into five groups by placing the sample of blood serum on an agarose gel, and then exposing the gel to an electric current. These groups are

- Albumin
- Alpha-1 globulin
- Alpha-2 globulin
- Beta globulin
- Gamma globulin (antibodies)

Hint

The healthcare provider may order urine protein electrophoresis. The healthcare provider may also order total serum protein test along with the SPE test.

Caution

Serum protein electrophoresis is not used to diagnose liver disorders, kidney disorders, or rheumatoid arthritis although abnormal protein level may be associated with these disorders.

WHAT IS BEING MEASURED?

- Albumin and globulin in blood

HOW IS THE TEST PERFORMED?

- See How to Collect Blood Specimen from a Vein.

RATIONALE FOR THE TEST

- Screen for
  - The underlying cause of hypogammaglobulinemia (HGG)
  - Amyloidosis
  - Multiple myeloma
  - Macroglobulinemia

NURSING IMPLICATIONS

- Assess if the patient
  - Has hyperlipidemia
  - Is pregnant
  - Has taken birth control pills, chlorpromazine, corticosteroids, aspirin, sulfonamides, neomycin, or isoniazid
  - Is undergoing chemotherapy

UNDERSTANDING THE RESULTS

- The SPE test results are available in 3 days. The laboratory determines normal values based on calibration of testing equipment with a control test. Test results are reported as high, normal, or low based on the laboratory's control test.
- Normal test results
  - Albumin: 58% to 74%
  - Alpha-1 globulin: 2% to 3.5%
  - Alpha-2 globulin: 5.4% to 10.6%
  - Beta globulin: 7% to 14%
  - Gamma globulin: 8% to 18%
- High level may indicate
  - Leukemia
  - Multiple myeloma
  - Lymphoma
  - Rheumatoid arthritis
  - Liver disease
  - Systemic lupus erythematosus
  - Pregnancy
  - Dehydration
  - Infection
  - Kidney disease
  - Heart disease
- Low level may indicate
  - Leukemia
  - Multiple myeloma
  - Lymphoma
  - Rheumatoid arthritis
  - Liver disease
  - Systemic lupus erythematosus
  - Pregnancy
  - Dehydration

- Infection
- Kidney disease
- Heart disease
- Hypothyroidism
- Burns
- Emphysema
- Celiac disease
- Crohn disease
- Malnutrition

TEACH THE PATIENT

- Explain
  - Why the sample is taken.
  - That the sample must be taken in the morning.
  - That the healthcare provider may ask the patient to refrain from taking birth control pills, chlorpromazine, corticosteroids, aspirin, sulfonamides, neomycin, or isoniazid prior to the test.

Summary

Hematology clinical laboratory tests examine blood and its components to assess for signs of disease. A sample of blood can be collected from a vein, a finger, or from the heel, depending on the amount of the blood sample that is required for the test.

The clinical laboratory then measures the sample and compares the results to a range of values that the clinical laboratory has determined to be normal. For many hematology tests, the clinical laboratory sets the normal range daily. The results contain the measured value of the sample, the clinical laboratory's normal range, and an indicator if the sample is higher or lower than the range.

A test result that is outside the normal range is assessed by the healthcare provider to determine if it is a significant sign of a disease.

Quiz

1. What does the total serum protein test assess?
   a. Albumin
   b. Globulin
   c. Total protein
   d. All of the above

2. What does the serum osmolality test measure?
   a. The volume of blood in the body
   b. Antidiuretic hormone
   c. The number of particles of substances that are dissolved in the serum
   d. The dose of a vasopressin

3. An elevated C-reactive protein (CRP) level prior to surgery may indicate
   a. Higher risk of infection following surgery
   b. Lower risk of infection following surgery
   c. Surgery is no longer necessary
   d. All the above

4. The mean corpuscular volume (MCV) indicates
   a. The size of white blood cells
   b. The size of red blood cells
   c. The concentration of hemoglobin
   d. The distribution of white blood cells

5. What is the purpose of a chemistry screen?
   a. Assess the sequential multichannel analysis with computer (SMAC) only.
   b. Assess blood-carrying capability of blood only.
   c. Assess blood serum only.
   d. Assess the patient's overall health.

6. Why would the prothrombin time (PT) test be ordered?
   a. Risk for bleeding
   b. Therapeutic level of Coumadin
   c. Vitamin K deficiency
   d. All of the above

7. Why is the reticulocyte count test ordered?
   a. Screen for anemia.
   b. Screen for risk of bleeding.
   c. Assess the vitamin B_{12} level.
   d. Assess the vitamin B_6 level.

8. What does the sedimentation rate (SR) measure?
   a. The number of erythrocytes that settle to the bottom of a test tube in an hour.
   b. Screen for inflammation.
   c. Assess treatment of inflammation.
   d. All of the above.

9. What is meant by type A blood?
   a. Has the A antigen and antibodies in plasma against B antigen
   b. Has the B antigen and antibodies in plasma against A antigen
   c. Has neither the A antigen nor the B antigen and antibodies in plasma against A antigen and B antigen
   d. Has the A antigen and the B antigen and no antibodies in plasma against A antigen and B antigen

10. Why would the uric acid blood test be ordered?
    a. Assess treatment of hypouricemia.
    b. Screen for uric acid kidney stones or gout.
    c. Assess for inflammation.
    d. Assess for infection.

Answers

1. d. All of the above.
2. c. The number of particles of substances that are dissolved in the serum.
3. a. Higher risk of infection following surgery.
4. b. The size of red blood cells.
5. d. Assess the patient's overall health.
6. d. All of the above.
7. a. Screen for anemia.
8. d. All of the above.
9. a. Has the A antigen and antibodies in plasma against B antigen.
10. b. Screen for uric acid kidney stones or gout.

CHAPTER 2

Electrolytes

Electrolytes are salts that are electrically charged ions used to maintain voltage across cell membranes and carry electrical impulses within the body. The concentration of electrolytes within the body is on constant change. However, the kidneys make adjustments to keep electrolytes in balance.

There are six key electrolytes that are monitored by clinical laboratory tests. These are calcium (Ca), magnesium (Mg), phosphate [contained in phosphorus (P)], potassium (K), sodium (Na), and chloride (Cl). All of these must be in balance to ensure proper physiology.

Electrolyte tests are referred to as an electrolyte panel, basic metabolic panel, or comprehensive metabolic panel. An electrolyte panel measures only electrolytes in a sample of blood. Basic metabolic and comprehensive metabolic panels measure electrolytes and other components.

In this chapter you will learn how each electrolyte is tested.

Learning Objectives

 Calcium

2 Magnesium

3 Phosphate
4 Potassium
5 Sodium
6 Chloride

Key Words

Aldosterone
Calcitonin
Ionized calcium test
Nonionized calcium test
Osteoclast
Parathyroid hormone (PTH)
Phosphorus
Sweat test
Vitamin D

Calcium 1

Calcium (Ca) is required for growth of bone and teeth and for muscle contraction and blood clotting. Nearly all calcium in the body is stored in bone with a minimum amount in blood. Calcium has a homeostasis relationship with phosphate. As calcium increases in blood, phosphate decreases and when phosphate increases in blood, calcium decreases.

The parathyroid keeps calcium and phosphate balanced. When there is too much phosphate (too little calcium) in blood, the parathyroid releases the parathyroid hormone (PTH) that stimulates osteoclast, breaking down bone to increase calcium level in blood. PTH also activates vitamin D to increase absorption of calcium in the gastrointestinal (GI) tract. Vitamin D is necessary for calcium absorption. And the kidneys retain calcium. Too much calcium in the blood causes the thyroid gland to release calcitonin, which moves calcium from blood to bone.

There are two kinds of calcium blood tests that are performed as part of a routine blood screening. The nonionized calcium test measures calcium attached to albumin in the blood. This test is affected by the amount of albumin in the blood. The ionized calcium test measures calcium not attached to albumin in the blood and therefore is not affected by the amount of albumin in the blood.

Hint

Calcium is also measured in urine.

WHAT IS BEING MEASURED?

- Calcium level in the blood

HOW IS THE TEST PERFORMED?

- See How to Collect Blood Specimen from a Vein in Chapter 1.

RATIONALE FOR THE TEST

- Screen for
  - Parathyroid gland function
  - Kidney function
  - Kidney stones
  - Pancreatitis
  - Bone disease
  - Underlying cause:
    - Muscle spasms, depression, confusion, tingling around the mouth and fingers, and muscle cramping and twitching are caused by low calcium level in blood.
    - Of an abnormal electrocardiographic (ECG) result.
    - Nausea, vomiting, bone pain, lack of appetite, weakness, abdominal pain, and constipation and increased urination are caused by a high calcium level in blood.

Hint

Symptoms appear only when calcium levels are either very high or very low.

Caution

The blood calcium test is not used to diagnose osteoporosis. The bone mineral density test is used for this purpose.

NURSING IMPLICATIONS

- Assess if the patient has
  - Taken milk, antacid, calcium salt, or calcium supplements 8 hours before administering the test, since this will affect the test result

- Eaten or drunk 12 hours before the test is administered
- Taken Diamox, estrogen, albuterol, corticosteroids, vitamin D, lithium, laxatives, aspirin, theophylline, or birth control pills, since these medications can affect the test result
- Has had a blood transfusion recently

UNDERSTANDING THE RESULTS

- Test results are available quickly. The laboratory determines normal values based on calibration of testing equipment with a control test. Test results are reported as high, normal, or low based on the laboratory's control test.
- Normal range
    - Nonionized calcium in adults: 9.0 to 10.5 mg/dL or 2.25 to 2.75 mmol/L
    - Nonionized calcium in children: 7.6 to 10.8 mg/dL or 1.9 to 2.7 mmol/L
    - Ionized calcium in adults: 4.65 to 5.28 mg/dL
- High-level values may indicate
    - Hyperparathyroidism
    - Cancer metastasized to the bone
    - Kidney disease
    - That the patient has been on bed rest for a long period
    - Tuberculosis
    - Sarcoidosis
    - Addison disease
    - Dehydration
    - Hyperthyroidism
    - Paget disease
    - Chronic liver problems
    - Ingesting too much calcium, vitamin D, or vitamin A
    - Decreased phosphate blood level
- Low-level values may indicate
    - Hypoparathyroidism
    - Malabsorption syndrome

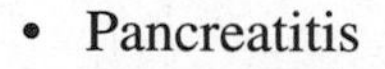

- Pancreatitis
- Hypoalbuminemia
- Low magnesium
- Kidney disease
- Increasing phosphate blood level

HINT

Calcium values are high in children who are growing. Calcium values are low in older men and pregnant women. If calcium levels in blood are abnormal, the healthcare provider may order additional tests for the parathyroid hormone, vitamin D, alkaline phosphatase, acid phosphatase, and chloride, which can affect the calcium levels in blood.

TEACH THE PATIENT

- Explain
    - Why the blood sample is taken.
    - That the patient should not eat or drink 12 hours before the test is administered.
    - That the healthcare provider may ask the patient to stop taking Diamox, estrogen, albuterol, corticosteroids, vitamin D, lithium, laxatives, aspirin, theophylline, or birth control pills 2 weeks before the test is administered, since these can affect the test result.

Magnesium 2

Magnesium (Mg), found mostly in bones and inside cells, is an electrolyte and is used to transfer potassium and sodium in and out of cells and to activate nerves, muscles, and enzymes. The magnesium blood test measures the level of magnesium in a blood sample and is tested along with other electrolytes.

HINT

Levels of calcium and magnesium have an inverse relationship.

WHAT IS BEING MEASURED?

- Magnesium level in the blood

HOW IS THE TEST PERFORMED?

- See How to Collect Blood Specimen from a Vein in Chapter 1.

RATIONALE FOR THE TEST

- Screen for
  - Underlying cause of
    - Muscle weakness, muscle twitches, and muscle irritability
    - Arrhythmia
    - The source of nausea, vomiting, low blood pressure, muscle weakness, and dizziness
  - The effects of medication that cause changes in the level of magnesium
  - The therapeutic treatment of high or low levels of magnesium

NURSING IMPLICATIONS

- Assess if the patient
  - Has taken antacids, laxatives, Milk of Magnesia, diuretics, Epsom salts, or magnesium supplements 3 days before administering the test, since these medications and supplements can affect the test result
  - Has taken insulin or antibiotics or received intravenous (IV) fluids before the test is administered
  - Is in the second or third trimester of pregnancy, since magnesium levels are normally low at this stage

UNDERSTANDING THE RESULTS

- Test results are available quickly. The laboratory determines normal values based on calibration of testing equipment with a control test. Test results are reported as high, normal, or low based on the laboratory's control test.
- Normal range
  - Adults: 1.3 to 2.1 mEq/L

- Children: 1.4 to 1.7 mEq/L
- Newborn: 1.4 to 2 mEq/L

- High values may indicate
  - Hyperparathyroidism
  - Addison disease
  - Hypothyroidism
  - Dehydration
  - Kidney failure
  - Overmedication with antacids and laxatives that contain magnesium
- Low values may indicate
  - Diabetic ketoacidosis
  - Pancreatitis
  - Hypercalcemia
  - Kidney disease
  - Alcohol abuse
  - Alcohol withdrawal
  - Sprue
  - Preeclampsia
  - Hyperthyroidism
  - Hypoparathyroidism
  - Burns
  - Starvation
  - Insufficient ingestion of magnesium
  - Diarrhea

TEACH THE PATIENT

- Explain
  - Why the blood sample is taken.
  - Not to take antacids, laxatives, Milk of Magnesia, diuretics, Epsom salts, or magnesium supplements 3 days before administering the test, since these medications can affect the test results

Phosphate 3

Phosphorus (P) is a mineral that contains a particle called phosphate, which is necessary for growth of bone and teeth and for contracting muscles. Most phosphate is in bone. Phosphate and calcium have an inverse relationship. The phosphate test measures the level of phosphate in blood.

HINT

Phosphate is measured along with other electrolytes.

WHAT IS BEING MEASURED?

- Phosphate level in blood

HOW IS THE TEST PERFORMED?

- See How to Collect Blood Specimen from a Vein in Chapter 1.
- See How to Collect Blood Specimen from a Heel Stick in Chapter 1.

RATIONALE FOR THE TEST

- Screen for
    - Bone disease
    - Kidney disease
    - Parathyroid glands function

NURSING IMPLICATIONS

- Assess if the patient has
    - Ingested alcohol, since this can affect the test results
    - Taken vitamin D supplements, antacids, insulin, epinephrine, acetazolamide, anabolic steroids, or phosphate-based enemas
    - Diabetes insipidus (DI)
    - Lymphoma
    - Type 2 diabetes

UNDERSTANDING THE RESULTS

- Test results are available in 2 hours. The laboratory determines normal values based on calibration of testing equipment with a control test. Test results are reported as high, normal, or low based on the laboratory's control test.
- Normal range
  - Adults: 3.0 to 4.5 mg/dL or 0.97 to 1.45 mmol/L
  - Children: 4.5 to 6.5 mg/dL or 1.45 to 2.1 mmol/L
  - Children younger than 1 year of age: 4.3 to 9.3 mg/dL or 1.4 to 3 mmol/L
- High values may indicate
  - Pregnancy
  - Acromegaly
  - Hypoparathyroidism
  - Kidney disease
  - Diabetic ketoacidosis
  - Bone fracture that is healing
  - Excess vitamin D
  - Low level of magnesium
- High values may indicate
  - Osteomalacia
  - Malnutrition
  - Hyperparathyroidism
  - Low level of vitamin D
  - Liver disease
  - Sprue
  - Alcohol abuse
  - High level of calcium
  - Burns

TEACH THE PATIENT

- Explain
  - Why the blood sample is taken.

- Not to ingest alcohol 12 hours before the test is administered.
- That the healthcare provider may ask the patient to stop taking vitamin D supplements, antacids, epinephrine, acetazolamide, anabolic steroids, or phosphate-based enemas 2 weeks prior to the test.

Potassium 4

Potassium (K) is a mineral stored inside the cell that has multiple functions, including muscle contractions, neural transmission, and fluid balance. Potassium is excreted by the kidneys, regulated by aldosterone hormone, and released by the adrenal glands. Potassium and sodium have an inverse relationship. The potassium test measures the level of potassium in blood.

HINT

Potassium is measured along with other electrolytes. The healthcare provider may order a urinalysis to determine the level of potassium that is being excreted by the kidneys.

WHAT IS BEING MEASURED?

- Potassium level in blood

HOW IS THE TEST PERFORMED?

- See How to Collect Blood Specimen from a Vein in Chapter 1.

RATIONALE FOR THE TEST

- Screen for cell lysis syndrome.
- Assess the
  - Effects of total parenteral nutrition (TPN) being administered to the patient
  - Adverse effect of diuretics, which can decrease potassium levels
  - Source of high blood pressure
  - Effect of kidney dialysis

NURSING IMPLICATIONS

- Assess if the patient has
  - Taken potassium supplements
  - Taken heparin, glucose, nonsteroidal anti-inflammatory drugs (NSAIDs), and antibiotics that contain potassium, natural licorice, corticosteroids, angiotensin-converting enzyme (ACE) inhibitors, or insulin
  - Experienced severe vomiting
  - Improperly used laxatives

UNDERSTANDING THE RESULTS

- Test results are available quickly. The laboratory determines normal values based on calibration of testing equipment with a control test. Test results are reported as high, normal, or low based on the laboratory's control test.
- Normal range
  - Adults: 3.5 to 5.0 mEq/L
  - Children: 3.4 to 4.7 mEq/L
  - Children younger than 1 year of age: 3.9 to 5.3 mEq/L
- High values may indicate
  - Myocardial infarction
  - Ingesting of too many potassium supplements
  - Intake of ACE inhibitors
  - Diabetic ketoacidosis
  - Kidney damage
- Low values may indicate
  - Alcoholism
  - Cystic fibrosis
  - Diarrhea
  - Hyperaldosteronism
  - Use of diuretics
  - Bartter syndrome
  - Vomiting
  - Burns

- Malnutrition
- Dehydration

TEACH THE PATIENT

- Explain
  - Why the blood sample is taken.
  - That the healthcare provider may ask the patient to stop taking for 2 weeks heparin, glucose, NSAIDs, and antibiotics that contain potassium, natural licorice, corticosteroids, or ACE inhibitors.

Sodium 5

Sodium (Na) is a mineral stored outside the cell in blood and lymph fluid that has multiple functions, including muscle contractions, neural transmission, and fluid balance. Sodium is excreted by the kidneys, regulated by aldosterone hormone, and released by the adrenal glands. Sodium and potassium have an inverse relationship. The sodium test measures the level of sodium in blood.

Hint

Sodium is measured along with other electrolytes. The healthcare provider may order a fractional excretion of sodium (FENa) urine test to determine the amount of sodium and creatinine in the urine.

WHAT IS BEING MEASURED?

- Sodium level in blood

HOW IS THE TEST PERFORMED?

- See How to Collect Blood Specimen from a Vein in Chapter 1.

RATIONALE FOR THE TEST

- Screen for
  - Adrenal gland diseases
  - Electrolyte balance

- Water balance
- Kidney disease

NURSING IMPLICATIONS

- Assess if the patient has
  - Elevated protein levels
  - Received IV fluid containing sodium
  - High triglyceride levels
  - Heparin, birth control pills, NSAIDs, antibiotics, tricyclic antidepressants (TCAs), corticosteroids, lithium, or estrogen

UNDERSTANDING THE RESULTS

- Test results are available quickly. The laboratory determines normal values based on calibration of testing equipment with a control test. Test results are reported as high, normal, or low based on the laboratory's control test.
- Normal range: 136 to 145 mEq/L
- High values may indicate
  - Diabetic ketoacidosis
  - Cushing syndrome
  - Dehydration
  - Kidney disease
  - Diabetes insipidus
  - High sodium intake
  - Vomiting
  - Diarrhea
  - Hyperaldosteronism
- Low values may indicate
  - Psychogenic polydipsia (drinking too much water)
  - Heart failure
  - Vomiting
  - Diarrhea
  - Burns

- Syndrome of inappropriate antidiuretic hormone (SIADH) secretion
- Cirrhosis
- Malnutrition
- Kidney disease

TEACH THE PATIENT

- Explain
  - Why the blood sample is taken.
  - That the healthcare provider may ask the patient to stop taking heparin, birth control pills, NSAIDs, antibiotics, TCAs, corticosteroids, lithium, or estrogen for 2 weeks.

Chloride 6

Chloride (Cl) is an electrolyte found outside the cell and is involved in fluid balance. The chloride test measures the level of chloride in blood.

Hint

The healthcare provider may order a chloride urine test or a skin sweat test. The healthcare provider will likely order the chloride test along with tests for other electrolytes.

WHAT IS BEING MEASURED?

- Chloride level in blood

HOW IS THE TEST PERFORMED?

- See How to Collect Blood Specimen from a Vein in Chapter 1.

RATIONALE FOR THE TEST

- Screen for the
  - Underlying cause of
    - Confusion, muscle spasm, muscle weakness, and difficulty breathing

- Metabolic alkalosis
- Kidney disorder
- Adrenal gland disorder

NURSING IMPLICATIONS

- Assess if the patient
  - Is pregnant
  - Has taken Coumadin, aspirin, NSAIDs, corticosteroids, cholestyramine, diuretics, androgens, or estrogens
  - Is dehydrated or overhydrated

UNDERSTANDING THE RESULTS

- The chloride test results are available quickly. The laboratory determines normal values based on calibration of testing equipment with a control test. Test results are reported as high, normal, or low based on the laboratory's control test.
- Normal test results
  - Adults: 98 to 106 mEq/L
  - Children: 90 to 110 mEq/L
  - Newborn: 96 to 106 mEq/L
  - Premature: 95 to 110 mEq/L
- Higher levels may indicate
  - Dehydration
  - Anemia
  - Kidney disease
  - Hyperparathyroidism
  - Ingestion of salt
- Lower levels may indicate
  - Kidney failure
  - Burns
  - Cushing syndrome
  - Heart failure

- Vomiting
- Diabetic ketoacidosis
- SIADH

TEACH THE PATIENT

- Explain
  - Why the sample is taken.
  - That the healthcare provider may ask the patient to refrain from taking Coumadin, aspirin, NSAIDs, corticosteroids, cholestyramine, diuretics, androgens, or estrogens prior to the test.

Summary

Electrical impulses within the body are conducted by electrically charged ions called electrolytes. The concentration of electrolytes must be within normal limits and are managed by the kidneys.

Diseases can cause one or more electrolytes to fall outside the normal range, resulting in signs and symptoms of the underlying disease. Healthcare providers order an electrolyte panel, a basic metabolic panel, or a comprehensive metabolic panel to measure a patient's electrolytes. These tests measure calcium, magnesium, phosphate, potassium, sodium, and chloride.

Knowing which electrolytes are outside the normal range enables the healthcare provider to compare the test results with other signs and symptoms to assist the healthcare provider in arriving at a medical diagnosis.

Quiz

1. What electrolyte decreases as calcium increases?
   a. Phosphate
   b. Potassium
   c. Chloride
   d. All of the above

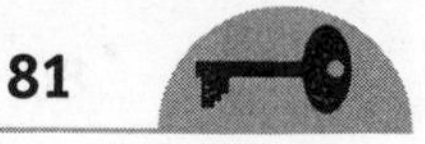

2. What can affect the results of an electrolyte test that measures calcium?
   a. Ingesting milk 8 hours before the test
   b. Taking vitamin D before the test
   c. Taking calcium supplements before the test
   d. All of the above

3. What should be explained to the patient before testing for magnesium?
   a. Avoid taking laxatives 3 days before the test.
   b. Exercise 2 hours before the test.
   c. Don't exercise 2 hours before the test.
   d. Avoid eating 3 days before the test.

4. Where is potassium stored in the body?
   a. Only in muscles
   b. Only in the kidneys
   c. Outside the cell
   d. Inside the cell

5. Why should you assess if the patient has taken antibiotics before testing for potassium?
   a. Taking antibiotics indicates that the patient has a bacterial infection.
   b. Some antibiotics contain potassium.
   c. Patients who take antibiotics always have higher than normal levels of potassium.
   d. Patients might vomit during the test.

6. Sodium has an inverse relationship with?
   a. Potassium
   b. Magnesium
   c. Phosphate
   d. Chloride

7. What does the fractional excretion of sodium test do?
   a. Determines the fractional amount of sodium in blood plasma
   b. Determines the fractional amount of sodium in urine
   c. Determines the fractional amount of potassium in blood plasma
   d. None of the above

8. What electrolyte is used to screen for cell lysis syndrome?
   a. Sodium
   b. Potassium
   c. Magnesium
   d. Phosphate

9. What are the two types of calcium blood tests?
   a. Ionized calcium test and the nonionized calcium test
   b. The parathyroid hormone (PTH) stimulation test and the vitamin D test
   c. The parathyroid test and the thyroid test
   d. None of the above

10. Calcium levels are
    a. High in growing children
    b. Low in older adults
    c. Low in pregnant women
    d. All of the above

Answers

1. a. Phosphate.
2. d. All of the above.
3. a. Avoid taking laxatives 3 days before the test.
4. d. Inside the cell.
5. b. Some antibiotics contain potassium.
6. a. Potassium.

7. b. Determines the fractional amount of sodium in urine.
8. b. Potassium.
9. a. Ionized calcium test and the nonionized calcium test.
10. d. All of the above.

CHAPTER 3

Arterial Blood Gases Test

Arterial blood gases test indicates how well the patient's lungs transfer oxygen and carbon dioxide to and from blood. Blood taken from an artery is analyzed to identify the concentration of oxygen, carbon dioxide, and bicarbonate in arterial blood. In addition, this test determines the pH value of the blood sample.

The concentration of blood is altered through the body depending on the patient's condition. The body is able to compensate for these changes by adjusting the levels of acid and bicarbonate in the blood.

In this chapter you'll learn about arterial blood gases test and how these test results determine if the patient is experiencing metabolic disorders.

Learning Objectives

1. Arterial Blood Gases
2. Total Carbon Dioxide (CO_2)
3. Carbon Monoxide (CO)

Key Words

Acidity range
Allen test
Arterial
Carboxyhemoglobin (COHb)

Arterial Blood Gases 1

Blood contains oxygen and carbon dioxide. Measuring the partial pressure of these gases indicates how well lungs exchange oxygen and carbon dioxide. In addition, measuring the oxygen saturation of the blood indicates the amount of hemoglobin (Hgb) that is carrying oxygen, and measuring the oxygen content of blood indicates the amount of oxygen in the blood.

Blood must be within acidity range. Acidity is measured using the pH scale, which measures the hydrogen ions (H^+). The pH of blood must be between 7.35 and 7.45 pH. Less than 7.35 and blood is considered too acidy. Greater than 7.45 and the blood is too alkaline (basic). Bicarbonate (HCO_3^-) is a chemical in the blood that ensures that the blood remains within the acceptable pH range. If blood becomes too acidic, the amount of bicarbonate is increased by increasing the absorption of bicarbonate by the kidneys.

WHAT IS BEING MEASURED?

- Partial pressure of oxygen (Pao_2)
- Partial pressure of carbon dioxide ($Paco_2$)
- Oxygen saturation (O_2Sat)
- Oxygen content (O_2CT)
- Blood pH
- Bicarbonate (HCO_3^-)

HOW IS THE TEST PERFORMED?

- See How to Collect Blood Specimen from a Vein in Chapter 1.

RATIONALE FOR THE TEST

- Screen for
  - Gas exchange capabilities of the lungs
  - Effectiveness of treatment of lung disease
  - Blood acidity level
  - Effectiveness of treatment of an imbalance of blood acidity
  - Kidney function

NURSING IMPLICATIONS

- Assess whether the patient has been taking anticoagulants (aspirin, Coumadin, nonsteroidal anti-inflammatory drugs [NSAIDs]). Ask the patient if any over-the-counter (OTC) medications have been taken since some contain aspirin. The test can still be performed; however, coagulation time following the test may be longer than 10 minutes.
- Assess for allergies especially to anesthetics.
- Discontinue oxygen therapy 20 minutes before the test unless the patient is unable to breathe without oxygen therapy. This ensures that gas exchange measurement is based on room air without supplemental oxygen.
- Assess for conditions that might affect the test and therefore may prevent the test from being performed:
  - Anemia
  - Polycythemia
  - Exposure to smoke (including second-hand smoke), paint removers, or carbon monoxide
  - Abnormally high or low body temperature

UNDERSTANDING THE RESULTS

- Results are available quickly. The laboratory determines normal values based on calibration of testing equipment with a control test. Test results are reported as high, normal, or low based on the laboratory's control test.

- Generally normal ranges are
  - Partial pressure of oxygen (Pao_2): 75 to 100 mm Hg
  - Partial pressure of carbon dioxide ($Paco_2$): 34 to 45 mm Hg
  - Oxygen saturation (O_2Sat): 95% to 100% (means 95 to 100 mL per 100 mL of blood)
  - Oxygen content (O_2CT): 15% to 22% (means 15 to 22 mL per 100 mL of blood)
  - Blood pH: 7.35 to 7.45 pH
  - Bicarbonate (HCO_3^-): 20 to 29 mEq/L or mmol/L
- High values can be a sign of
  - Chronic obstructive pulmonary disease (COPD)
  - Pulmonary edema
  - Cushing disease
  - Conn syndrome
  - Kidney failure
  - Sepsis
  - Dehydration
  - Drug overdose
  - Diabetes
  - The patient who is taking hydrocortisone, prednisone, or diuretics
- Low value can indicate
  - Liver disease (low bicarbonate)
  - Aspirin overdose (low bicarbonate)
  - Alcohol overdose (low bicarbonate)
  - Severe malnutrition (low bicarbonate)
  - Diarrhea (low bicarbonate)
  - Dehydration (low bicarbonate)
  - Hyperventilation (low bicarbonate)
  - Kidney disease (low bicarbonate)
  - Hyperthyroidism (low bicarbonate)
  - Shock (low bicarbonate)
  - Severe burns (low bicarbonate)

TEACH THE PATIENT

- Explain
  - Why the blood sample is taken.
  - The function of blood gases and the meaning of abnormal values.
  - That pressure may be applied to the site for 10 minutes or more especially if the patient is taking anticoagulants.
  - That oxygen therapy will be discontinued for 20 minutes before the test unless the patient is unable to breathe without oxygen therapy.

Total Carbon Dioxide 2

Carbon dioxide is a gaseous by-product of metabolism that is transported to the lungs where carbon dioxide is exhaled. Blood contains three forms of carbon dioxide. These are bicarbonate (HCO_3^-), carbonic acid (H_2CO_3), and dissolved carbon dioxide (CO_2). Most of the carbon dioxide is in the form of bicarbonate. Levels of these types of carbon dioxide are balanced by the lungs and kidneys. The total carbon dioxide test measures the level of all three types of carbon dioxide and is administered for the same amount of time as the arterial blood gas is administered.

Hint

The total carbon dioxide test is part of the chemistry screen.

WHAT IS BEING MEASURED?

- Total carbon dioxide level in the blood

HOW IS THE TEST PERFORMED?

- See How to Collect Blood Specimen from a Vein in Chapter 1.

RATIONALE FOR THE TEST

- Assess
  - Lungs
  - Kidney function

NURSING IMPLICATIONS

- Assess if the patient
    - Has taken corticosteroids, antibiotics, antacids, diuretics, aspirin, sodium bicarbonate, or barbiturates
    - Has hyperthermia

UNDERSTANDING THE RESULTS

- Test results are available quickly. The laboratory determines normal values based on calibration of testing equipment with a control test. Test results are reported as high, normal, or low based on the laboratory's control test.
- Normal range is
    - Adult : 23 to 29 mmol/L
    - Children: 20 to 28 mmol/L
    - Younger than 1 year old: 13 to 22 mmol/L
- High-level values may indicate
    - Vomiting
    - Pneumonia (respiratory acidosis)
    - COPD (respiratory acidosis)
    - Conn syndrome (metabolic acidosis)
    - Alcoholism (metabolic acidosis)
    - Cushing syndrome (metabolic acidosis)
- Low-level values may indicate
    - Hyperventilation (respiratory alkalosis)
    - Cirrhosis (respiratory alkalosis)
    - Liver failure (respiratory alkalosis)
    - Pneumonia (respiratory alkalosis)
    - Diabetes (metabolic alkalosis)
    - Aspirin overdose (metabolic alkalosis)
    - Diarrhea (metabolic alkalosis)
    - Heart failure (metabolic alkalosis)
    - Kidney failure (metabolic alkalosis)
    - Chronic starvation (metabolic alkalosis)

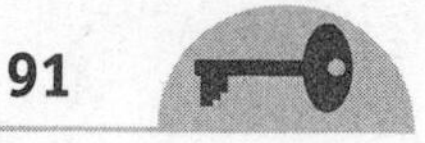

- Ingestion of antifreeze (metabolic alkalosis)
- Ingestion of methanol (metabolic alkalosis)

TEACH THE PATIENT

- Explain
  - Why blood sample is taken.
  - That the healthcare provider may ask the patient to stop taking for 2 weeks prior to the test corticosteroids, antibiotics, antacids, diuretics, aspirin, sodium bicarbonate, or barbiturates.

Carbon Monoxide

Carbon monoxide (CO) is a colorless, odorless gas that replaces oxygen attached to the hemoglobin (red blood cells), creating a compound called carboxyhemoglobin that decreases oxygenation of blood and can result in death. The carbon monoxide blood test measures the level of carboxyhemoglobin in blood.

WHAT IS BEING MEASURED?

- Carboxyhemoglobin level in the blood

HOW IS THE TEST PERFORMED?

- See How to Collect Blood Specimen from a Vein in Chapter 1.

RATIONALE FOR THE TEST

- Screen for
  - Exposure to breathing carbon monoxide
  - Underlying cause of headache, dizziness, vision problem, muscle weakness, confusion, extreme sleepiness, and nausea or vomiting which is caused by carbon monoxide

NURSING IMPLICATIONS

- Assess
  - If the patient has smoked before administering the test
  - If the patient has been exposed to carbon monoxide from automobiles and gas burning machine such as an old heating system

- Quick assessment
  - 20% to 30%
    - Nausea
    - Vomiting
    - Headache
    - Poor judgment
  - 30% to 40%
    - Vision problems
    - Dizziness
    - Confusion
    - Increasing heart rate (HR)
    - Increasing respiratory rate (RR)
    - Muscle weakness
  - 50% to 60%
    - Loss of consciousness
  - Greater than 60%
    - Seizures
    - Risk of death

UNDERSTANDING THE RESULTS

- Test results are available right away. The laboratory determines normal values based on calibration of testing equipment with a control test. Test results are reported as high, normal, or low based on the laboratory's control test.
- Normal range is
  - Nonsmoker: Less than 3% of total hemoglobin
  - Smoker: 2% to 10% of total hemoglobin
- High-level values may indicate carbon monoxide poisoning.

Hint

Patients exposed to automobile traffic have normal carbon monoxide levels of 8% to 12%. A patient with 20% carbon monoxide level will show no symptoms of carbon monoxide poisoning. Women and children have fewer red blood cells than men and therefore have more severe symptoms at lower carbon monoxide levels. The healthcare provider will likely order arterial blood gases and complete blood count tests if carbon monoxide levels are high.

TEACH THE PATIENT

- Explain
  - Why the blood sample is taken.
  - That the patient must stop smoking before the test is administered.

Summary

The effective transfer of oxygen and carbon dioxide by the lungs is determined by the arterial blood gases tests. A sample of arterial blood is taken from the patient and is analyzed for the concentration of oxygen, carbon dioxide, bicarbonate, and pH value.

Based on the test results, the healthcare provider can assess gas exchange and the patient's metabolic status. If the blood sample is too acidic, the body compensates by removing carbon dioxide, which is an acid, and by retaining bicarbonate. If the blood sample is too alkaline, then the body retains carbon dioxide and removes bicarbonate. This is referred to as compensation where the body automatically makes adjustments to bring blood within an acceptable range.

Quiz

1. What test is performed before drawing an arterial blood sample?
   a. Allen test
   b. Collection test
   c. Acid-alkaline test
   d. None of the above

2. What should you tell a patient after taking an arterial blood sample?
   a. There may be a small bruise at the puncture site.
   b. The patient may feel light-headed.
   c. The patient should not carry objects for 24 hours after the test.
   d. All of the above.

3. What is the potential result if the patient has been taking Coumadin prior to drawing the arterial blood sample?
   a. Inaccurate test results.
   b. The patient must take a baby aspirin following the test.
   c. Coagulation time may be longer than 10 minutes.
   d. The test should not be performed.

4. What test result would you expect to find in a patient with kidney failure?
   a. Respiratory alkalosis
   b. Metabolic alkalosis
   c. Metabolic acidosis
   d. Respiratory acidosis

5. A patient with 20% carbon monoxide blood level
   a. Will show symptoms of carbon monoxide poisoning
   b. Will not show symptoms of carbon monoxide poisoning
   c. Will show slight symptoms of carbon monoxide poisoning
   d. None of the above

6. Diarrhea can result in
   a. Respiratory alkalosis
   b. Metabolic alkalosis
   c. Metabolic acidosis
   d. Respiratory acidosis

7. A sign of low bicarbonate is
   a. Low blood pressure
   b. High blood pressure
   c. Hyperventilation
   d. Hypoventilation

8. What conditions might affect the blood gases test?
   a. Anemia
   b. High body temperature
   c. Low body temperature
   d. All of the above

9. Alcoholism can results in
   a. Respiratory alkalosis
   b. Metabolic alkalosis
   c. Metabolic acidosis
   d. Respiratory acidosis

10. What test measures carboxyhemoglobin level in the blood?
    a. Carbon monoxide test
    b. Total carbon dioxide test
    c. Arterial blood gases
    d. None of the above

Answers

1. a. Allen test.
2. d. All of the above.
3. c. Coagulation time may be longer than 10 minutes.
4. b. Metabolic alkalosis.
5. b. Will not show symptoms of carbon monoxide poisoning.
6. b. Metabolic alkalosis.
7. c. Hyperventilation.
8. d. All of the above
9. c. Metabolic acidosis.
10. a. Carbon monoxide test.

CHAPTER 4

Liver Tests

The liver is the largest gland in the body that makes and secretes substances. The liver synthesizes albumin, which maintains blood volume and clotting factors. The liver also synthesizes, stores, and metabolizes fatty acids and cholesterol. Fatty acids are used for energy by the body. The liver stores and metabolizes carbohydrates. Carbohydrates are converted into glucose for energy.

The liver forms and secretes bile. Bile contains acids that help the intestines absorb fats and vitamins A, D, E, and K, which are fat-soluble vitamins. In addition, the liver clears the body of medication and harmful chemicals, such as bilirubin, which is the result of the metabolism of aged red blood cells, and ammonia, which is the result of the metabolism of proteins. The liver transforms these chemicals into components that are easily excreted by the body in urine or stool.

In this chapter you will learn about blood tests that determine how well the liver is performing.

Learning Objectives

1. Hepatitis A Virus (HAV) Test
2. Hepatitis B Virus (HBV) Tests
3. Alanine Aminotransferase (ALT)
4. Alkaline Phosphatase (ALP)
5. Ammonia
6. Aspartate Aminotransferase (AST)
7. Bilirubin

Key Words

Conjugated bilirubin
Direct bilirubin
Gamma glutamyl transferase (GGT)
Gamma glutamyl transpeptidase
IgG anti-HAV
IgM anti-HAV
Indirect bilirubin
Lactulose
Serum glutamate oxaloacetate transaminase (SGOT)
Serum glutamate pyruvate transaminase (SGPT)
Total bilirubin
Unconjugated bilirubin

Hepatitis A Virus Test 1

A patient who is or has been infected with the hepatitis A virus (HAV) will have hepatitis A antibodies in their blood. A patient who received the HAV will also have these antibodies, indicating the effectiveness of the vaccine. There are two types of antibodies:

- IgM anti-HAV: Presence of this antibody indicates that the patient was recently infected. This antibody is detectable 2 weeks after being infected and remains in the blood for 3 to 12 months.
- IgG anti-HAV: Presence of this antibody indicates that the patient has at some point been infected. This antibody is detectable between 8 and 12 weeks following the infection and remains in the blood.

WHAT IS BEING MEASURED?

- The presence or absence of the IgM anti-HAV and IgG anti-HAV antibodies

HOW IS THE TEST PERFORMED?

- See How to Collect Blood Specimen from a Vein in Chapter 1.

RATIONALE FOR THE TEST

- Screen for
  - HAV infection
- Assess
  - The effectiveness of the hepatitis A vaccine
  - If the patient is protected against the hepatitis A vaccine
  - If the infection from hepatitis A is the source of abnormal liver function tests

NURSING IMPLICATIONS

- Assess
  - Whether the patient has been taking anticoagulants (aspirin, Coumadin, nonsteroidal anti-inflammatory drugs [NSAIDs]). Ask the patient if any over-the-counter medications have been taken since some contain aspirin. The test can still be performed; however, coagulation time following the test may be longer than 10 minutes.
  - If the vein is swollen after the test (phlebitis). If so, apply a warm compress several times a day.
- Consult the state's regulations to determine if cases of hepatitis A infection must be reported to the health department.

UNDERSTANDING THE RESULTS

- Results are available in 5 to 7 days and reported as positive (the presence of the antibody) or negative (the antibody was not found in the blood sample).
- A negative test result does not mean that the patient is not infected with HAV. It can take 2 weeks or more before the antibodies are detectable. This is referred to as a false negative.
- A positive test result in the absence of the patient previously receiving the hepatitis A vaccine may be followed with liver function tests.

TEACH THE PATIENT

- Explain
  - Why the test is administered.
  - A negative test result does not mean that the patient is infection free. It takes 2 weeks or more to develop antibodies.
  - A positive test result indicates the presence of antibodies and not the presence of the HAV.

Hepatitis B Virus Tests 2

Hepatitis B is a virus that can cause an infection. There are several hepatitis B virus (HBV) tests used to determine if blood has signs of an HBV. There are three signs:

- HBV antibodies: HBV antibodies are produced as part of the immune response to the presence of the HBV in the patient and may remain in the patient's blood long after the HBV is destroyed.
- HBV antigens: HBV antigens are markers created by the HBV when the HBV infects the patient.
- HBV DNA: HBV DNA is present when HBV infects the patient.

There are seven types of HBV tests:

- Hepatitis B surface antigen (HBsAg): This is the first test that detects HBV antigen even before symptoms are present. It is also used to detect if the patient will be an HBV carrier if the HBsAg level is elevated for more than 6 months.
- Hepatitis B surface antibody (HBsAb): This test detects HBV antibodies, which are elevated 4 weeks after HBsAg is no longer detectable, and is used to determine if the patient requires an HBV vaccination.
- Hepatitis B e-antigen (HBeAg): This test detects the HBeAg antigen, which is present if the patient is currently infected and is used to monitor HBV treatment.
- Hepatitis DNA test: This test determines the level of HBV DNA in the patient's blood and is used to monitor treatment of chronic HBV infection.
- Hepatitis B core antibody (HBcAb): This test detects the HBcAb antibody a month after the HBV infection and is used to screen transfused blood for hepatitis B.

- Hepatitis B core antibody IgM (HBcAbIgM): This test detects the HBcAbIgM antibody within 6 months of the patient becoming infected with HBV.
- Hepatitis B e-antibody (HBeAb): This test detects the HBeAb antibody, indicating that the patient has almost recovered from an acute HBV infection.

Hint

The healthcare provider may also order tests for alanine aminotransferase (ALT), alkaline phosphatase (ALP), bilirubin, and aspartate aminotransferase (AST) to assess the patient's liver function.

WHAT IS BEING MEASURED?

- HBV antibody levels in blood
- HBV antigen levels in blood
- HBV DNA level in blood

HOW IS THE TEST PERFORMED?

- See How to Collect Blood Specimen from a Vein in Chapter 1.

RATIONALE FOR THE TEST

- Assess
  - For HBV infection
  - If transfused blood is or was infected with HBV
  - The effect of hepatitis B vaccination
  - The treatment of HBV

NURSING IMPLICATIONS

- A negative test result does not rule out HBV since signs of HBV can take weeks to develop.
- A patient who received an HBV vaccination may have HBsAb antibodies.
- A patient who received an HBV vaccination is protected against hepatitis D infection.

UNDERSTANDING THE RESULTS

- HBV test results are available quickly.
- Normal HBV level is (negative): No presence of antigens, antibodies, or DNA.
- Abnormal HBV level is (positive): Antigens, antibodies, or DNA are present.

CAUTION

Consult your local laws to determine if HBV infections must be reported to the health department.

TEACH THE PATIENT

- Explain why sample is taken.

Alanine Aminotransferase 3

Alanine aminotransferase (ALT), formerly called serum glutamate pyruvate transaminase (SGPT), is an enzyme mainly found in the liver, and is also in the heart, pancreas, muscles, and kidneys in small amounts. Damage to the liver caused by injury or disease results in the release of ALT in the blood. The ALT test measures the level of ALT in the blood as a way to detect liver disease. The ALT test is typically performed with the aspartate aminotransferase (AST) test to detect liver damage. Other tests that detect liver problems are the lactate dehydrogenase (LDH) test and the bilirubin test.

WHAT IS BEING MEASURED?

- The level of the ALT in the blood

HOW IS THE TEST PERFORMED?

- See How to Collect Blood Specimen from a Vein in Chapter 1.

RATIONALE FOR THE TEST

- Screen for liver disorder.
- Assess

- The effects of medications that can cause liver damage
- The underlying cause of jaundice

NURSING IMPLICATIONS

- Assess if the patient
  - Has performed strenuous exercise before the test is administered. Strenuous exercise can affect the test results.
  - Is pregnant. This can affect the test results.
  - Has taken Echinacea or valerian and other herbal and natural treatments. If so, the patient should refrain from taking these for several days before the test since they can affect the test results.
  - Is allergic to any medications.
  - Received any intramuscular (IM) injections, injured a muscle, cardiac catheterization, or surgery recently. These can release ALT from injured muscle and affect the results of the test.
- Identify prescription and nonprescription medications taken by the patient. Medications such as narcotics, barbiturates, antibiotics, chemotherapy, and statins affect the test results and should be stopped several days before taking the test.

UNDERSTANDING THE RESULTS

- Test results are available in 12 hours. The laboratory determines normal values based on calibration of testing equipment with a control test. Test results are reported as high, normal, or low based on the laboratory's control test.
- Generally the normal range is 4 to 36 U/L or 0.07 to 0.62 microkat/L.
- Slightly high levels may indicate
  - Chronic diseases that affect the liver
  - Cirrhosis
  - Fat deposits in the liver
  - Medications such as narcotics, barbiturates, antibiotics, chemotherapy, and statins
- Moderately high levels may indicate
  - Excessive amounts of acetaminophen.
  - Alcohol abuse.

    - Mononucleosis.
    - Hepatitis.
    - Young child is growing quickly.
- Very high levels may indicate
    - Necrosis
    - Lead poisoning
    - Shock
    - Carbon tetrachloride exposure
    - Reaction to medication
    - Severe liver damage
    - Recent liver damage
    - Viral hepatitis
    - Rapid progression of acute lymphocytic leukemia (ALL) in a child

TEACH THE PATIENT

- Explain
    - Why blood sample is taken. ALT testing is part of routine blood screening.
    - That the ALT enzyme can be abnormal for a number of reasons and therefore other tests are necessary to confirm an abnormal ALT test result.
    - That the patient should make it known before the test is performed if the patient
        - Had strenuous exercise
        - Has taken narcotics, barbiturates, antibiotics, chemotherapy, and statins
        - Takes Echinacea or valerian and other herbal and natural treatments
        - Experienced anything that could cause a muscle injury such as IM injections, injured a muscle, cardiac catheterization, or surgery

Alkaline Phosphatase 4

Alkaline phosphatase (ALP) is an enzyme produced mainly in the liver and is also produced by bones, kidneys, intestines, and placenta. The ALP test measures the level of ALP in blood.

Hint

An ALP isoenzymes test is likely to be ordered if the ALP level is high. The healthcare provider may order an ultrasound or CT scan. The gamma glutamyl transferase (GGT) test, 5-nucleotidase, or the gamma glutamyl transpeptidase test might be ordered if the ALP level is high to differentiate between bone ALP and liver ALP.

WHAT IS BEING MEASURED?

- ALP level in blood

HOW IS THE TEST PERFORMED?

- See How to Collect Blood Specimen from a Vein in Chapter 1.

RATIONALE FOR THE TEST

- Screen for
  - Liver disease
  - Side effects of medication on liver
  - Rickets
  - Osteomalacia
  - Paget disease
  - Bone tumor
- Assess
  - For treatment of Paget disease, rickets, and osteomalacia
  - The underlying cause of high calcium level in blood

NURSING IMPLICATIONS

- The ALP test is given as part of routine blood screening.
- There is no special preparation required for this test.
- If ALP is high, the healthcare provider may order a second test and ask the patient to refrain from eating or drinking 10 hours before the test. ALP increases after eating.
- Assess if the patient is taking antibiotics, birth control pills, oral diabetes medication, or is on aspirin therapy since these can affect the test results.

- Assess if the patient is pregnant and determine the gestational age of the fetus. The placenta produces ALP in the third trimester.
- Assess if the patient
  - Is postmenopausal. The patient will have a high level of ALP.
  - Is a child, since rapid bone growth increases ALP level.
  - Uses alcohol, since this affects the test result.

UNDERSTANDING THE RESULTS

- Test results are available between 2 and 5 days. The laboratory determines normal values based on calibration of testing equipment with a control test.
- Test results are reported as high, normal, or low based on the laboratory's control test.
- Generally the normal range is
  - Adult: 30 to 126 U/L or 0.5 to 2.0 microkat/L
  - Children: 30 to 300 U/L or 0.5 to 5.0 microkat/L
- Very high level may indicate
  - Gallstones
  - Hepatitis
  - Obstructive jaundice
  - Cancer metastasized to the liver
  - Liver disease
- High level may indicate
  - Rickets
  - Paget disease
  - Osteomalacia
  - Cancer metastasized to the bone
  - Bone tumors
  - Hyperparathyroidism
  - Mononucleosis
  - Kidney cancer
  - Heart failure
  - Myocardial infarction

    - Sepsis
    - Bone healing
- Low level may indicate
    - Malnutrition
    - Scurvy
    - Celiac disease

Hint

It is normal for a pregnant woman in the third trimester to have an increased ALP because the placenta makes ALP. It is normal for a child who is experiencing rapid bone growth to have an increased ALP.

TEACH THE PATIENT

Explain

- Why blood sample is taken.
- The function of ALP and the meaning of abnormal values.
- That the patient may be asked to take the ALP test a second time if the test result indicates a high level of ALP. For the second test, the patient will be asked to refrain from eating or drinking 10 hours before the test. ALP increases after eating.

Ammonia 5

Ammonia is formed when bacteria in the intestines breaks down protein. Ammonia is then converted into urea by the liver, which is excreted by the kidneys in urine. The ammonia test measures the ammonia level in the blood. If the liver is unable to convert ammonia to urea, ammonia levels in the blood increase, indicating that there may be a liver function problem.

WHAT IS BEING MEASURED?

- Ammonia level in blood

HOW IS THE TEST PERFORMED?

- See How to Collect Blood Specimen from a Vein in Chapter 1.

RATIONALE FOR THE TEST

- Screen for
  - Liver function
  - Reye syndrome
  - Hyperalimentation
  - Cirrhosis
  - Acute liver failure
- Assess for treatment of liver disease.

NURSING IMPLICATIONS

- High ammonia levels can result in symptoms of confusion, sleepiness, hand tremors, or coma.
- Assess if the patient
  - Has eaten or drunk other than water 8 hours before the test.
  - Smoked for 8 hours before the test.
  - Performed strenuous exercise before the test.
  - Is constipated.
  - Has eaten a high- or low-protein diet.
  - Is taking furosemide, heparin, valproic acid, acetazolamide, neomycin, tetracycline, lactulose, Marplan, Nardil, diphenhydramine, or Parnate. The patient will be asked to refrain from taking these medications several days before the test.

UNDERSTANDING THE RESULTS

- Test results are available in about 12 hours. The laboratory determines normal values based on calibration of testing equipment with a control test. Test results are reported as high, normal, or low based on the laboratory's control test.
- Generally the normal range is
  - Adult: 15 to 45 mcg/dL or 11 to 32 mcmol/L
  - Children: 40 to 80 mcg/dL or 28 to 57 mcmol/L
  - Newborn: 90 to 150 mcg/dL or 64 to 107 mcmol/L

- High level may indicate
  - Cirrhosis
  - Hepatitis
  - Reye syndrome
  - Heart failure
  - Bleeding from intestines or stomach
  - Hemolytic disease of the newborn (incompatible blood types of mother and fetus)
  - Kidney failure

Hint

Lactulose is a laxative that reduces intestinal bacteria production of ammonia. Ammonia levels might be slightly elevated in a patient who has severe cirrhosis. Newborns normally have high levels of ammonia that is temporary.

TEACH THE PATIENT

- Explain
  - Why blood sample is taken.
  - The function of ammonia and the meaning of abnormal values.
  - That the patient must refrain from smoking, eating, or drinking, except water, for 8 hours before the test and avoid strenuous exercise before the test.
  - That the patient should consult the healthcare provider whether he/she should stop taking furosemide, heparin, valproic acid, acetazolamide, neomycin, tetracycline, lactulose, Marplan, Nardil, diphenhydramine, or Parnate several days before the test.

Aspartate Aminotransferase 6

Aspartate aminotransferase (AST), previously known as serum glutamate oxaloacetate transaminase (SGOT), is an enzyme in the liver, heart, pancreas, kidneys, red blood cells, and muscle tissues. When these tissues or cells are damaged, there is an increase of AST in the blood 6 to 10 hours after the damage and remains for 4 days. The AST test measures the level of AST in the blood.

Hint

The healthcare provider orders tests to measure AST and ALT in a normal screen for liver damage and liver disease. The AST test is more effective in detecting liver damage caused by alcohol abuse than the ALT test. As the patient recovers from tissue damage, the AST level in the blood decreases.

WHAT IS BEING MEASURED?

- AST level in blood

HOW IS THE TEST PERFORMED?

- See How to Collect Blood Specimen from a Vein in Chapter 1.

RATIONALE FOR THE TEST

- Screen for
  - Liver damage
  - Hepatitis
  - Cirrhosis
- Assess
  - The treatment of liver disease
  - Underlying cause of jaundice
  - If medication is causing liver damage

NURSING IMPLICATIONS

- Assess if the patient
  - Performed strenuous exercise before administration of the test since this can affect the test result.
  - Is taking Echinacea, valerian, or other herbs since this can affect the test result. The patient may be required to stop taking these for several days before the test is administered.
  - Is pregnant.
  - Takes mega doses of vitamin A.

- Has had surgery, cardiac catheterization, IM injections, or injury to muscle recently, since these can cause an increase in AST level in blood.
- Has recently taken antibiotics, narcotics, statins, and barbiturates or is undergoing chemotherapy.

UNDERSTANDING THE RESULTS

- Test results are available in about 12 hours. The laboratory determines normal values based on calibration of testing equipment with a control test. Test results are reported as high, normal, or low based on the laboratory's control test.
- Generally the normal range is 8 to 35 U/L or 0.14 to 0.58 microkat/L
- Slightly high level may indicate
  - Fatty deposits in the liver
  - Alcohol abuse
  - Excessive acetaminophen
  - The patient has taken antibiotics, narcotics, statins, barbiturates, or is undergoing chemotherapy
- Moderately high level may indicate
  - Cirrhosis
  - Chronic disease that directly or indirectly affects the liver
  - Kidney damage
  - Lung damage
  - Duchenne muscular dystrophy
  - Mega dose of vitamin A
  - Myocardial infraction
  - Heart failure
  - Mononucleosis
  - Cancer
  - Myositis
  - Alcohol abuse
- Very high level may indicate
  - Viral hepatitis
  - Necrosis

- Shock
- Drug reaction
- Burns
- Pulmonary embolism
- Trauma
- Heatstroke
- Poison

TEACH THE PATIENT

- Explain
  - Why blood sample is taken.
  - The function of AST and the meaning of abnormal values.
  - That the patient must refrain from taking Echinacea, valerian, or other herbs, vitamin A.
  - That the patient must notify the healthcare provider if he/she is taking antibiotics, narcotics, statins, and barbiturates or is undergoing chemotherapy.

Bilirubin 7

The liver breaks down old red blood cells into bilirubin, which becomes the brownish yellow component of bile. Bilirubin is excreted through feces and gives feces its brown color. Indirect (unconjugated) bilirubin, which is insoluble in water, is carried by the blood to the liver, where indirect bilirubin is transformed into direct bilirubin, which is soluble in water. The bilirubin test measures the total and the direct bilirubin levels in the blood. Indirect bilirubin is measured by subtracting direct bilirubin level from the total bilirubin level.

HINT

Bilirubin can also be measured in urine.

WHAT IS BEING MEASURED?

- Total bilirubin level in the blood
- Direct bilirubin level in the blood

HOW IS THE TEST PERFORMED?

- See How to Collect Blood Specimen from a Vein in Chapter 1.
- See How to Collect Blood Specimen from a Heel Stick in Chapter 1.
- A transcutaneous bilirubin meter may be used to measure the bilirubin level in a newborn. This handheld meter is placed against the skin. The skin is not punctured. If the bilirubin level is high, then the heel stick test is used to collect a blood sample from the newborn.

RATIONALE FOR THE TEST

- Screen for
  - Liver function
  - Hepatitis
  - Cirrhosis
  - Block bile duct from gallstones or pancreatic tumor
  - Hemolytic anemia
  - Hemolytic disease
  - Neonatal jaundice
  - Liver damage caused by medication

NURSING IMPLICATIONS

- Assess if the patient
  - Has eaten or drunk 4 hours before the test.
  - Has taken aspirin or Coumadin.
  - Is pregnant.
  - Has taken caffeine. This can lower bilirubin levels.
  - Has fasted for a long period. This increases indirect bilirubin levels.
  - Has taken Dilantin, Valium, Indocin, birth control pills, or Dalmane, since these medications can increase bilirubin levels.
  - Has taken theophylline, phenobarbital, or ascorbic acid (vitamin C) since these can lower bilirubin levels.

UNDERSTANDING THE RESULTS

- Test results are available in 2 hours. The laboratory determines normal values based on calibration of testing equipment with a control test.

Test results are reported as high, normal, or low based on the laboratory's control test.

- Normal range is
  - Total bilirubin: 0.3 to 1.0 mg/dL or 5.1 to 17.0 mmol/L
  - Direct bilirubin: 0.1 to 0.3 mg/dL or 1.0 to 5.1 mmol/L
  - Indirect bilirubin: 0.2 to 0.7 mg/dL or 3.4 to 11.9 mmol/L
- Normal range for newborn
  - Younger than 24 hours old
    - Full term: Less than 6.0 mg/dL or less than 103 mmol/L
    - Premature: Less than 8.0 mg/dL or less than 137 mmol/L
  - Younger than 48 hours old
    - Full term: Less than 10.0 mg/dL or less than 170 mmol/L
    - Premature: Less than 12.0 mg/dL or less than 205 mmol/L
  - 3 days to 5 days old
    - Full term: Less than 12.0 mg/dL or less than 205 mmol/L
    - Premature: Less than 15.0 mg/dL or less than 256 mmol/L
  - 7 days or older
    - Full term: Less than 10.0 mg/dL or less than 170 mmol/L
    - Premature: Less than 15.0 mg/dL or less than 256 mmol/L
- High level may indicate
  - Cholecystitis
  - Infected gallbladder
  - Hepatitis
  - Cirrhosis
  - Mononucleosis
  - Blocked bile duct
  - Gallstones
  - Pancreatic cancer
  - Gilbert syndrome
  - Sickle cell disease
  - Transfusion reaction
  - That the patient has taken Dilantin, Valium, Indocin, birth control pills, Dalmane

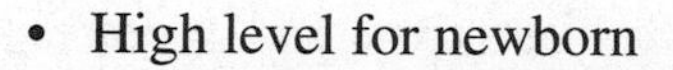

- High level for newborn
  - 24 hours old or younger: More than 10 mg/dL or more than 170 mmol/L
  - 25 to 48 hours old: More than 15 mg/dL or more than 256 mmol/L
  - 49 to 72 hours old: More than 18 mg/dL or more than 205 mmol/L
  - Older than 72 hours: More than 20 mg/dL or more than 340 mmol/L
- Low level may indicate
  - That the patient has taken theophylline, phenobarbital, or ascorbic acid

---

HINT

Newborns can experience physiologic jaundice between 1 and 3 days old because the newborn is unable to break down red blood cells. This usually resolves in a week. The healthcare provider may order phototherapy to prevent collateral problems from developing.

TEACH THE PATIENT

- Explain
  - Why blood sample is taken.
  - The function of bilirubin and the meaning of abnormal values.
  - That the patient must refrain from eating Echinacea, valerian, or other herbs.
  - That the patient should refrain from eating or drinking anything for 4 hours before the test is administered.
  - That the healthcare provider may want the patient to stop taking aspirin, Coumadin, Dilantin, Valium, Indocin, birth control pills, Dalmane, theophylline, phenobarbital, or ascorbic acid and caffeine several days before the test is administered.

Summary

The liver plays a critical role in maintaining various bodily functions. It is this organ that produces the clotting factor that stops bleeding. The liver also synthesizes albumin that is used to maintain blood volume.

The intestines are able to absorb fat and fat-soluble vitamins because the liver synthesizes and secretes bile. Bile contains acids that metabolize fat and fat-soluble vitamins into elements that are easily absorbed into the body.

The liver also provides an energy store for the body by storing carbohydrates and metabolizing carbohydrates into glucose and synthesizes, stores, and metabolizes fatty acids and cholesterol.

The liver can also be thought of as the garbage disposal because the liver metabolizes harmful biochemicals and medications into elements that can safely be excreted by the body in urine and stool.

Quiz

1. What does the presence of the IgM anti-HAV mean?
   a. The patient was recently infected with hepatitis A virus.
   b. The patient at some point was infected with hepatitis A virus.
   c. The patient was recently infected with hepatitis B virus.
   d. The patient at some point was infected with hepatitis B virus.

2. What is alanine aminotransferase?
   a. An enzyme only found in the liver
   b. An enzyme mainly found in the liver
   c. The test used to determine the amount of carbohydrates stored in the liver
   d. None of the above

3. Why is the ammonia blood test performed?
   a. Estimates the amount of protein in the liver.
   b. Estimates the level of bacteria in the liver.
   c. Estimates the level of urea in the liver.
   d. If the liver is unable to convert ammonia to urea, ammonia levels in blood increase indicating liver function problem.

4. Why is it important to know if the patient has taken lactulose prior to the ammonia test?
   a. Lactulose is a laxative that reduces intestinal bacteria production of ammonia; therefore the test results will not accurately reflect the function of the liver.
   b. Lactulose increases intestinal bacteria production of ammonia.
   c. Lactulose interrupts the liver's metabolism of ammonia.
   d. All of the above.

5. The bilirubin test measures
   a. Only total bilirubin
   b. Total bilirubin and direct bilirubin
   c. Total bilirubin and indirect bilirubin
   d. Only indirect bilirubin

6. What is bilirubin?
   a. Bilirubin gives feces its brown color.
   b. Bilirubin is the brownish yellow component of bile.
   c. Bilirubin is metabolized older red blood cells.
   d. All of the above.

7. Why do newborns sometimes experience physiologic jaundice?
   a. The newborn is unable to break down red blood cells due to physiologic immaturity of the liver. This is usually resolved in a week.
   b. The newborn is unable to break down red blood cells due to physiologic immaturity of the liver. This is resolved with a blood transfusion.
   c. The newborn is unable to break down red blood cells due to physiologic immaturity of the liver. This is resolved with a liver transplant.
   d. The mother has hepatitis C.

8. What does the transcutaneous bilirubin meter measure?
   a. It is a handheld meter that helps determine if the newborn has jaundice.
   b. Measures the level of bilirubin by being placed on the skin.
   c. The level of bilirubin in a newborn.
   d. All of the above.

9. Why is the ammonia test administered?
   a. To screen for Reye syndrome
   b. To screen for hepatitis A
   c. To screen for hepatitis B
   d. To screen for hepatitis C

10. Why is the ALT test administered?
    a. To screen for liver disorder
    b. To assess the underlying cause of jaundice
    c. To assess liver damage caused by medication
    d. All of the above

Answers

1. a. The patient was recently infected with hepatitis A virus.
2. b. An enzyme mainly found in the liver.
3. d. If the liver is unable to convert ammonia to urea, ammonia levels in blood increase indicating liver function problem.
4. a. Lactulose is a laxative that reduces intestinal bacteria production of ammonia; therefore the test results will not accurately reflect the function of the liver.
5. b. Total bilirubin and direct bilirubin.
6. d. All of the above.
7. a. The newborn is unable to break down red blood cells due to physiologic immaturity of the liver. This is resolved in a week.
8. d. All of the above.
9. a. To screen for Reye syndrome.
10. d. All of the above.

CHAPTER 5

Cardiac Enzymes and Markers Tests

Cardiac muscle contains enzymes. In a myocardial infarction, cardiac muscle is damaged, causing the release of cardiac enzymes into the blood stream. When a patient is suspected of having a myocardial infarction, the healthcare provider will order the cardiac enzymes and cardiac markers tests to determine if cardiac muscle enzymes appear in the patient's blood.

It can take between 2 and 24 hours for cardiac muscle enzymes to reach a detectable level in blood. Therefore, healthcare providers typically use an electrocardiogram (ECG) to diagnose the acute phase of a myocardial infarction. The cardiac enzymes and cardiac markers tests are used to confirm a previous acute myocardial infarction.

There are several cardiac enzymes and cardiac markers tests commonly used by healthcare providers to confirm a myocardial infarction diagnosis. You'll learn about these tests in this chapter.

Learning Objectives

1. Brain Natriuretic Peptide (BNP)
2. Cardiac Enzyme Studies
3. Homocysteine
4. Renin Assay

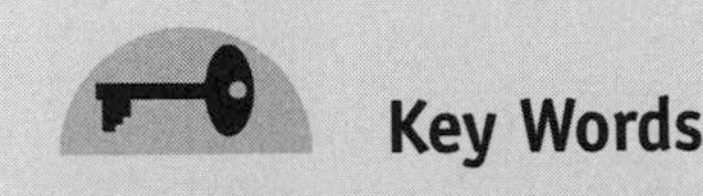

Key Words

Aldosterone
Angiotensin
Creatine phosphokinase
N-terminal pro-brain natriuretic peptide
Protein troponin
Renin-angiotensin system

Brain Natriuretic Peptide Test 1

The heart produces the hormone brain natriuretic peptide (BNP). Low level of BNP is normally found in blood. However, BNP level increases when the heart works harder for long periods such as in heart failure. The BNP test measures the level of BNP in the blood.

Hint

The healthcare provider may order the N-terminal pro-brain natriuretic peptide (NT-proBNP) test that measures the NT-proBNP hormone. This test provides similar diagnostic results as the BNP test.

WHAT IS BEING MEASURED?

- BNP level in blood

HOW IS THE TEST PERFORMED?

- See How to Collect Blood Specimen from a Vein in Chapter 1.

RATIONALE FOR THE TEST

- Screen for heart failure.
- Assess the treatment of heart failure.

NURSING IMPLICATIONS

- Assess if the patient
  - Has eaten or drunk anything except water for 12 hours before the test is administered.
  - Is taking diuretics or cardiac glycosides, which affect the test results.
  - Has myocardial infarction, kidney disease and is on dialysis, and chronic obstructive pulmonary disease (COPD).
  - Is taking Echinacea, valerian, or other herbs, since these herbs can affect the test results. The patient may be required to stop taking these herbs for several days before the test is administered.

UNDERSTANDING THE RESULTS

- Test results are available quickly. The laboratory determines normal values based on calibration of testing equipment with a control test. Test results are reported as high, normal, or low based on the laboratory's control test.
- Generally the normal range: 0 to 99 pg/mL or 0 to 99 ng/L
- Slightly high level (100 to 300 pg/mL or 100 to 300 ng/L) may indicate
  - Minimum heart failure
- Moderately high level (300 pg/mL or 300 ng/L) may indicate
  - Mild heart failure
- High level (600 pg/mL or 600 ng/L) may indicate
  - Moderate heart failure
- Very high level (900 pg/mL or 900 ng/L) may indicate
  - Severe heart failure

HINT

An abnormal level of BNP indicates there is increased pressure inside the heart and may indicate heart failure if the patient is on kidney dialysis.

TEACH THE PATIENT

- Explain
  - Why the blood sample is taken.
  - The function of BNP and the meaning of abnormal values.

- That the patient must refrain from taking Echinacea, valerian, or other herbs.
- That the patient should refrain from eating or drinking anything except water for 12 hours before the test is administered.

Cardiac Enzyme Studies 2

The cells of heart muscles and other tissues contain the enzyme creatine phosphokinase (CK, CPK) and the protein troponin (TnT, TnI). Creatine phosphokinase and troponin enter the blood when heart muscle and other tissues are damaged. If levels of creatine phosphokinase and troponin are elevated, the healthcare provider orders an ECG to differentiate between heart muscle damage and other tissue damage. Troponin and CPK-MB are mostly found in cardiac muscle.

HINT

Blood samples are taken every 12 hours for 2 days following a suspected myocardial infarction. It takes 6 hours for troponin levels to rise after a myocardial infarction. The healthcare provider may order a myoglobin test along with the cardiac enzymes test to help diagnose a myocardial infarction.

WHAT IS BEING MEASURED?

- Creatine phosphokinase and troponin levels in blood

HOW IS THE TEST PERFORMED?

- See How to Collect Blood Specimen from a Vein in Chapter 1.

RATIONALE FOR THE TEST

- Screen for
  - A myocardial infarction
  - Heart muscle injury following bypass surgery
  - Unstable angina
- Assess the results of percutaneous coronary intervention (PCI) or thrombolytic medication to restore blood flow through the coronary artery.

NURSING IMPLICATIONS

- Assess if the patient
    - Has Reye syndrome, muscular dystrophy, myocarditis, autoimmune diseases, or cardiac disease
    - Has undergone cardioversion or defibrillation
    - Has recently received an intramuscular (IM) injection
    - Has had recent surgery
    - Has undergone strenuous exercise before the test is administered
    - Is taking statins
    - Abuses alcohol
    - Has kidney failure

UNDERSTANDING THE RESULTS

- Test results are available in 1 hour. The laboratory determines normal values based on calibration of testing equipment with a control test. Test results are reported as high, normal, or low based on the laboratory's control test.
- Normal troponin range is
    - TnI: Less than 0.3 mcg/L
    - TnT: Less than 0.1 mcg/L
- Normal total creatine phosphokinase range is
    - Men: 55 to 170 IU/L
    - Women: 30 to 135 IU/L
- Normal CPK-MB range: Less than 3.0 ng/mL
- High troponin values may indicate
    - Cardiac muscle injury. Troponin level increases in 4 hours and reaches the highest level in 24 hours after the cardiac muscle injury. Troponin level returns to normal levels in 10 days following the cardiac muscle injury.
- High total creatine phosphokinase values may indicate
    - Cardiac muscle injury or tissue injury. Total creatine phosphokinase increases in 4 hours and reaches the highest level in 24 hours after the cardiac muscle injury. Total creatine phosphokinase level returns to normal levels in 3 days following the cardiac muscle injury.

- High CPK-MB values may indicate
  - Cardiac muscle injury. CPK-MB increases within 2 hours and reaches the highest level within 24 hours after the cardiac muscle injury. CPK-MB level returns to normal levels in 3 days following the cardiac muscle injury. If levels are high after 3 days, then additional cardiac muscle is being damaged and the myocardial infarction is continuing.

TEACH THE PATIENT

- Explain
  - Why the blood sample is taken.
  - That the patient should not exercise before the test is administered.
  - That the patient should not receive an IM injection before the test is administered.
  - That the healthcare provider may request that the patient stop taking statins for 2 weeks prior to the test.

Homocysteine 3

Homocysteine is an amino acid found in blood. Homocysteine levels increase along with increases in levels of cholesterol that can lead to a risk of stroke, deep venous thrombosis (DVT), myocardial infarction, and pulmonary embolism. The homocysteine test measures the level of homocysteine in blood. The healthcare provider may order the homocysteine test for a patient who has a family history of heart disease, but who has not exhibited other risk factors.

HINT

The healthcare provider may order a urine homocysteine test.

WHAT IS BEING MEASURED?

- Homocysteine level in blood

HOW IS THE TEST PERFORMED?

- Avoid eating and drinking anything except water for 12 hours before the test is administered.
- See How to Collect Blood Specimen from a Vein in Chapter 1.

RATIONALE FOR THE TEST

- Assess risk for stroke and myocardial infarction.

NURSING IMPLICATIONS

- Assess if the patient
  - Has eaten or drunk anything except water for 12 hours before the test
  - Is currently hypertensive
  - Smokes
  - Has psoriasis, leukemia, kidney disease, or homocystinuria
  - Is experiencing menopause
  - Is deficient of vitamin B_{12}, vitamin B_6, and folic acid
  - Drinks three or more cups of coffee daily
  - Is taking birth control pills, theophylline, tamoxifen, antibiotics, or anticonvulsants
  - Has a family history of elevated homocysteine
  - Is a diabetic

UNDERSTANDING THE RESULTS

- The homocysteine test results are available in 24 hours. The laboratory determines normal values based on calibration of testing equipment with a control test. Test results are reported as high, normal, or low based on the laboratory's control test.
- Normal homocysteine test results: 4 to 14 mcmol/L
- High homocysteine level may indicate
  - Homocystinuria
  - Hypothyroidism
  - Kidney disease
  - Alzheimer disease
  - Inflammatory bowel disease
  - Diet deficient in vitamin B_{12}, vitamin B_6, or folic acid
- Low homocysteine level may indicate
  - Diabetes

Hint

Homocysteine level naturally increases with age.

TEACH THE PATIENT

- Explain
  - Why the sample is taken.
  - That the patient cannot eat or drink anything except water for 12 hours before the test.
  - That the healthcare provider may ask the patient to refrain from taking birth control pills, theophylline, tamoxifen, antibiotics, or anticonvulsants 2 weeks prior to the test.

Renin Assay

Blood pressure is regulated by the renin-angiotensin system (RAS). Low blood pressure causes the secretion of the renin enzyme by the kidneys, which increases angiotensin production that constricts blood vessels, resulting in an increase in blood pressure. Angiotensin causes the adrenal cortex to replace aldosterone, causing the kidneys to retain water and sodium, resulting in an increase in fluid volume and blood pressure. The renin assay test measures the level of renin in blood.

Hint

The healthcare provider may also order the aldosterone test. The renin stimulation test may be ordered if the renin level is low.

WHAT IS BEING MEASURED?

- Renin level in blood

HOW IS THE TEST PERFORMED?

- Avoid
  - Eating natural black licorice for 2 weeks before the test
  - Ingesting caffeine 24 hours before the test
  - Eating or drinking 8 hours before the test
  - Taking β-blockers, corticostoroids, angiotensin-converting enzyme (ACE) inhibitors, estrogen, aspirin, or diuretics 4 weeks before the test

- Maintain a low-sodium diet for 3 days before the test.
- The patient must
  - Relax for 2 hours before the first blood sample is taken
  - Ambulate for 2 hours after the first blood sample is taken and before the second blood sample is taken
- The patient should be upright when the sample is taken.
- See How to Collect Blood Specimen from a Vein in Chapter 1.

RATIONALE FOR THE TEST

- Assess the underlying cause of hypertension.

NURSING IMPLICATIONS

- Assess if the patient
  - Has eaten natural black licorice 2 weeks before the test
  - Has ingested caffeine 24 hours before the test
  - Has eaten or drunk 8 hours before the test
  - Maintained a low-sodium diet 3 days before the test
  - Has taken β-blockers, corticosteroids, ACE inhibitors, estrogen, aspirin, or diuretics 4 weeks before the test
  - Has relaxed 2 hours before the first blood sample is taken
  - Has ambulated for 2 hours after the first blood sample is taken and before the second blood sample is taken
  - Is pregnant
  - Was upright when the sample was taken

UNDERSTANDING THE RESULTS

- The renin assay test results are available quickly. The laboratory determines normal values based on calibration of testing equipment with a control test. Test results are reported as high, normal, or low based on the laboratory's control test.
- Normal renin assay test results (patient in upright position)
  - Age 20 to 39 years (normal sodium diet): 0.1 to 4.3 ng/mL/h
  - Age 20 to 39 years (low-sodium diet): 2.9 to 24 ng/mL/h

    - Age 40 and older (normal sodium diet): 0.1 to 3.0 ng/mL/h
    - Age 40 and older (low-sodium diet): 2.9 to 10.8 ng/mL/h
- High renin assay test results may indicate
    - Malignant high blood pressure
    - Kidney disease
    - Blocked artery
    - Cirrhosis
    - Addison disease
    - Hemorrhage
- Low renin assay test results may indicate
    - Conn syndrome

TEACH THE PATIENT

- Explain
    - Why the sample is taken.
    - That two samples are taken—one after resting for 2 hours and another after ambulating for 2 hours.
    - That the healthcare provider may ask the patient to refrain from taking β-blockers, corticosteroids, ACE inhibitors, estrogen, aspirin, or diuretics 4 weeks prior to the test.
- Avoid
    - Eating natural black licorice for 2 weeks before the test
    - Ingesting caffeine 24 hours before the test
- Maintain a low-sodium diet for 3 days before the test.
- Avoid eating or drinking 8 hours before the test.

Summary

Healthcare providers use cardiac enzymes and cardiac markers tests to detect if a patient has experienced a myocardial infarction. Cardiac enzymes and cardiac markers tests determine the level of cardiac enzymes in the patient's blood.

Cardiac enzymes are normally contained inside cardiac muscle. In a myocardial infarction, some cardiac muscles rupture, releasing cardiac enzymes into the blood stream. It can take between 2 and 24 hours for cardiac muscle enzymes to reach a detectable level. It is during this time period when the cardiac enzymes can be detected by the cardiac enzymes and cardiac markers tests.

Healthcare providers use an ECG to diagnose the acute phase of a myocardial infarction and use the cardiac enzymes and cardiac markers tests to confirm the diagnosis.

Quiz

1. What hormone increases when the heart works harder for long periods?
   a. Brain natriuretic peptide (BNP)
   b. Creatine
   c. Phosphokinase
   d. None of the above

2. How long does it take for the troponin level to rise after a myocardial infarction?
   a. 6 days.
   b. 6 hours.
   c. Troponin levels decrease in 6 hours.
   d. 2 hours.

3. Cardiac muscle is the only source of creatine phosphokinase.
   a. True
   b. False

4. What amino acid in blood indicates a risk for myocardial infarction if its level increases in blood?
   a. Polycystin
   b. Cysteine
   c. Homocysteine
   d. None of the above

5. What must a patient do for the renin assay test?
    a. Relax for 2 hours before the first blood sample is taken.
    b. Ambulate for 2 hours after the first blood sample is taken.
    c. Sit upright when blood samples are taken.
    d. All of the above.

6. Before the homocysteine test, the patient should
    a. Avoid eating and drinking except for water 12 hours before the test is administered
    b. Walk 10 blocks before the test is administered
    c. Avoid walking
    d. Avoid driving

7. What can cause misleading results of a cardiac enzyme study?
    a. The patient has Reye syndrome.
    b. The patient recently received an intramuscular (IM) injection.
    c. The patient has undergone strenuous exercise before the test is administered.
    d. All of the above.

8. What are the enzymes mostly found in cardiac muscles?
    a. Troponin and CPK-MB
    b. CPK and creatine phosphokinase
    c. CK and creatine phosphokinase
    d. CKB and creatine phosphokinase

9. What can negatively affect the hormone BNP test?
    a. The patient has kidney dialysis.
    b. The patient is taking diuretics.
    c. The patient is taking Echinacea.
    d. All of the above.

10. Why would a recent surgery have a negative effect on the cardiac enzyme studies?
    a. The patient must be NPO (nothing by mouth) when blood sample for the cardiac enzyme studies is taken from the patient.
    b. Surgery places stress on the cardiac muscle.
    c. Surgery causes disruption of muscle, resulting in the release of enzymes contained in muscle tissue. Some of those same enzymes are contained in cardiac muscle. Therefore, those enzymes will be reported as a high level; however, the cause is not a myocardial infarction.
    d. Surgery weakens the patient's immune system.

Answers

1. a. Brain natriuretic peptide (BNP).
2. b. 6 hours.
3. b. False.
4. c. Homocysteine.
5. d. All of the above.
6. a. Avoid eating and drinking except for water 12 hours before the test is administered.
7. d. All of the above.
8. a. Troponin and CPK-MB.
9. d. All of the above.
10. c. Surgery causes disruption of muscle, resulting in the release of enzymes contained in muscle tissue. Some of the same enzymes are contained in cardiac muscle. Therefore, those enzymes will be reported as a high level; however, the cause is not a myocardial infarction.

CHAPTER 6

Serologic Tests

The presence of foreign protein in the body from a microorganism or from mismatched donated blood causes a reaction of the body's immune system. This reaction produces antibodies that destroy the foreign protein by metabolizing it into components that can be excreted safely by the body.

Serology tests examine the patient's blood serum for antibodies. Healthcare providers order serologic tests for a number of purposes. These purposes include to diagnose an infection, to determine if the patient has developed immunity to specific antigens, to determine a patient's blood type, and to determine if the patient has an autoimmune disorder.

An autoimmune disorder occurs when the patient's immune system identifies the patient's own protein as a foreign protein, resulting in the patient's immune system creating antibodies to that protein.

In this chapter you'll learn about common serologic tests.

Learning Objectives

1. Immunoglobulins
2. Antinuclear Antibodies (ANA)
3. Human Immunodeficiency Virus (HIV) Tests
4. CD4+ Count
5. Viral Load Measurement
6. Rheumatoid Factor (RF)

Key Words

Agglutination test
Autoimmune
Branched DNA (bDNA)
Enzyme-linked immunosorbent assay (ELISA)
HIV RNA
Home blood test kits
Human immunodeficiency virus
IgA
IgD
IgE
IgG
IgM
Indirect fluorescent antibody (IFA)
Macroglobulinemia
Nephelometry test
Nucleic acid sequence–based amplification (NASBA)
Polymerase chain reaction (PCR)
Rapid test kits
Reverse-transcriptase polymerase chain reaction (RT-PCR)
Saliva test
T-helper cells
Western blot

Immunoglobulins 1

Immunoglobulins are antibodies made by the immune system in response to microorganisms that enter the body, an allergen, and abnormal cells such as cancer cells. An antibody is specific to an antigen. The immunoglobulin test measures the level of an immunoglobulin in the patient's blood. A low level of a specific immunoglobulin increases the risk of repeated infections from antigen. There are five major types of immunoglobulins:

- IgA: This is found in tears and saliva and protects the ears, eyes, breathing passages, digestive tract, and vagina that are exposed to outside antigens. Body produces 10% of this type of immunoglobulin.
- IgD: This is found in abdominal and chest tissues. Body produces less than 5% of this type of immunoglobulin.
- IgE: This is found on mucous membranes, lungs, and skin and defends against allergens. A high level of IgE immunoglobulins is common in patients who are hypoallergenic. Body produces less than 5% of this type of immunoglobulin.
- IgG: This is found in all body fluids and defends the body against viruses and bacteria. This immunoglobulin crosses the placenta. Body produces 80% of this type of immunoglobulin.
- IgM: This is found in blood and lymph fluid and is the first response to infection. Body produces 5% of this type of immunoglobulin, which form when an infection occurs for the first time.

HINT

The immunoglobulin test is commonly administered as a follow-up to an abnormal result from the total blood protein test or from the blood protein electrophoresis test.

WHAT IS BEING MEASURED?

- IgA immunoglobulins
- IgD immunoglobulins
- IgE immunoglobulins
- IgG immunoglobulins
- IgM immunoglobulins

HOW IS THE TEST PERFORMED?

See How to Collect Blood Specimen from a Vein in Chapter 1.

RATIONALE FOR THE TEST

- Assess
    - Levels of immunoglobulins to determine the strengths of the patient's immune system

- The patient's immunization
- The effectiveness of treatment of infection and bone marrow cancer
- Screen for
  - Allergies
  - Autoimmune diseases
  - Multiple myeloma and macroglobulinemia cancers

NURSING IMPLICATIONS

- Assess if there is reason for the patient not to take the test:
  - Has taken hydralazine (Apresoline), phenylbutazone, birth control pills, anticonvulsants (phenytoin), methotrexate, aminophenazone, asparaginase, and corticosteroids
  - Received a blood transfusion within 6 months before the test
  - Used alcohol or illegal medication
  - Received boosters of a vaccination or new vaccinations within 6 months before the test
  - Received a radioactive scan within 3 days before the test
  - Has undergone radiation and chemotherapy treatments

UNDERSTANDING THE RESULTS

- Results are available in 2 days. The laboratory determines normal values based on calibration of testing equipment with a control test. Test results are reported as high, normal, or low based on the laboratory's control test.
- Generally normal ranges are
  - IgA: 85 to 385 mg/dL
  - IgD: Less than 8 mg/dL or 5 to 30 mcg/L
  - IgE: 10 to 1,421 mcg/L
  - IgG: 565 to 1,765 mg/dL
  - IgM: 55 to 375 mg/dL
- Abnormally high values are not used to diagnose a condition. However, they can be a sign of
  - IgA
    - Systemic lupus erythematosus (SLE)

- Rheumatoid arthritis (RA)
- Multiple myeloma
- Cirrhosis
- Chronic hepatitis
- IgD
  - Multiple myeloma
- IgE
  - Asthma
  - Allergic reaction
  - Parasitic infection
  - Atopic dermatitis
  - Multiple myeloma
- IgG
  - Chronic infection
  - Multiple sclerosis (MS)
  - Multiple myeloma
  - Chronic hepatitis
- IgM
  - Nephrotic syndrome
  - RA
  - Macroglobulinemia
  - Mononucleosis
  - Viral hepatitis
  - Parasite infection
  - Infection that started in a newborn before delivery
- Abnormally low values are not used to diagnose a condition. However, they can be a sign of
- IgA
  - Enteropathy
  - Leukemia
  - Nephrotic syndrome
  - Ataxia/telangiectasia

- IgD
    - Not significant
- IgE
    - Ataxia/telangiectasia
- IgG
    - Leukemia
    - Nephrotic syndrome
    - Macroglobulinemia
- IgM
    - Multiple myeloma
    - Leukemia

TEACH THE PATIENT

- No special preparation is needed for the test.
- Explain why the test is ordered. An immunoglobulin test can help the healthcare provider diagnose the infection.

Antinuclear Antibodies 2

In autoimmune diseases, the body produces antibodies that attach and destroy body's own cells. The antinuclear antibody (ANA) test measures the pattern and amount of these antibodies.

WHAT IS BEING MEASURED?

- ANA

HOW IS THE TEST PERFORMED?

See How to Collect Blood Specimen from a Vein in Chapter 1.

RATIONALE FOR THE TEST

- Screen for
    - RA
    - SLE

- Polymyositis
- Scleroderma

NURSING IMPLICATIONS

- No special patient preparation is necessary for the test.
- Ask the patient for a list of prescription and nonprescription medications he/she has taken. Consult the laboratory to determine if these medications might affect the test results. Medication for high blood pressure, heart disease, and tuberculosis (TB) can cause the results to be high.

UNDERSTANDING THE RESULTS

- Results are available in 1 week. The laboratory determines normal values based on calibration of testing equipment with a control test. Test results are reported as high, normal, or low based on the laboratory's control test.
- Results are reported as a titer. A titer specifies how much of the sample of blood is diluted with saline before antibodies are no longer detected. The titer is reported as a ratio of parts of the blood sample and saline. The higher the second number of the ratio, the greater the number of antibodies in the blood sample.
- Generally normal range is
  - 1:40 or less than 40
- Abnormally high values are not used to diagnose a condition. However, they can be a sign of
  - RA
  - SLE
  - Scleroderma
  - Polymyositis
  - Raynaud syndrome
  - Addison disease
  - Hashimoto thyroiditis
  - Hemolytic anemia
  - Vitamin B_{12} deficiency

- Hepatitis
- Idiopathic thrombocytopenia

Hint
An elevated result can be caused by medication for high blood pressure, heart disease, and tuberculosis and a viral infection. Older adults also show an elevated result caused by aging. A patient who has a family history of autoimmune disease may have an elevated result.

TEACH THE PATIENT

- No special preparation is needed for the test.
- Explain
  - Why the test is ordered.
  - That the test result is a sign and not a definitive diagnosis. Additional assessment must be made before a diagnosis is reached.
  - The conditions that can cause an elevated test result other than disease.

Human Immunodeficiency Virus Tests 3

The human immunodeficiency virus (HIV) infects the CD4+ white blood cells that are the body's defense against infection. There are two types of HIV. HIV-1 is common in nearly all AIDS cases. HIV-2 is associated with West Africa. HIV causes AIDS, which is incurable. There is a period when the HIV infection is not detectable in a patient. This is called the seroconversion period and also known as the window period. The patient can spread HIV during this period. The seroconversion period can be up to 2 weeks and as long as 6 months. After the seroconversion period, the HIV test is able to detect HIV antibodies or the HIV RNA in the patient's blood.

There are several HIV tests:

- Enzyme-linked immunosorbent assay (ELISA): The first test administered to screen a patient. This looks for HIV antibodies in the blood. If the test result is negative, then other tests are not performed. If the test result is positive, another ELISA test is performed. Other tests are performed if there are two positive ELISA tests because the ELISA test can produce a false positive.

- Western blot: This tests for HIV antibodies and is more difficult to perform than the ELISA test.
- Polymerase chain reaction (PCR): This looks for HIV RNA in the blood. PCR is used to identify a very recent infection and is administered to screen blood and organs before using them for donations and when HIV antibody tests are inconclusive. The PCR test is not performed often because of expense.
- Indirect fluorescent antibody (IFA): This tests for HIV antibodies and is performed secondary to two positive ELISA tests.
- Saliva test: This tests the presence of HIV in saliva. Results must be confirmed by the Western blot test.
- Rapid test kits: These test results are available in a half hour. Results must be confirmed by the Western blot test.
- Home blood test kits: Available without prescription. Samples are sent through the mail to the laboratory. Results are provided over the phone using an anonymous code.

HINT

Detecting HIV in a newborn is difficult because the newborn's blood still has HIV antibodies from either HIV positive parent for 18 months.

Testing is performed when the patient

- Exhibits symptoms of HIV.
- Has HIV risk factors.
- Is donating blood or organs.
- Is pregnant. An infected mother can be treated, decreasing the likelihood that HIV will be passed to the fetus.

WHAT IS BEING MEASURED?

- HIV antibodies in blood
- HIV RNA in blood

HOW IS THE TEST PERFORMED?

- The patient must consent to HIV testing.
- Tests are performed at 6 weeks, 3 months, and 6 months after the patient is exposed to HIV.

- The PCR test is performed days or a few weeks following exposure to HIV.
- There are no special preparations for this test.
- See How to Collect Blood Specimen from a Vein in Chapter 1.

RATIONALE FOR THE TEST

- Assess if the patient has been infected with HIV.

NURSING IMPLICATIONS

- If the patient is diagnosed with HIV, then the healthcare provider will order regularly CD4+ count and viral load measurement tests. The CD4+ count determines the impact HIV has on the immune system. The viral load measurement test determines the amount of HIV in the blood.
- Consult with the healthcare institution's policy on reporting positive test results.
- Assess
  - For use of corticosteroids since this can affect the test results.
  - If the patient has an autoimmune disease, syphilis, or leukemia. These diseases can affect the test results.
  - If the patient consumes alcohol. This can affect the test results.
- Arrange for counselors to be available to assist the patient with the emotional, financial, and social repercussions of being HIV positive should there be positive test results.

UNDERSTANDING THE RESULTS

- ELISA test results are available in 2 to 4 days. Results of other tests are available within 2 weeks.
- Normal (negative) is
  - No HIV antibodies detected.
  - The test is repeated for 6 months if the test is negative. The immune system can take up to 6 months to produce detectable HIV antibodies.
- Uncertain (indeterminate) is
  - Results are not clear if the patient is infected with HIV because another factor is possibly interfering with the results.

- The PCR test is typically ordered.
- After 6 months, the patient is said to be stable indeterminate and not infected with HIV.

- Abnormal (positive) is
  - HIV antibodies are found in the patient's blood.
- If the ELISA test is positive, the same blood sample is tested again. If the second ELISA test is positive, then the Western blot or IFA test is performed.
- A patient is considered HIV positive if the Western blot, IFA, or PCR tests are positive, confirming two positive ELISA test results.

TEACH THE PATIENT

- Explain
  - That HIV testing cannot be performed without the patient's consent.
  - Why blood sample is taken. HIV testing is part of routine blood screening.
  - That HIV testing does not determine if the person has AIDS.
  - That a negative test results will require retesting at 6 weeks, 3 months, and 6 months after the patient is exposed to HIV because of the seroconversion period.
  - That it is difficult to detect HIV in a newborn for 18 months because the newborn has antibodies from the mother.
  - That a positive ELISA test result does not mean that the patient is HIV positive. Additional testing is necessary to confirm this result.
  - That counselors will be available to assist the patient with the emotional, financial, and social repercussions of being HIV positive.
  - The benefits of early detection and treatment.
  - That treatment can be given to an infected mother, decreasing the likelihood that HIV will be passed on to the fetus.

CD4+ Count 4

Three types of leukocytes (WBC) important in fighting infection are T-lymphocytes, T-cells, and T-helper cells. The CD4+ count test measures the level of these leukocytes to assess the patient's immune system. Patients who have a low CD4+ count are at risk for opportunistic infections.

HINT

The CD4+ count test is used to assess the immune system of patients who have HIV. The result of the CD4+ count test and the viral load test determines when antiretroviral treatment of HIV is to be started. The healthcare provider may also order the CD4+ percentage test that determines the percentage of CD4+ cells in the total number of lymphocytes and the CD8 count test, which determines the level of T-suppressor cells.

WHAT IS BEING MEASURED?

- The CD4+ cells level in blood

HOW IS THE TEST PERFORMED?

- See How to Collect Blood Specimen from a Vein in Chapter 1.

RATIONALE FOR THE TEST

- Assess
  - The patient's immune system
  - The progress of HIV
  - The treatment of HIV
- Assist in the diagnosis of AIDS.
- Develop a baseline of CD4+ count. The healthcare provider compares CD4+ values over time to identify changes in levels.

NURSING IMPLICATIONS

- Note the time when the CD4+ blood sample is taken since CD4+ cells are normally lower in the morning.
- Be sure that the blood sample is refrigerated.
- Assess if the patient has influenza, pneumonia, or herpes simplex since these diseases can affect the test results.
- Assess if the patient has taken corticosteroids or is undergoing chemotherapy.

UNDERSTANDING THE RESULTS

- Test results are available in 3 days.
- Normal CD4+ cell count range is
  - Not infected with HIV: 600 to 1,200 cells/μL
  - Low risk for opportunistic infection: Greater than 350 cells/μL
- Low CD4+ cell count may indicate
  - 200 to 350 cells/μL
    - Weak immune system
    - Risk for opportunistic infection
    - Start antiretroviral treatment
  - Under 200 cells/μL
    - High risk for opportunistic infection
    - AIDS
    - Start antiretroviral treatment

TEACH THE PATIENT

- Explain
  - Why blood sample is taken.
  - That the healthcare provider may request that the patient stop taking corticosteroids for 2 weeks prior to the test.

Viral Load Measurement 5

The HIV in a patient's blood is determined by presence of HIV RNA. The amount of HIV RNA indicates if the infection is decreasing, stabilized, or increasing. The viral load measurement test determines the amount of HIV RNA in the patient's blood. A viral load measurement test is administered when the patient is diagnosed with HIV and becomes the baseline. Results of subsequent viral load measurement tests are compared to the baseline to determine the infection's progress.

There are three types of viral load measurement tests:

- Branched DNA (bDNA) test
- Nucleic acid sequence–based amplification (NASBA) test
- Reverse-transcriptase polymerase chain reaction (RT-PCR) test

---

HINT

Although the viral load measurement test is more accurate, the healthcare provider may also order the CD4+ count test, which assesses the progress of the HIV infection. The viral load measurement test is not used to diagnose HIV.

CAUTION

A patient diagnosed with HIV who has a negative viral load measurement test result can infect another person.

WHAT IS BEING MEASURED?

- HIV RNA in blood

HOW IS THE TEST PERFORMED?

- Baseline every 4 months if the patient is not receiving highly active antiretroviral therapy (HAART).
- Baseline 8 weeks after the start of HAART, and then every 4 months if the patient is receiving HAART.
- See How to Collect Blood Specimen from a Vein in Chapter 1.

RATIONALE FOR THE TEST

- Assess treatment of HIV infection.
- Assess progress of HIV infection.

NURSING IMPLICATIONS

- Assess if the patient has
  - Another infection
  - Received immunizations recently

UNDERSTANDING THE RESULTS

- The viral load measurement test results take 2 weeks.
- Normal viral load measurement test result is (negative): No HIV RNA is found in blood.
- Abnormal viral load measurement test result: HIV RNA is found and is reported as copies per milliliter where a copy is an HIV RNA.
- The viral load measurement test result is compared to the baseline test result and to previous tests results to determine the trend of the infection.

TEACH THE PATIENT

- Explain
  - Why the sample is taken.
  - That a negative test result does not mean that the patient is cured. The patient can still spread the infection.

Rheumatoid Factor

The rheumatoid factor (RF) is an autoantibody that destroys the patient's own tissues, resulting in stiffness, joint pain, and inflammation. The RF test measures the amount of the RF in the blood sample and is used to differentiate rheumatoid arthritis (RA) from other forms of arthritis. The two types of RF tests are the agglutination test and the nephelometry test.

Hint

A patient may have a high level of RF but no symptoms of RA. However, the patient is likely to develop RA in the future. A patient may have a normal level of RF and symptoms of RA will likely require a second RF test.

Caution

The RF is one of other signs and symptoms used to diagnose RA.

WHAT IS BEING MEASURED?

- Measures the RF in blood

HOW IS THE TEST PERFORMED?

- See How to Collect Blood Specimen from a Vein in Chapter 1.

RATIONALE FOR THE TEST

- Screen for RA.

NURSING IMPLICATIONS

- Assess the patient's age. Patients older than 65 years of age normally have a higher RF level.

UNDERSTANDING THE RESULTS

- The RF test results are available quickly.
- Results are reported as a titer. A titer specifies how much of the sample of blood is diluted with saline before antibodies are no longer detected. The titer is reported as a ratio of parts of the blood sample and saline. The higher the second number of the ratio, the greater number of antibodies in the blood sample.
- Normal test results: 1:20 to 1:40 or less.
- Higher levels may indicate
  - RA
  - SLE
  - Vasculitis
  - Scleroderma
  - Cirrhosis
  - Hepatitis
  - Leukemia
  - Endocarditis
  - Mononucleosis
  - Malaria
  - Tuberculosis
  - Syphilis

TEACH THE PATIENT

- Explain
  - Why the sample is taken.
  - That the healthcare provider may ask for a second RF test if the first test is normal and the patient shows signs of RA.

Summary

When a foreign protein enters the body, the body's immune system is alerted and goes on the defensive by creating antibodies that seek out the foreign protein. Once found, the antibody kills the foreign protein by metabolizing it into components that can be safely excreted by the body.

Once the antibody is produced, it remains in the body for a long time, providing protection against the related foreign protein. The patient is considered immune to that foreign protein. In this chapter, you learned that healthcare providers order serology tests to examine antibodies in the patient's blood serum.

The presence of an antibody can mean that the patient is currently infected with foreign protein or has been exposed to the foreign protein in the past either through an invasion by a microorganism or from a vaccination. The serology test is used to determine the status of a patient's immunization and can be used to determine if the patient has an autoimmune disorder.

Quiz

1. What are immunoglobulins?
   a. Antibodies by the immune system in response to a microorganism
   b. Antibodies by the immune system in response to an allergen
   c. Antibodies by the immune system in response to abnormal cells
   d. All of the above

2. What is IgM?
   a. An immunoglobulin that forms when an infection occurs for the first time
   b. An immunoglobulin that crosses the placenta
   c. An immunoglobulin found on mucous membranes
   d. An immunoglobulin found in tears

3. What can negatively affect the immunoglobulin test?
   a. Recent blood transfusion
   b. Radioactive scan 3 days before the test is administered
   c. Recent vaccination
   d. All of the above

4. What does the antinuclear antibodies (ANA) test measure?
   a. Antibodies that destroy bacteria
   b. Antibodies that destroy the body's own cells
   c. Antibodies that destroy viruses
   d. Antibodies that destroy microorganisms

5. What test can be used to screen for rheumatoid arthritis (RA)?
   a. Immunoglobulin test
   b. IgM test
   c. ANA test
   d. CD4+ count test

6. What does the CD4+ count test measure?
   a. T-lymphocytes
   b. T-cells
   c. T-helper cells
   d. All of the above

7. Why would a healthcare provider order the CD4+ count test?
   a. To assess the patient's immune system
   b. To assist in the diagnosis of AIDS
   c. To assess the treatment of HIV
   d. All of the above

8. Why is the viral load measurement test administered?
   a. To confirm AIDS
   b. To determine if HIV RNA is decreasing, stabilized, or increasing
   c. To confirm HIV
   d. None of the above

9. What does the rheumatoid factor (RF) test measure?
   a. The presence of the autoantibody rheumatoid factor
   b. The presence of the rheumatoid antigen
   c. The presence of the rheumatoid bacteria
   d. The presence of the rheumatoid virus

10. What happens if the ELISA test is positive?
    a. The patient has a confirmed diagnosis of HIV.
    b. The patient has a confirmed diagnosis of AIDS.
    c. A second ELISA test is administered.
    d. None of the above.

Answers

1. d. All of the above.
2. a. An immunoglobulin that forms when an infection occurs for the first time.
3. d. All of the above.
4. b. Antibodies that destroy the body's own cells.
5. c. ANA test.

6. d. All of the above.
7. d. All of the above.
8. b. To determine if HIV RNA is decreasing, stabilized, or increasing.
9. a. The presence of the autoantibody rheumatoid factor.
10. c. A second ELISA test is administered.

CHAPTER 7

Endocrine Tests

The endocrine system sends hormones via blood vessels to regulate bodily functions, including metabolism, growth, mood, and tissue function. Hormones are created, stored, and released by glands and act as messengers, signaling other glands and organs to react in a specific manner.

Hormones are released based on existing hormone levels in the blood in order to keep hormonal levels in balance. For example, an excess amount of a hormone may cause the release of a different hormone that causes the gland to stop or reduce excretion of the hormone, thereby bringing hormones in balance.

Diseases can dysregulate the release of hormones, resulting in under- or overproduction of one or more hormones. Endocrine tests are administered to assess if the patient is experiencing an endocrine disease. You'll learn about endocrine tests in this chapter.

Learning Objectives

1. Adrenocorticotropic Hormone (ACTH) and Cortisol
2. Overnight Dexamethasone Suppression Test

3 Aldosterone in Blood
4 Cortisol in Blood
5 Estrogens
6 Growth Hormone
7 Luteinizing Hormone (LH)
8 Parathyroid Hormone (PTH)
9 Thyroid Hormone Tests
10 Thyroid-Stimulating Hormone (TSH)
11 Testosterone

Key Words

Addison disease
Adrenal gland tumor
Adrenal glands
Aldactone
α-Fetoprotein (AFP)
Corticotropin-releasing hormone (CRH)
Cushing disease
Diabetes
Estradiol
Estriol
Estrone
Human chorionic gonadotropin (hCG)
Hyperaldosteronism
Hyperthyroidism
Hypoglycemic
Hypothyroidism
Inferior petrosal sinus
Insulinlike growth factor 1 (IGF-1)
Pituitary gland
Renin
Sheehan syndrome
Thyroxine (T4)
Triiodothyronine (T3)

Adrenocorticotropic Hormone and Cortisol 1

The hypothalamus releases corticotropin-releasing hormone (CRH), which causes the pituitary gland to release the adrenocorticotropic hormone (ACTH). ACTH causes the adrenal gland to release cortisol. Cortisol increases blood pressure and glucose and reduces the immune response. The ACTH test measures the level of

ACTH in the blood. ACTH levels fall when cortisol levels rise and ACTH levels rise when cortisol levels fall.

HINT
The healthcare provider may request that a sample be taken from the inferior petrosal sinus near the pituitary gland to determine if the pituitary gland is producing ACTH or if ACTH is being produced elsewhere in the patient's body.

WHAT IS BEING MEASURED?

- The level of ACTH in the patient's blood
- The level of cortisol in the patient's blood

HOW IS THE TEST PERFORMED?

- A blood specimen is taken either in the morning if the healthcare provider wants a peak level or in the evening if a trough level is required. ACTH is highest between 6 AM and 8 AM and lowest between 6 PM and 11 PM.
- See How to Collect Blood Specimen from a Vein in Chapter 1.
- The sample must be placed in a special test tube, placed on ice and immediately processed.
- Note on the test tube the day and time when the sample was taken.

RATIONALE FOR THE TEST

- Assess
    - The pituitary gland function
    - The adrenal gland function

NURSING IMPLICATIONS

- Assess the patient for conditions that might affect the test results
    - Intoxication
    - Pregnancy
    - Menstruation
    - Emotional or physical stress or severe injury

    - Undergoing a radioactive tracer medical test the previous week before the ACTH test
    - Taking corticosteroids acts like cortisol
    - Taking amphetamines, insulin, and lithium carbonate, which causes the release of cortisol
- Assess whether or not the patient
    - Ate or drank 12 hours before the test
    - Has exercised within the previous 12 hours
- The patient may experience bruising if either the ACTH or cortisol levels are elevated.

UNDERSTANDING THE RESULTS

- The result is available in 6 days. The laboratory determines normal values based on calibration of testing equipment with a control test. Test results are reported as high, normal, or low based on the laboratory's control test.
- Generally normal ranges for ACTH levels are
    - 6 AM to 8 AM: Less than 80 pg/mL or less than 18 pmol/L
    - 6 PM to 11 PM: Less than 50 pg/mL or less than 11 pmol/L
- Generally normal ranges for cortisol levels are
    - 6 AM to 8 AM: 5 to 23 mcg/dL or 138 to 635 nmol/L
    - 6 PM to 11 PM: 3 to 13 mcg/dL or 83 to 359 nmol/L
- High levels may indicate
    - Physical or emotional stress
    - Addison disease
    - Cushing disease
    - Adrenal gland tumor
    - Pituitary gland tumor
- Low levels might indicate
    - Head injury
    - Stroke
    - Pituitary gland damaged from radiation or surgery
    - Pituitary gland tumor
    - Adrenal gland tumor—Cushing syndrome (increased release of cortisol)

- Comparing levels of ACTH and cortisol can give a sign for the following diseases/conditions
  - High ACTH high cortisol: ACTH made outside the pituitary gland; Cushing disease
  - Low ACTH high cortisol: Adrenal tumor (Cushing syndrome)
  - High ACTH low cortisol: Addison disease
  - Low ACTH low cortisol: Hypopituitarism

Hint

Levels of ACTH vary by the minute, making test results difficult to interpret.

TEACH THE PATIENT

- Don't exercise 12 hours before the test.
- Avoid stressful situations 12 hours before the test.
- Explain
  - That there might be bruising if the ACTH and cortisol levels are high.
  - That the test cannot be taken if the patient is pregnant, menstruating, or intoxicated or if the patient has taken corticosteroids, amphetamines, insulin, and lithium carbonate. If the patient is taking these medications, then the patient must stop taking them for 2 days before the test.
  - When and why blood samples are taken.
  - That the healthcare provider may ask the patient not to eat or drink 12 hours before the test. The patient may be permitted to eat low-carbohydrate foods for 2 days before the test.

Overnight Dexamethasone Suppression Test 2

The pituitary gland releases ACTH whenever there is a low level of cortisol in the blood. ACTH signals the adrenal glands to release cortisol. An adrenal gland tumor causes the release of cortisol in the absence of ACTH.

Dexamethasone is medication similar to cortisol in that dexamethasone signals the pituitary gland that there is a high level of cortisol in the blood, causing the pituitary gland to suppress the release of ACTH.

The overnight dexamethasone suppression test examines the patient's cortisol level after dexamethasone has been administered. The cortisol level should be lower because there is no ACTH in the blood to signal the adrenal glands to release cortisol. If cortisol levels remain high, then this is a sign of Cushing syndrome as a result of an adrenal gland tumor, which is producing cortisol.

WHAT IS BEING MEASURED?

- The level of ACTH in the patient's blood
- The level of cortisol in the patient's blood

HOW IS THE TEST PERFORMED?

- Dexamethasone is administered at 11 PM the night before the test.
- Administer dexamethasone with antacid or milk to reduce the risk of heartburn or stomach upset.
- Patient is not permitted to eat or drink anything 12 hours before the specimen is drawn.
- A blood specimen is taken at 8 AM.
- See How to Collect Blood Specimen from a Vein in Chapter 1.

RATIONALE FOR THE TEST

- Assess
  - The adrenal gland function
  - For Cushing syndrome

NURSING IMPLICATIONS

- Assess the patient for conditions that might affect the test results by preventing a decrease in cortisol levels
  - Severe obesity
  - Pregnancy
  - Dehydration
  - Uncontrolled diabetes
  - Severe weight loss

- Acute alcohol withdrawal
- Severe injury
- Stress
- Taking birth control pills, phenytoin (Dilantin), monoamine oxidase inhibitors (MAOIs), lithium, morphine, aspirin, barbiturates, methadone, diuretics, or Aldactone

- Assess the patient's metabolism. Some patients metabolize dexamethasone quickly, resulting in a lower than expected blood level when the blood specimen is taken. In this case, the healthcare provider will increase the dose of dexamethasone. The patient may be given up to 8 doses of dexamethasone over 2 days.
- The patient should refrain from taking birth control pills, phenytoin (Dilantin), MAOIs, lithium, morphine, aspirin, barbiturates, methadone, diuretics or Aldactone for 48 hours before the blood specimen is drawn.
- A patient with high ACTH and cortisol levels may experience bruising.

UNDERSTANDING THE RESULTS

- The result is available in 3 days. The laboratory determines normal values based on calibration of testing equipment with a control test. Test results are reported as high, normal, or low based on the laboratory's control test.
- Generally normal ranges for cortisol levels
    - Less than 5 mcg/dL or less than 150 nmol/L
- High levels may indicate
    - Cushing syndrome
    - Hyperthyroidism
    - Uncontrolled diabetes
    - Heart failure
    - Myocardial infarction
    - Poor diet
    - Fever
    - Alcoholism
    - Depression
    - Anorexia nervosa
    - Lung cancer

Hint

Further testing is indicated if there is a high cortisol level. Also, a normal result does not rule out Cushing syndrome. The healthcare provider may prefer conducting a 24-hour urine free cortisol test, since this is not influenced by the timing of the blood specimen.

TEACH THE PATIENT

- Don't exercise 12 hours before the test.
- Avoid stressful situations 12 hours before the test.
- Explain
  - That there might be bruising if the ACTH and cortisol levels are high.
  - That the test cannot be administered if the patient has taken birth control pills, phenytoin (Dilantin), MAOIs, lithium, morphine, aspirin, barbiturates, methadone, diuretics, or Aldactone. If the patient is taking these medications, he/she must stop taking them for 2 days before the test.
  - When and why the blood sample is taken.
- Don't eat or drink 12 hours before the test.

Aldosterone in Blood 3

Kidneys release renin. Renin is a hormone that signals the adrenal glands to release aldosterone to control blood pressure and fluids and electrolyte balance by retaining fluid and sodium. The aldosterone test determines the level of aldosterone in the blood. This test is typically performed with the renin activity test.

WHAT IS BEING MEASURED?

- Aldosterone level in blood

HOW IS THE TEST PERFORMED?

- See How to Collect Blood Specimen from a Vein in Chapter 1.

Hint

The healthcare provider may choose to use a 24-hour urine test in place of measuring aldosterone levels in the blood if the patient is unable to change position.

RATIONALE FOR THE TEST

- Assess
  - For an adrenal gland tumor
  - The cause of high blood pressure
  - The cause of a low potassium level in the blood
  - For an overactive adrenal gland

NURSING IMPLICATIONS

- The aldosterone test is administered as part of routine blood screening.
- There is no special preparation required for this test.
- If the test result is abnormal, the healthcare provider may order subsequent aldosterone tests. For subsequent aldosterone tests, the patient should
  - Refrain from eating natural black licorice 2 weeks prior to the test.
  - Eat 3 g/d of sodium for 2 weeks prior to the test.
  - Avoid a low-salt diet, since this increases aldosterone levels.
  - Avoid a high-salt diet, since this decreases aldosterone levels. Do not eat olives, soy sauces, pretzels, potato chips, canned soups, bacon, and vegetables.
  - Might be asked to stand, sit, or lie when the test is administered, since the patient's position may affect the test results. Standing or sitting for 2 hours before the test increases the aldosterone level.
- Assess
  - Prescription and nonprescription medication that the patient takes. Diuretics, estrogen, corticosteroids, progesterone, opiates, nonsteroidal anti-inflammatory drugs (NSAIDs), heparin, and Aldactone affect the test and should be stopped 2 weeks before the test.
  - If the patient is in the third trimester of pregnancy, since aldosterone level can be high at this stage of pregnancy.

- If the patient is under emotional stress or experiences strenuous exercise prior to the test. This can affect the test results.
- The patient's age, since aldosterone level decreases with age.

HINT

A patient who has muscle cramps, muscle weakness, tingling in the hands, or high blood pressure may have a high level of aldosterone.

UNDERSTANDING THE RESULTS

- Test results are available between 2 and 5 days. The laboratory determines normal values based on calibration of testing equipment with a control test. Test results are reported as high, normal, or low based on the laboratory's control test.
- The healthcare provider will likely order subsequent aldosterone tests if there is an abnormal result.
- Generally the normal range is
  - Lying down: 3 to 10 ng/dL or 0.08 to 0.3 nmol/L
  - Standing or sitting: 5 to 30 ng/dL or 0.14 to 0.8 nmol/L
  - Child 1 year and older: 5 to 80 ng/dL or 0.14 to 2.13 nmol/L
  - Child younger than 1 year: 1 to 160 ng/dL or 0.03 to 2.26 nmol/L
- High level may indicate
  - Primary hyperaldosteronism directly affects the adrenal glands—renin activity test is low.
    - Adrenal hyperplasia
    - Conn syndrome/adrenal gland tumor
  - Secondary hyperaldosteronism does not directly affect the adrenal glands—renin activity test is high.
    - Cirrhosis
    - Heart failure
    - Kidney disease
    - Preeclampsia
- Low level may indicate
  - Addison disease

HINT

The healthcare provider may order a potassium blood test if there is abnormal adrenal growth or if overactive adrenal glands are suspected as the cause of an abnormal aldosterone test result.

TEACH THE PATIENT

- Explain
  - Why blood sample is taken.
  - The function of aldosterone and the meaning of abnormal values.
  - That the patient may be asked to stand, sit, or lie during the test.
  - That the healthcare provider may ask the patient to stop taking diuretics, estrogen, corticosteroids, progesterone, opiates, NSAIDs, heparin, or Aldactone 2 weeks prior to the test, since these medications affect the test results.
- Ask the patient to refrain from
  - Eating natural black licorice 2 weeks prior to the test.
  - Keeping a low- or high-salt diet.

Cortisol in Blood 4

Cortisol, produced by the adrenal glands, is a hormone that causes an increase in blood pressure and glucose while decreasing the immune response. Cortisol levels reach the highest at 7 AM and the lowest 3 hours after sleep based on the diurnal rhythm. However, the diurnal rhythm reverses if the patient works at night and sleeps during the day.

The pituitary gland releases ACTH whenever there is a low level of cortisol in the blood. ACTH signals the adrenal glands to release cortisol.

HINT

The healthcare provider may also order the dexamethasone suppression test, adrenocorticotropic hormone test, or a 24-hour urine test.

WHAT IS BEING MEASURED?

- Cortisol level in the blood

HOW IS THE TEST PERFORMED?

- A blood sample is taken in the morning to measure peak level.
- See How to Collect Blood Specimen from a Vein in Chapter 1.

RATIONALE FOR THE TEST

- Assess the function of the
  - Pituitary gland
  - Adrenal glands

NURSING IMPLICATIONS

- Ask the patient to lie down for 30 minutes before administering the test to reduce stress levels.
- Assess if the patient
  - Has had strenuous physical activity 24 hours before the test is administered
  - Is pregnant, since this test can increase cortisol levels
  - Is hypoglycemic
  - Has had a radioactive scan a week before the test
  - Has taken amphetamines, estrogen, birth control pills, or corticosteroids, since these medications affect the test results
  - Has eaten or drunk before the test, since this can affect the test results

UNDERSTANDING THE RESULTS

- Test results are available quickly. The laboratory determines normal values based on calibration of testing equipment with a control test. Test results are reported as high, normal, or low based on the laboratory's control test.
- Normal range
  - Adults (morning): 5 to 23 mcg/dL or 138 to 635 nmol/L
  - Adults (afternoon): 3 to 13 mcg/dL or 83 to 359 nmol/L
  - Children (morning): 3 to 21 mcg/dL or 83 to 580 nmol/L
  - Children (afternoon): 3 to 10 mcg/dL or 83 to 276 nmol/L
  - Newborn: 1 to 24 mcg/dL or 27 to 663 nmol/L

- High level values may indicate
  - Adrenal gland tumor
  - Overactive adrenal gland
  - Cushing syndrome
  - Cushing disease
  - Adenoma
  - Depression
  - Severe liver disease
  - Obesity
  - Hyperthyroidism
  - Severe kidney disease
  - Surgery
  - Sepsis
- Low level values may indicate
  - Addison disease
  - Shock
  - Sheehan syndrome
  - Autoimmune disease

TEACH THE PATIENT

- Explain
  - Why blood sample is taken.
  - That the healthcare provider may ask the patient to stop taking amphetamines, estrogen, birth control pills, or corticosteroids 2 weeks prior to the test, since these medications affect the test results.
  - That the healthcare provider may ask the patient not to eat or drink 12 hours before the test.

Estrogens 5

Estrogen is a hormone produced in the ovaries, placenta, muscle tissue, adipose tissue, adrenal glands, and testicles in men. There are three types of estrogen hormones:

- Estradiol: This is an estrogen found in nonpregnant women that varies with the menstrual cycle.
- Estriol: This is an estrogen that is produced by the placenta and is measured in pregnant women who are in at least the ninth week of pregnancy.
- Estrone: This is an estrogen measured in women who have finished menopause and in both men and women suggestive of having testicular cancer, ovarian cancer, or an adrenal gland tumor.

HINT

Estrogen levels can also be measured in urine. The healthcare provider may order a triple test or quad marker screen to assess for fetal birth defects. The maternal serum triple test measures levels of estrogen, α-fetoprotein (AFP), and human chorionic gonadotropin (hCG). The quad marker screen measures the same as the triple test and also tests inhibin A hormone.

WHAT IS BEING MEASURED?

- Estrogen levels in blood

HOW IS THE TEST PERFORMED?

- See How to Collect Blood Specimen from a Vein in Chapter 1.

RATIONALE FOR THE TEST

- Assess
  - The effect of fertility therapy
  - For abnormal sexual characteristics in men
  - For estrogen-producing tumors
  - For fetal birth defects

NURSING IMPLICATIONS

- Assess
  - The stage of the patient's menstrual cycle
  - If the patient is pregnant
  - If the patient has completed menopause

- If the patient is undergoing hormone replacement therapy
- Type of birth control that is used by the patient
- If the patient has undergone a radioactive scan in the week prior to administering the test
- If the patient has taken prednisone, Clomid, or Serophene, since these medications might affect the test result

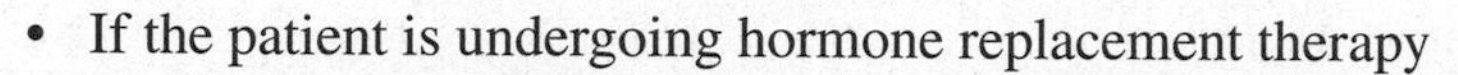

UNDERSTANDING THE RESULTS

- Estrogen test results are available in 1 day. The laboratory determines normal values based on calibration of testing equipment with a control test. Test results are reported as high, normal, or low based on the laboratory's control test.
- Normal estradiol range
  - Before menopause: 30 to 400 pg/mL
  - After menopause: 0 to 30 pg/mL
  - Men: 10 to 50 pg/mL
  - Children: 0 to 15 pg/mL
- Normal estriol range in pregnant women
  - First trimester: Less than 38 ng/mL
  - Second trimester: 38 to 140 ng/mL
  - Third trimester: 31 to 460 ng/mL
- High estrogen values may indicate
  - Cirrhosis
  - Multiple fetuses
  - Early puberty
  - Infertility treatment is working
  - Ovarian cancer
  - Testicular cancer
  - Adrenal gland tumor
- Low estrogen values may indicate
  - Turner syndrome
  - Anorexia nervosa
  - Dysfunctional pituitary gland

- Dysfunctional placenta
- Fetal disorder

TEACH THE PATIENT

- Explain
    - Why blood sample is taken.
    - That the healthcare provider may ask the patient to stop taking prednisone, Clomid, or Serophene for 2 weeks prior to the test.

Growth Hormone 6

The human growth hormone (HGH), secreted by the pituitary gland, stimulates cell growth and reproduction and insulinlike growth factor 1 (IGF-1). The growth hormone test measures the level of growth hormone in blood.

HINT

The healthcare provider may order the IGF-1 test, the growth hormone (GH) suppression test (glucose loading test), and the GH stimulation test (insulin tolerance test) along with the GH test.

WHAT IS BEING MEASURED?

- Growth hormone level in blood

HOW IS THE TEST PERFORMED?

- Avoid eating and drinking for 10 hours before the test.
- Avoid exercising 10 hours before the test.
- Rest for 30 minutes before the test.
- See How to Collect Blood Specimen from a Vein in Chapter 1.

RATIONALE FOR THE TEST

- Assess for
    - Abnormal growth

- Treatment of abnormal growth
- A pituitary gland tumor

NURSING IMPLICATIONS

- Assess if the patient
  - Has eaten or drunk 10 hours before the test
  - Has exercised 10 hours before the test
  - Rested for 30 minutes before the test
  - Is taking St. John's wort, insulin, estrogen, amphetamines, birth control pills, or corticosteroids, since these medications can affect the test result
  - Is hypoglycemic
  - Is obese

UNDERSTANDING THE RESULTS

- Normal growth hormone test results are available quickly. The laboratory determines normal values based on calibration of testing equipment with a control test. Test results are reported as high, normal, or low based on the laboratory's control test.
- Normal growth hormone level range
  - Men: Less than 5 ng/mL
  - Women: Less than 10 ng/mL
  - Children: 0 to 10 ng/mL
  - Newborn: 10 to 40 ng/mL
- High growth hormone levels may indicate
  - Acromegaly
  - Gigantism
  - Pituitary gland tumor
  - Starvation
  - Diabetes
  - Kidney disease
- Low growth hormone levels may indicate
  - Dwarfism

- Hypopituitarism
- Sarcoidosis

TEACH THE PATIENT

- Explain
  - Why sample is taken.
  - That the patient must avoid eating and drinking for 10 hours before the test.
  - That the patient must avoid exercising for 10 hours before the test.
  - That the patient must rest for 30 minutes before the test.
  - That the healthcare provider may want the patient to refrain from taking St. John's wort, insulin, estrogen, amphetamines, birth control pills, or corticosteroids 2 weeks before the test.

Luteinizing Hormone 7

Luteinizing hormone (LH) is produced by the pituitary gland that stimulates production of testosterone and ovulation and regulates the menstrual cycle.

HINT

Home ovulation testing kits detect LH levels in urine.

WHAT IS BEING MEASURED?

- Luteinizing level in blood

HOW IS THE TEST PERFORMED?

- The healthcare provider may order several samples taken on 1 day or samples taken over several days.
- See How to Collect Blood Specimen from a Vein in Chapter 1.

RATIONALE FOR THE TEST

- Assess for
  - Underlying cause of infertility
  - Infertility treatment

- Underlying cause of irregular menstrual period or amenorrhea
- Menopause
- Precocious puberty and delayed puberty
- Erectile dysfunction

NURSING IMPLICATIONS

- Assess
  - First day of the patient's last menstrual period
  - If the patient experiences light or heavy bleeding on the first day of the menstrual period
  - If the patient has hyperthyroidism
  - If the patient has undergone a radioactive tracer test 1 week prior to the test
  - If the patient is taking phenothiazine, cimetidine, clomiphene, spironolactone, digitalis, naloxone, anticonvulsants, levodopa, or birth control pills a month before the test is administered
  - If the patient has liver disease

HINT

Age, menstrual cycle, and stage of sexual development affect the test results.

UNDERSTANDING THE RESULTS

- The LH test results are available quickly. The laboratory determines normal values based on calibration of testing equipment with a control test. Test results are reported as high, normal, or low based on the laboratory's control test.
- Normal LH test results
  - Menstruating
    - Follicular phase: 1 to 18 IU/L
    - Midcycle phase: 8.7 to 80 IU/L
    - Luteal phase: 0.5 to 18 IU/L
    - Postmenopause: 12 to 55 IU/L
  - Men: 1 to 9 IU/L
  - Before puberty: 0 to 1 IU/L

    - Male puberty: 0.4 to 7 IU/L
    - Female puberty: 0.4 to 12 IU/L
- High LH level may indicate
    - Women
        - Polycystic ovary syndrome
        - Early puberty
        - No ovaries
    - Men
        - Klinefelter syndrome
        - No testicles
        - Malfunctioning testicles
- Low LH level may indicate
    - Malfunctioning pituitary gland
    - Malfunctioning hypothalamus
    - Anorexia nervosa
    - Underweight
    - That the patient is stressed

TEACH THE PATIENT

- Explain
    - Why the sample is taken.
    - That the healthcare provider may order several samples taken on 1 day or samples taken over several days.
    - That the healthcare provider may ask the patient to refrain from taking phenothiazine, cimetidine, clomiphene, spironolactone, digitalis, naloxone, anticonvulsants, levodopa, or birth control pills a month before the test is administered.

Parathyroid Hormone 8

The parathyroid glands release parathyroid hormone (PTH) when calcium level in the blood is low, causing the kidneys to retain calcium and bone to release calcium into the blood. PTH converts vitamin D to an active form, resulting in an increase

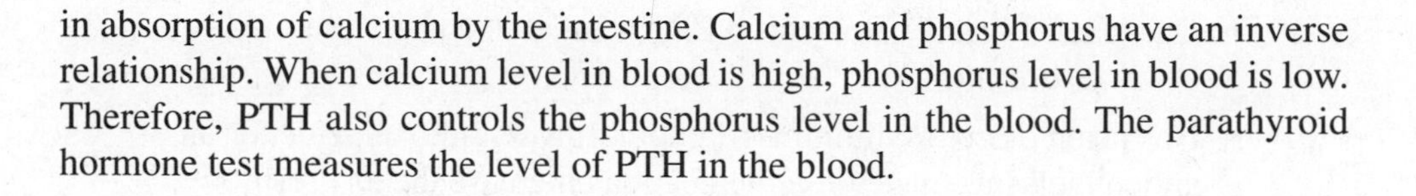

in absorption of calcium by the intestine. Calcium and phosphorus have an inverse relationship. When calcium level in blood is high, phosphorus level in blood is low. Therefore, PTH also controls the phosphorus level in the blood. The parathyroid hormone test measures the level of PTH in the blood.

Hint

The healthcare provider may also order tests for calcium and phosphorus levels in the blood and creatinine tests to assess kidney function.

WHAT IS BEING MEASURED?

- Parathyroid hormone level in blood

HOW IS THE TEST PERFORMED?

- Avoid eating or drinking anything except water 10 hours before the test.
- Take the blood sample shortly after the patient awakens.
- See How to Collect Blood Specimen from a Vein in Chapter 1.

RATIONALE FOR THE TEST

- Assess for hyperparathyroidism.
- Assess the underlying cause of abnormal calcium level in blood.

NURSING IMPLICATIONS

- Assess if the patient
  - Works days or nights to determine the patient's sleep habits
  - Has high cholesterol levels
  - Has high triglyceride levels
  - Is pregnant
  - Is breast-feeding
  - Has eaten or drunk anything except water 10 hours before the test
  - Has taken calcium supplements or is on a high-calcium diet
  - Has taken phosphorus supplements or is on a high-phosphorus diet
  - Has undergone a radioactive tracer test in a week prior to the test

- Has taken Tagamet, Betachron ER, or Inderal, since these medications lower test results
- Has taken diuretics, lithium, rifampin, furosemide, thiazide, or an anticonvulsant, since these medications increase the test results

UNDERSTANDING THE RESULTS

- The PTH test results are available in 2 days. The laboratory determines normal values based on calibration of testing equipment with a control test. Test results are reported as high, normal, or low based on the laboratory's control test.
- Normal PTH test results: 10 to 65 pg/mL
- High PTH test results may indicate
  - Hyperplasia
  - Parathyroid tumor
  - Low calcium level in blood
  - Kidney disease
  - Pancreatic cancer
  - Lung cancer
  - Ovarian cancer
  - Absorption disorder of the intestines
  - Vitamin D deficiency
- Low PTH test results may indicate
  - Lymphoma
  - Low magnesium level in blood
  - Excess calcium intake
  - Malfunctioning parathyroid gland
  - Multiple myeloma
  - Sarcoidosis

TEACH THE PATIENT

- Explain
  - Why the sample is taken.
  - To avoid eating or drinking anything except water 10 hours before the test.

- That the sample will be taken shortly after the patient awakens.
- That the healthcare provide may ask the patient to refrain from taking Tagamet, Betachron ER, Inderal, diuretics, lithium, rifampin, furosemide, thiazide, or anticonvulsant medication 2 weeks prior to the test.

Thyroid Hormone Tests 9

The hypothalamus gland releases thyrotropin-releasing hormone (TRH), which stimulates the anterior pituitary gland to release thyroid-stimulating hormone (TSH). TSH causes the thyroid gland to release thyroxine (T4) and triiodothyronine (T3), both of which regulate metabolism. T4 and T3 are produced only if there is sufficient iodine in the thyroid gland. T4 and T3 are transported in blood either freely or bound to globulin. Free T4 and T3 affect metabolism. The thyroid gland also releases calcitonin when the patient has hypercalcemia. Calcitonin regulates calcium levels in blood by moving calcium from the blood to bone. The thyroid hormone tests measure the level of T4 and T3 in the blood. There are four thyroid hormone tests:

- Total thyroxine (T4) test: This test measures the amount of T4 hormone that is bound to globulin and the amount of unbound T4 hormone, called free thyroxine in the blood.
- Free thyroxine (FT4) test: This test measures the amount of unbound T4 hormone.
- Free thyroxine index (FTI) test: This test compares the amount of bound thyroxine to total thyroxine and thereby indirectly measures unbound thyroxine.
- Triiodothyronine (T3) test: This test measures T3 hormone that is bound to globulin and the amount of unbound T3 hormone, called free triiodothyronine.

WHAT IS BEING MEASURED?

- T4 level in blood
- T3 level in blood

HOW IS THE TEST PERFORMED?

- See How to Collect Blood Specimen from a Vein in Chapter 1.
- See How to Collect Blood Specimen from a Heel Stick in Chapter 1.

RATIONALE FOR THE TEST

- Assess
  - The underlying cause for an abnormal TSH test
  - For hyperthyroidism
  - For treatment of hyperthyroidism
  - For hypothyroidism
  - For treatment of hypothyroidism

NURSING IMPLICATIONS

- Assess if the patient
  - Is taking Dilantin, Tegretol, Coumadin, aspirin, heparin, propranolol, lithium, amiodarone, estrogen, birth control pills, corticosteroids, or progesterone
  - Is pregnant
  - Has had an iodine dye X-ray 4 weeks before the test

UNDERSTANDING THE RESULTS

- Thyroid hormone test results are available in 3 days. The laboratory determines normal values based on calibration of testing equipment with a control test. Test results are reported as high, normal, or low based on the laboratory's control test.
- Normal test results
  - Total thyroxine
    - Adults: 5 to 14 mcg/dL
    - Children: 5.6 to 16.6 mcg/dL
    - Newborn: 9.8 to 22.6 mcg/dL
  - Total triiodothyronine
    - Adults: 80 to 230 ng/dL
    - Children: 83 to 280 ng/dL
    - Newborn: 32 to 250 ng/dL
  - Free thyroxine: 0.8 to 2.4 ng/dL
  - Free thyroxine index
    - Adults: 4.2 to 13
    - Children: 5 to 12.8

    - Newborn: 7.5 to 17.5
    - Free triiodothyronine: 0.2 to 0.6 ng/dL
- High levels may indicate
    - Hyperthyroidism
    - Graves disease
    - Goiter
    - Thyroiditis
    - Overdose of thyroxine replacement therapy
- Low levels may indicate
    - Hypothyroidism
    - Thyroid radiation treatment
    - Thyroid surgery
    - Pituitary gland disorder
    - Thyroiditis

TEACH THE PATIENT

- Explain
    - Why the sample is taken.
    - That the healthcare provider may ask the patient to refrain from taking Dilantin, Tegretol, Coumadin, aspirin, heparin, propranolol, lithium, amiodarone, estrogen, birth control pills, corticosteroids, or progesterone before the test.

Thyroid-Stimulating Hormone 10

The hypothalamus gland releases thyrotropin-releasing hormone (TRH), which stimulates the anterior pituitary gland to release thyroid-stimulating hormone (TSH). TSH causes the thyroid gland to release thyroxine (T4) and triiodothyronine (T3) both of which regulate metabolism. The TSH test measures the level of TSH in blood.

Hint

The healthcare provider will likely order the thyroid hormone tests along with the TSH test.

WHAT IS BEING MEASURED?

- TSH level in blood

HOW IS THE TEST PERFORMED?

- See How to Collect Blood Specimen from a Vein in Chapter 1.

RATIONALE FOR THE TEST

- Assess
  - For hyperpituitarism
  - Treatment of hyperpituitarism
  - For hypopituitarism
  - Treatment of hypopituitarism
  - For hypothalamus disorder
  - The underlying cause of hyperthyroidism
  - The underlying cause of hypothyroidism

NURSING IMPLICATIONS

- Assess if the patient
  - Is taking Dopamine, Tegretol, Coumadin, aspirin, heparin, corticosteroids, lithium, levodopa, propylthiouracil, or Tapazole
  - Is pregnant
  - Is stressed
  - Has had an iodine dye X-ray 4 weeks before the test

UNDERSTANDING THE RESULTS

- The thyroid hormone test results are available in 3 days. The laboratory determines normal values based on calibration of testing equipment with a control test. Test results are reported as high, normal, or low based on the laboratory's control test.
- Normal test results
  - Adults: 0.4 to 4.5 mIU/L
  - Newborn: 3 to 18 mIU/L

- High levels may indicate
  - Hypothyroidism
  - Pituitary gland tumor
  - Hashimoto thyroiditis
- Low levels may indicate
  - Hyperthyroidism
  - Goiter
  - Toxic nodule tumor
  - Graves disease
  - Pituitary gland damage

TEACH THE PATIENT

- Explain
  - Why sample is taken.
  - That the healthcare provider may ask the patient to refrain from taking Dopamine, Tegretol, Coumadin, aspirin, heparin, corticosteroids, lithium, levodopa, propylthiouracil, or Tapazole before the test.

Testosterone 11

The pituitary gland releases LH that stimulates the release of testosterone by the adrenal glands, testes, and ovaries. Testosterone in blood is unbound (called free) or bound to the sex hormone-binding globulin (SHBG) protein in blood. The testosterone test measures the level of testosterone in blood.

WHAT IS BEING MEASURED?

- TSH level in blood

HOW IS THE TEST PERFORMED?

- See How to Collect Blood Specimen from a Vein in Chapter 1.

RATIONALE FOR THE TEST

- Assess
  - The underlying cause of infertility

- The underlying cause of erectile dysfunction
- The underlying cause of osteoporosis in men
- For precocious puberty (boys)
- For the underlying cause of hirsutism in women
- For the underlying cause of irregular menstruation
- Treatment of prostate cancer

NURSING IMPLICATIONS

- Assess if the patient
    - Is taking birth control pills, digoxin, Aldactone, corticosteroids, testosterone, estrogen, barbiturates, or seizure medication
    - Has hyperthyroidism or hypothyroidism
    - Is obese

UNDERSTANDING THE RESULTS

- The testosterone hormone test results are available in 2 days. The laboratory determines normal values based on calibration of testing equipment with a control test. Test results are reported as high, normal, or low based on the laboratory's control test.
- Normal test results
    - Bound testosterone
        - Men
            - Older than 19 years: 280 to 1080 ng/dL
            - 16 to 19 years: 250 to 910 ng/dL
            - 14 to 15 years: 170 to 540 ng/dL
            - 10 to 13 years: Less than 300 ng/dL
            - 7 months to 9 years: Less than 30 ng/dL
        - Women
            - Older than 16 years: Less than 70 ng/dL
            - 14 to 15 years: Less than 60 ng/dL
            - 10 to 13 years: Less than 40 ng/dL
            - 7 months to 9 years: Less than 30 ng/dL
    - Unbound testosterone: 0.3 to 2.0 pg/mL

- High levels may indicate
  - Testicular cancer
  - Adrenal gland cancer
  - Ovarian cancer
  - Polycystic ovary syndrome (PCOS)
  - Precocious puberty (boys)
- Low levels may indicate
  - Klinefelter syndrome
  - Cirrhosis
  - Alcohol abuse
  - Testicle disorder
  - Pituitary gland damage

TEACH THE PATIENT

- Explain
  - Why the sample is taken.
  - That the healthcare provider may ask the patient to refrain from taking birth control pills, digoxin, Aldactone, corticosteroids, testosterone, estrogen, barbiturates, or seizure medication before the test.

Summary

Hormones are biochemical messengers that signal glands and organs how to behave. Hormones are created, stored, and excreted by glands in response to the amount of hormones in the blood stream.

Generally, there is a balance of hormones. At times, hormones can become imbalanced caused by a gland excreting too much of a hormone. When this occurs, a different hormone is excreted into the blood stream by a different gland, causing a decrease in the production and excretion of the initial hormone.

There are times when an imbalance of hormones is caused by a disease. The disease causes an under- or overproduction of specific hormones. Endocrine tests are ordered by the healthcare provider to assist in determining the source of the disease.

Quiz

1. What does the adrenocorticotropic hormone (ACTH) test measure?
   a. Adrenocorticotropic hormone
   b. Corticotropin-releasing hormone
   c. Cortisol
   d. All of the above

2. What is the purpose of cortisol?
   a. Increases blood pressure
   b. Increases glucose
   c. Decreases immune response
   d. All of the above

3. In the absence of ACTH, what can cause release of cortisol?
   a. CTH
   b. Dexamethasone
   c. Adrenal gland tumor
   d. ACT

4. What is the function of dexamethasone?
   a. Suppresses ACTH
   b. Increases cortisol levels
   c. Increases ACTH
   d. Tests for Addison disease

5. Which of the following hormone signals the adrenal glands to release aldosterone?
   a. Renin
   b. Cortisol
   c. Phenytoin
   d. Dexamethasone

6. What is the purpose of aldosterone?
   a. Controls blood pressure
   b. Controls fluid and electrolyte balance
   c. Retains sodium
   d. All of the above

7. When is estriol measured?
   a. In women who have finished menopause
   b. In nonpregnant women during the menstrual cycle
   c. In pregnant women who are in at least the ninth week of pregnancy
   d. In women who are in menopause

8. What is the purpose of the luteinizing hormone (LH) test?
   a. To assess the cause of infertility
   b. To assess menopause
   c. To assess precocious puberty
   d. All of the above

9. Why is the parathyroid hormone (PTH) test administered?
   a. To assess the cause of hyperparathyroidism
   b. To assess cholesterol levels
   c. To assess thyroid hormone levels
   d. To assess insulinlike growth factor 1 (IGF-1) levels

10. What is the reason for ordering the thyroid-stimulating hormone (TSH) test?
    a. To assess hypothalamus disorder
    b. To assess hyperpituitarism
    c. To assess hypopituitarism
    d. All of the above

Answers

1. a. Adrenocorticotropic hormone.
2. d. All of the above.
3. c. Adrenal gland tumor.
4. a. Suppresses ACTH.
5. a. Renin.
6. d. All of the above.
7. c. In pregnant women who are in at least the ninth week of pregnancy.
8. d. All of the above.
9. a. To assess the cause of hyperparathyroidism.
10. d. All of the above.

CHAPTER 8

Glucose Tests

There are two pancreatic endocrine hormones secreted by the islet cells in the pancreas. These are insulin and glucagon. Both are secreted based on blood glucose levels. When blood glucose is elevated, the pancreas secretes insulin, which causes glucose to cross the cell membrane allowing it to be used for energy, resulting in a decrease in blood glucose.

When blood glucose levels are low, the pancreas secretes glucagon, which signals the liver to release stored glucose into the blood, resulting in an increase in blood glucose. Other cells such as muscles also release glucose in response to glucagon.

Blood glucose must be maintained within a narrow range, which occurs naturally with the secretion of insulin and glucagon. However, failure of islet cells to properly function, caused by diseases such as diabetes, can result in high levels of blood glucose (hyperglycemia) or low levels of blood glucose (hypoglycemia).

Healthcare providers order glucose tests to monitor the blood glucose level. Based on the results of these tests, the healthcare provider may administer insulin or glucose to the patient. You'll learn about glucose tests in this chapter.

Learning Objectives

1. C-peptide
2. D-xylose Absorption Test
3. Blood Glucose
4. Glycohemoglobin (GHb)

Key Words

D-xylose
Fasting blood glucose (FBG)
Insulinoma
Malabsorption
Oral glucose tolerance test
Proinsulin
Random blood glucose (RBG)

C-peptide 1

Proinsulin is split into C-peptide and insulin. The level of C-peptide is considered equal to the amount of insulin, indicating the amount of insulin made by the pancreas. The C-peptide test measures the level of C-peptide in blood and is used to differentiate between types 1 and 2 diabetes.

HINT

The healthcare provider will order a blood glucose test along with the C-peptide test. The healthcare provider may order a C-peptide stimulation test to differentiate between types 1 and 2 diabetes.

WHAT IS BEING MEASURED?

- C-peptide level in blood

HOW IS THE TEST PERFORMED?

- See How to Collect Blood Specimen from a Vein in Chapter 1.

RATIONALE FOR THE TEST

- Differentiate between types 1 and 2 diabetes.
- Assess the result of removal of an insulinoma from the pancreas.
- Assess the underlying cause of hypoglycemia.

NURSING IMPLICATIONS

- Assess if the patient
  - Ate or drank 8 hours before the test is administered.
  - Has taken insulin or oral type 2 diabetes medication, since this can affect the test results.
  - Is taking alcohol, insulin, meglitinides or sulfonylureas, birth control pills, diuretics, or corticosteroids, since these medications can affect the test results.
  - Has kidney failure.
- Note if the patient is obese.

UNDERSTANDING THE RESULTS

- Test results are available quickly. The laboratory determines normal values based on calibration of testing equipment with a control test. Test results are reported as high, normal, or low based on the laboratory's control test.
- The blood glucose test is performed along with the C-peptide test and the results of both tests are compared.
- Normal C-peptide range: 0.78 to 1.89 ng/mL
- High C-peptide and high blood glucose values may indicate
  - Type 2 diabetes
  - Cushing syndrome
  - Insulin resistance
- High C-peptide and low blood glucose values may indicate
  - Insulinoma/pancreatic insulin-producing tumor
  - That the patient is taking sulfonylureas
  - That the patient is taking meglitinides

- Metastasized insulinoma if the insulinoma was removed
- Low C-peptide and low blood glucose values may indicate
  - Addison disease
  - Liver disease
  - That the patient has taken insulin
  - Infection
- Low C-peptide and high blood glucose values may indicate
  - Type 1 diabetes
  - Pancreatectomy

HINT

A patient with type 2 diabetes can develop low C-peptide level over time.

TEACH THE PATIENT

- Explain
  - Why blood sample is taken.
  - That the patient should not eat or drink 8 hours before the test is administered.
  - That the healthcare provider may ask the patient to stop taking alcohol, insulin, meglitinides or sulfonylureas, birth control pills, diuretics, or corticosteroids for 2 weeks prior to the test.

D-xylose Absorption Test 2

D-xylose is a simple sugar that is absorbed by the intestines. The D-xylose absorption test measures the level of D-xylose in blood.

HINT

D-xylose absorption can also be tested with a urine sample. A urine test is less accurate than the blood D-xylose absorption test for patients younger than 12 years of age. The healthcare provider may order an upper gastrointestinal (GI) series if the D-xylose test is positive.

WHAT IS BEING MEASURED?

- D-xylose level in blood

HOW IS THE TEST PERFORMED?

- The patient drinks 8 oz of water that contains 25 g of D-xylose.
- A blood sample is taken after 2 hours and then again 5 hours after drinking the water.
- See How to Collect Blood Specimen from a Vein in Chapter 1.

RATIONALE FOR THE TEST

- Screen for malabsorption syndrome.

NURSING IMPLICATIONS

- The patient should avoid fruits, jams, pastries, and jellies 1 day before administration of the test.
- Adult patients should avoid eating or drinking, except water, 12 hours before the test.
- Patients younger than 9 years of age should avoid eating or drinking, except water, 4 hours before the test.
- Patients are not permitted to eat or drink until the last blood sample is taken.
- The patient should rest quietly until the last blood sample is taken.
- Assess if the patient is
    - Undergoing radiation treatment.
    - Taking aspirin, antibiotics, indomethacin, or cardiac medication, since these medications can affect the test result.

UNDERSTANDING THE RESULTS

- Test results are available quickly. The laboratory determines normal values based on calibration of testing equipment with a control test. Test results are reported as high, normal, or low based on the laboratory's control test.
- Normal D-xylose range
    - Adult: 21 to 57 mg/dL

- Younger than 13 years of age: Greater than 20 mg/dL
- Younger than 6 months of age: Greater than 15 mg/dL
- High D-xylose levels may indicate
  - That the patient is undergoing radiation treatment
  - Hodgkin disease
  - Scleroderma
- Low D-xylose levels may indicate
  - Celiac disease
  - Hookworm
  - Malabsorption syndrome
  - Inflammation of the intestine
  - Crohn disease
  - Giardiasis
  - Whipple disease

TEACH THE PATIENT

- Explain
  - Why the blood sample is taken.
  - That the patient should avoid fruits, jams, pastries, and jellies 1 day before administering the test.
  - That the adult patient should avoid eating or drinking, except water, 12 hours before the test and patients younger than 9 years of age for 4 hours.
  - That the patient will not be permitted to eat or drink until after the last blood sample is taken.
  - That the patient will be asked to rest quietly until the last blood sample is taken.

Blood Glucose 3

Glucose is the source of energy for cells and is transported into cells by insulin. As blood glucose levels rise following ingestion of food, the pancreas releases insulin to move the glucose from blood into cells. The blood glucose tests measure the level of glucose in the blood. There are four blood glucose tests:

- Oral glucose tolerance test (OGTT): This test measures the blood glucose levels at specific time intervals after the patient ingests a glucose drink. The OGTT is ordered to screen for gestational diabetes and to confirm positive results of other blood glucose tests.
- Two-hour postprandial blood sugar test: This test measures blood glucose levels 2 hours after the patient ingests food.
- Fasting blood glucose (FBG) test: This test measures blood glucose levels after the patient has fasted for 8 hours. The FBG test is the initial test for diabetes.
- Random blood glucose (RBG): This test measures blood glucose levels several times a day regardless of food intake.

Hint

Glucose levels can also be measured in urine; however, this is not used to diagnose or monitor glucose levels. The healthcare provider may order the glycohemoglobin (GHb) blood test, which is used to monitor blood glucose levels for the previous 120 days.

WHAT IS BEING MEASURED?

- Glucose levels in blood

HOW IS THE TEST PERFORMED?

- FBG test requires the patient to fast for 8 hours before the test is administered. Insulin and other diabetes medication are withheld until the test is administered.
- Two-hour postprandial test requires the patient to eat a normal meal 2 hours before the test is administered.
- Oral glucose tolerance test requires the patient to fast 8 hours before the test. Ingesting water is permitted. A blood sample is taken to be used as a baseline. The patient ingests a glucose drink. A blood sample is taken in hourly intervals for 3 hours.
- See How to Collect Blood Specimen from a Vein in Chapter 1.

Caution

The blood glucose test for diagnosing diabetes must be repeated on two different days.

RATIONALE FOR THE TEST

- Screen for
  - Diabetes
  - Hypoglycemia
- Assessment of the treatment for diabetes.

NURSING IMPLICATIONS

- Assess if the patient
  - Is ill. Illness can increase blood glucose levels and affect the test result.
  - Smokes, since this can affect the test result.
  - Uses caffeine, since this can affect the test result.
  - Is experiencing stress, since this can affect the test result.
  - Has eaten or fasted according to procedures for the blood glucose test.
  - Has taken Dyrenium, Dyazide, Dilantin, Lasix, niacin, Inderal, Esidrix, Hydro Par, Oretic, prednisone, or birth control pills, since these medications may affect the test result.

UNDERSTANDING THE RESULTS

- Blood glucose test results are available in 2 hours. The laboratory determines normal values based on calibration of testing equipment with a control test. Test results are reported as high, normal, or low based on the laboratory's control test.
- Normal blood glucose range
  - FBG test: 70 to 99 mg/dL
  - Two-hour postprandial blood sugar test: 70 to 145 mg/dL
  - RBG test: 70 to 125 mg/dL
- High blood glucose may indicate
  - Diabetes
    - FBG test: Greater than 125 mg/dL
    - Two-hour postprandial blood sugar test: Greater than 199 mg/dL
    - RBG test: Greater than 199 mg/dL and the presence of symptoms of diabetes

- Prediabetes
  - FBG test: 100 to 125 mg/dL
- Acromegaly
- Cushing syndrome
- Stroke
- Myocardial infarction
- That the patient has taken corticosteroid medication

- Low blood glucose may indicate
  - Hypoglycemia (FBG less than 40 mg/dL)
  - Insulinoma/insulin-producing tumor
  - Hypothyroidism
  - Addison disease
  - Kidney failure
  - Malnutrition
  - Cirrhosis
  - Anorexia
  - That the patient has taken insulin or other diabetes medication

TEACH THE PATIENT

- Explain
  - Why the blood sample is taken.
  - The dietary restriction required by the test if any.
  - That the healthcare provider may ask the patient to stop taking Dyrenium, Dyazide, Dilantin, Lasix, niacin, Inderal, Esidrix, Hydro Par, Oretic, prednisone, or birth control pills for 2 weeks prior to the test.

Glycohemoglobin 4

Glucose binds to hemoglobin in red blood cells, which has a life span of 120 days. The glycohemoglobin (GHb), commonly known as the HgbAIC test, measures the level of glucose bound to hemoglobin. This differs from the blood glucose test, which measures the level of glucose in plasma. The healthcare provider orders the

GHb test to assess if treatment is controlling diabetes and if the patient is adhering to the treatment over a 120-day period.

Hint

The GHb test can be administered at any time; however, the healthcare provider is likely to order the test four times a year. The GHb test is not a replacement for the blood glucose test and cannot dictate hypoglycemia.

WHAT IS BEING MEASURED?

- Glucose that adheres to hemoglobin

HOW IS THE TEST PERFORMED?

- See How to Collect Blood Specimen from a Vein in Chapter 1.

RATIONALE FOR THE TEST

- Assess
  - Treatment for diabetes
  - If the patient is adhering to treatment for diabetes

NURSING IMPLICATIONS

- Assess if the patient has
  - Had a blood transfusion in the past 120 days
  - Hemolytic anemia, thalassemia, sickle cell anemia, polycystic ovary syndrome, pheochromocytoma, Cushing syndrome, or kidney disease
  - Had his/her spleen removed
  - Taken corticosteroids

UNDERSTANDING THE RESULTS

- Normal GHb test results are available in 2 days. The laboratory determines normal values based on calibration of testing equipment with a control test.

Test results are reported as high, normal, or low based on the laboratory's control test.

- Normal GHb level range
  - GHb A1c: 4.5% to 5.7%
  - Total GHb: 5.3% to 7.5%
- High GHb levels (8% or greater) may indicate
  - Poorly controlled diabetes
  - The patient is undergoing corticosteroid treatment
  - Pheochromocytoma
  - Polycystic ovary syndrome (PCOS)
  - Cushing syndrome

TEACH THE PATIENT

- Explain
  - Why the sample is taken.
  - Why the test is administered several times during the year.
  - That the GHb test is not an alternative to daily blood glucose tests.

Summary

The pancreas secretes insulin and glucagon based on the level of glucose in the blood. A high level of blood glucose signals the pancreas to secrete insulin, which assists glucose to cross cell membranes where glucose is used for energy.

A low level of blood glucose causes the pancreas to secrete glucagon. Glucagon is a hormone that signals the liver, muscles, and other organs to release stored glucose into the blood, resulting in an increase in blood glucose levels.

The balance between insulin and glucagon secretions maintains a narrow range of blood glucose. However, diseases such as diabetes can cause an imbalance, resulting in high levels of blood glucose (hyperglycemia) or lower levels of blood glucose (hypoglycemia) that cannot be brought into range naturally.

Glucose tests are used by healthcare providers to monitor a patient's blood glucose level to determine if the patient has a disease that affects insulin and/or glucagon. These tests are also used to monitor treatment of these diseases.

Quiz

1. What is the importance of the C-peptide level?
   a. The C-peptide level determines the level of glucose in arterial blood.
   b. The C-peptide level is considered equal to the amount of insulin produced by the pancreas.
   c. The C-peptide level determines the level of glucose in venous blood.
   d. None of the above.

2. What can affect the C-peptide level?
   a. The patient has kidney failure.
   b. The patient has taken insulin.
   c. The patient has taken oral type 2 diabetes medication.
   d. All of the above.

3. Why is the D-xylose absorption test administered?
   a. To screen malabsorption syndrome
   b. To screen diabetes mellitus (DM)
   c. To screen insulin production
   d. To screen type 2 DM

4. What is the initial test for diabetes?
   a. Fasting blood glucose (FBG)
   b. Random blood glucose (RBG)
   c. Oral glucose tolerance test (OGTT)
   d. None of the above

5. Why is the glycohemoglobin (GHb) test administered?
   a. To determine if the DM patient has been maintaining an adequate blood glucose level for the previous 60 days
   b. To determine if the DM patient has been maintaining an adequate blood glucose level for the previous 120 days
   c. To determine if the DM patient has been maintaining an adequate blood glucose level for the previous 30 days
   d. None of the above

6. What can affect blood glucose levels?
   a. Illness
   b. Smoking
   c. Stress
   d. All of the above

7. What does the GHb test determine?
   a. Two positive GHb tests confirm the diagnosis of DM.
   b. Glucose binds to hemoglobin in RBC. RBC has a 120-day life span and therefore indicates the amount of glucose in blood.
   c. One positive GHb test confirms the diagnosis of DM.
   d. One positive GHb test confirms the diagnosis of type 2 DM.

8. What would a low C-peptide level indicate?
   a. Liver disease
   b. Infection
   c. Addison disease
   d. All of the above

9. How is the FBG test administered?
   a. The patient fasts for 8 hours. All DM medications are withheld until the test is completed.
   b. The patient eats a regular meal but refrains from ingesting sugar-sweetened beverages.
   c. The patient eats a regular meal including ingesting sugar-sweetened beverages.
   d. The patient eats a bland meal but refrains from ingesting sugar-sweetened beverages.

10. What should you do if the patient tells you he/she ate a jelly donut the morning of the D-xylose absorption test?
    a. Postpone the test for 24 hours.
    b. Administer the test.
    c. Administer the test but tell the lab tech that the patient ate a jelly donut.
    d. Postpone the test for 12 hours.

Answers

1. b. The C-peptide level is considered equal to the amount of insulin produced by the pancreas.
2. d. All of the above.
3. a. Screen for malabsorption syndrome.
4. a. Fasting blood glucose (FBG).
5. b. To determine if the DM patient has been maintaining an adequate blood glucose level for the previous 120 days.
6. d. All of the above.
7. b. Glucose binds to hemoglobin in RBC. RBC has a 120-day life span and therefore indicates the amount of glucose in blood.
8. d. All of the above.
9. a. The patient fasts for 8 hours. All DM medications are withheld until the test is completed.
10. a. Postpone the test for 24 hours.

CHAPTER 9

Tumor Markers

A tumor is uncontrollable growth of cells that may be malignant (cancerous) or benign (noncancerous). Blood tests are performed to detect the presence of tumor markers. A tumor marker is a substance, which is usually a protein, which is produced either by tumor cells or by other cells in response to the presence of the tumor.

The presence of a tumor marker does not mean that the patient has cancer. Conditions other than cancer can also generate the tumor marker. Likewise, the absence of a tumor marker does not mean that the patient is cancer free because many times early stages of cancer do not produce a tumor marker.

In addition to tumor markers, a patient's blood can also be tested for cancer genomics. Cancer genomics are mutation of specific genes, which is called a risk marker. A risk marker indicates that a patient is at higher than normal risk for developing cancer, however, there isn't a sign of cancer as yet. Conversely, the presence of the tumor marker indicates a possibility that a tumor is present.

In this chapter you'll learn about these tests.

Learning Objectives

1. Cancer Antigen 125 (CA-125)
2. Carcinoembryonic Antigen (CEA)
3. Prostate-Specific Antigen (PSA)

Key Words

Complex prostate-specific antigen (cPSA)
Cystoscopy
Metastasized
Prostate-specific antigen density (PSAD)
Prostatitis
Transrectal ultrasound (TRUS)

Cancer Antigen 125 (CA-125) 1

Cancer antigen 125 (CA-125) is a protein attached to the ovarian cancer cells and other cancer cells. The CA-125 test measures the level of the CA-125 in the blood.

HINT

The CA-125 test is not used to screen patients for ovarian cancer. This test cannot differentiate between a benign or malignant ovarian tumor. The healthcare provider may order testing of peritoneal fluid and the chest to assess if CA-125 is present.

WHAT IS BEING MEASURED?

- CA-125 protein level in blood

HOW IS THE TEST PERFORMED?

- See How to Collect Blood Specimen from a Vein in Chapter 1.

RATIONALE FOR THE TEST

- Screen for
  - Ovarian cancer
  - Other types of cancers
- Assess the effectiveness of cancer treatment.

NURSING IMPLICATIONS

- Assess if the patient
  - Has undergone abdominal surgery in the previous 3 weeks.
  - Has undergone chemotherapy recently.
  - Has undergone a radioactive scan recently.
  - Is pregnant.
  - Is menstruating.

UNDERSTANDING THE RESULTS

- Test results are available in 3 days. The laboratory determines normal values based on calibration of testing equipment with a control test. Test results are reported as high, normal, or low based on the laboratory's control test.
- Normal CA-125 range: Less than 35 U/mL
- High CA-125 values may indicate
  - Ovarian cancer
  - Cancer of the endometrium
  - Cancer of the fallopian tubes
  - Liver cancer
  - Breast cancer
  - Lung cancer
  - Colon cancer
  - Pancreatic cancer
  - Esophageal cancer
  - Stomach cancer
  - Cancer of the peritoneum

- Lymphoma
- Pancreatitis
- Hepatitis
- Cirrhosis
- Lupus
- Endometriosis
- Pelvic inflammatory disease
- Uterine fibroids
- Menstruation
- Pregnancy

HINT

A high level of CA-125 might be found months before other diagnostic tests, indicating the return of previously treated ovarian cancer.

TEACH THE PATIENT

- Explain
  - Why blood sample is taken.
  - That this test is not used to screen for ovarian cancer.
  - That a positive result does not differentiate between a benign or malignant ovarian tumor.

Carcinoembryonic Antigen 2

The carcinoembryonic antigen (CEA) is a protein present during fetal development that terminates at birth. The CEA is produced in certain types of cancers. The CEA test measures the level of CEA antigen in blood.

HINT

The CEA is not used to screen for early detection of a cancer and is not used to diagnose cancer. Most cancers do not cause high levels of the CEA. The healthcare provider may order a test of peritoneal fluid and cerebrospinal fluid to determine if the cancer has metastasized.

WHAT IS BEING MEASURED?

- CEA protein level in blood

HOW IS THE TEST PERFORMED?

- See How to Collect Blood Specimen from a Vein in Chapter 1.

RATIONALE FOR THE TEST

- Screen for cancer
- Assess
  - Effectiveness of cancer treatment
  - Success of surgery to remove the cancer

NURSING IMPLICATIONS

- Assess if the patient
  - Smokes, since it can affect the test result.
  - Has undergone chemotherapy recently.
  - Has undergone a radioactive scan recently.
  - Is pregnant.
  - Is menstruating.

UNDERSTANDING THE RESULTS

- Test results are available in 3 days. The laboratory determines normal values based on calibration of testing equipment with a control test. Test results are reported as high, normal, or low based on the laboratory's control test.
- Normal CEA range
  - Nonsmoker: Less than 3 ng/mL
  - Smoker: Less than 5 ng/mL
- High CEA values may indicate
  - Colon cancer
  - Rectal cancer

- Lung cancer
- Breast cancer
- Pancreatic cancer
- Ovarian cancer
- Cancer has returned following treatment. Successful treatment results in CEA levels returning to normal within 6 weeks of treatment.
- Cancer has metastasized
- Pancreatitis
- Kidney failure
- Cirrhosis
- Peptic ulcer
- Chronic obstructive pulmonary disease (COPD)
- Inflammatory bowel disease
- Crohn disease

HINT

CEA maybe normal although the patient has cancer.

TEACH THE PATIENT

- Explain
  - Why the blood sample is taken.
  - That further testing is likely if there is a positive result.
  - That a negative result does not rule out cancer.

Prostate-Specific Antigen 3

The prostate gland releases prostate-specific antigen (PSA) in low amounts. Increased amounts are released with an enlarged prostate gland, prostatitis, prostate cancer, and from injury resulting from a digital rectal examination, cystoscopy, or sexual activity. The PSA test measures the level of PSA blood.

Hint

PSA test is performed in conjunction with a digital rectal examination. The healthcare provider may also order the prostate-specific antigen density (PSAD) test to compare the PSA value to the prostate gland size, the prostate-specific antigen velocity (PSAV) test to determine the if the PSA increases over time, the complex PSA (cPSA) to detect prostate cancer, and the transrectal ultrasound (TRUS) to measure the size of the prostate gland.

Caution

The PSA level can be normal in a patient who has prostate cancer.

WHAT IS BEING MEASURED?

- PSA level in blood

HOW IS THE TEST PERFORMED?

- Do not ejaculate 48 hours before the test.
- See How to Collect Blood Specimen from a Vein in Chapter 1.

RATIONALE FOR THE TEST

- Screen for prostate cancer.
- Assess the treatment of prostate cancer.

NURSING IMPLICATIONS

- Assess if the patient
  - Has ejaculated 48 hours before the test
  - Has a urinary tract infection (UTI)
  - Has been catheterized within the past 8 weeks
  - Has prostatitis
  - Has undergone a prostate biopsy in the past 8 weeks
  - Is taking Avodart, finasteride, Cytoxan, Neosar, methotrexate, or diethylstilbestrol.

UNDERSTANDING THE RESULTS

- The PSA test results are available quickly.
- Normal test results (maximum level):

- Younger than 50 years of age: 2.5 ng/mL
- 51 to 60 years of age: 3.5 ng/mL
- 61 to 70 years of age: 4.5 ng/mL
- Older than 70 years of age: 6.5 ng/mL

- Higher levels may indicate
  - Prostate cancer
  - Benign prostatic hypertrophy (BPH)
  - Prostatitis
  - Recent catheterization
  - Injury

TEACH THE PATIENT

- Explain
  - Why sample is taken.
  - That further testing is necessary if there is a high level of PSA in the blood. This does not mean that the patient has prostate cancer.
  - That a normal level of PSA does not mean that the patient is free from prostate cancer.

Summary

Tumor marker tests assess the patient's blood for proteins that are frequently present if a tumor is present. These substances are secreted either by the tumor or by other cells in response to the presence of the tumor.

The existence of a tumor marker in the patient's blood does not necessarily mean that the patient has a tumor since the same proteins can be produced by other disorders. The existence of a tumor marker simply indicates that the healthcare provider needs to examine the patient more closely for other signs of tumor.

Conversely, the absence of a tumor marker in the patient's blood does not necessarily mean that a tumor is absent. A tumor might be present, but its growth is at an early stage when tumor markers have yet to be produced.

Tests are also performed on the patient's blood to identify risk markers. A risk marker is a mutation of specific genes that can later result in a tumor. The patient with a risk marker may not develop a tumor, but it is a risk for possibly developing a tumor.

Quiz

1. The cancer antigen 125 (CA-125) test is used to screen patients for ovarian cancer.

   a. True

   b. False

2. Why is the carcinoembryonic antigen (CEA) test used to screen for cancer?

   a. The CEA is normally present during fetal development and is terminated at birth. It is present if there is a tumor.

   b. The CEA is excellent for diagnosing cancer.

   c. The CEA is used for early detection of cancer.

   d. None of the above.

3. Why is the prostate-specific antigen (PSA) test administered?

   a. A low level always indicates prostatitis.

   b. A low level might indicate prostate cancer.

   c. A high level might indicate an enlarged prostate gland.

   d. None of the above.

4. A positive PSA test means that the patient may have

   a. Prostate cancer

   b. Had sexual activity

   c. An enlarged prostate gland

   d. All of the above

5. A patient who has a normal PSA level is free from prostate cancer.

   a. True

   b. False

6. The CA-125 test can differentiate between benign and malignant tumors.

   a. True

   b. False

7. A high level of CEA can indicate
   a. Peptic ulcer
   b. Crohn disease
   c. Pancreatitis
   d. All of the above

8. What should you tell a patient who has a positive CEA test result?
   a. A positive result will require further tests.
   b. The patient will undergo chemotherapy.
   c. The patient will undergo radiation treatments.
   d. The patient will be admitted to the hospital.

9. Give three reasons for administering the CEA test.
   a. Screening for cancer, assessing cancer treatment, and identifying the location of the tumor for surgery.
   b. Screening for cancer, assessing cancer treatment, and for diagnosing cancer.
   c. Screening for cancer, assessing cancer treatment, and assessing for the success of surgery to remove the tumor.
   d. None of the above.

10. Why would the prostate-specific antigen density (PSAD) test be ordered?
    a. To diagnose prostate cancer
    b. To compare the PSA value to the prostate gland size.
    c. To confirm the magnetic resonance imaging (MRI) diagnosis of prostate cancer
    d. None of the above

Answers

1. b. False.
2. a. The CEA is normally present during fetal development and is terminated at birth. It is present if there is a tumor.

3. c. A high level might indicate an enlarged prostate gland.
4. d. All of the above.
5. b. False.
6. b. False.
7. d. All of the above.
8. a. A positive result will require further tests.
9. c. Screening for cancer, assessing cancer treatment, and assessing the success of surgery to remove the tumor.
10. b. To compare the PSA value to the prostate gland size.

CHAPTER 10

Pregnancy and Genetic Tests

Healthcare providers have an arsenal of tests that provide clues to the underlying cause of infertility and risk for genetic disorders that can affect the fetus or newborn. These tests are ordered to screen patients for disorders when there are no telltale signs or symptoms, and other tests are ordered when the healthcare provider is looking to confirm a sign or symptoms that fall outside the normal parameters.

These tests look for certain components in blood, such as the presence of antibodies or certain enzymes, or the levels of protein and hormones. Scientific research has determined that the absence or existence of these components in blood correlates to the presence or absence of a specific disorder.

In this chapter you will be introduced to these tests and learn what they are, how they are performed, how to prepare the patient for the test, and how to interpret the test results.

Learning Objectives

1. Antisperm Antibody Test
2. α-Fetoprotein (AFP)
3. Follicle-Stimulating Hormone (FSH)
4. Human Chorionic Gonadotropin (hCG) Hormone
5. Inhibin A Test
6. Prolactin
7. Phenylketonuria (PKU) Test
8. Tay-Sachs Test
9. Sickle Cell Test
10. Hemochromatosis (HFE) Gene Test

Key Words

Amniocentesis
Down syndrome
Lactogenesis
Luteinizing hormone
Menstrual cycle
Multiple of median
Tube defect
Unconjugated estriol (uE3)

Antisperm Antibody Test 1

An immune system response can be caused by semen, resulting in antibodies attaching to and killing sperm causing immunologic infertility. These antibodies can be in blood, vaginal fluids, or semen. Antibodies can be made in a man if his sperm comes in contact with his immune system as a result of testicular injury, prostate gland infection, vasectomy, or other surgeries that expose sperm to the immune system. Semen from her partner's sperm can cause an allergic reaction in a woman.

WHAT IS BEING MEASURED?

- Antisperm antibodies

HOW IS THE TEST PERFORMED?

- A blood sample is taken from a female patient (See How to Collect Blood Specimen from a Vein in Chapter 1).
- A semen sample is obtained.
  - Ejaculation should not occur for a 2-day period before and should not have taken place any more than 5 days before the test.
  - Collect semen samples after blood and vaginal fluid samples are taken.
  - Samples cannot be taken by engaging in sexual intercourse and withdrawal during ejaculation. Vaginal fluid may mix with the sperm.
  - A semen sample is collected through masturbation.
    - Urinate.
    - Wash and rinse your hands and penis.
    - Collect the semen in a sterile cup.
    - Do not use lubricants or condoms.
    - Take sample to the laboratory within 30 minutes.

RATIONALE FOR THE TEST

- Assess if
  - Antisperm antibodies are in blood, vaginal fluid, or semen.
  - Infertility maybe caused by immunologic infertility.

HINT

Some healthcare providers disagree about the usefulness of the test since treatment is the same regardless of the test results.

NURSING IMPLICATIONS

- Male patients may be embarrassed about collecting a sample.
- Confer with the healthcare provider if masturbation is against the patient's religious beliefs.

UNDERSTANDING THE RESULTS

- Negative test results: Antisperm antibodies are not found.
- Positive test results: Antisperm antibodies are found.

TEACH THE PATIENT

- Explain why the test is ordered.
- Explain to male patients
  - Do not ejaculate for 2 days before the test, and ejaculation should not have taken place any more than 5 days before the test.
  - How to collect a semen sample.
  - That a sample is collected through masturbation and not by engaging in sexual intercourse.
  - Do not use lubricants or condoms when collecting the sample.
  - Ask the patient to consult his healthcare provider if masturbation is against his religious beliefs.
  - Take the sample to the laboratory within 30 minutes.

α-Fetoprotein 2

α-Fetoprotein (AFP) is produced by the fetal liver and is detectable in a pregnant woman's blood. The level of AFP rises gradually in 14 weeks of gestation and continues to rise until around the 34th week of gestation when the AFP level gradually decreases.

A high or low level of AFP is a sign that there may be a problem with the fetal development. The AFP test is commonly administered as part of a maternal serum triple or quadruple screening test, which, along with other factors including the pregnant woman's age, is used to estimate the chances of birth defects.

The maternal serum triple screening test examines levels of

- AFP
- β-hCG
- uE3 (estrogen)

The maternal serum quadruple screening test examines levels of

- The same substance as the maternal serum triple screening test
- The hormone inhibin A

A healthcare provider may also administer the AFP test in nonpregnant women and in children and men to assess for a number of other diseases that cause a high level of AFP in the blood. These diseases include lymphoma, Hodgkin disease, ovarian cancer, testicular cancer, renal cell cancer, and pancreatic cancer. One-half of patients diagnosed with these cancers have normal AFP test results.

WHAT IS BEING MEASURED?

- The level of AFP in the blood

HOW IS THE TEST PERFORMED?

- The test is performed between 15 and 20 weeks of gestation.
- The patient is weighed before the blood sample is drawn. The results of the test are evaluated based on the patient's weight.
- Estimate the gestation.
- Assess if the patient is a diabetic who requires insulin injections to manage the diabetes.
- The patient's race is noted. Normal values of AFP are higher in African-American women and lower in Asian women.
- See How to Collect Blood Specimen from a Vein in Chapter 1.

RATIONALE FOR THE TEST

- Assess for
    - Fetal neural tube defects
    - Spinal bifida
    - Anencephaly
    - Edward syndrome (trisomy 18)
    - Down syndrome (trisomy 21)
    - Omphalocele
    - Hepatoma in patients who have chronic hepatitis B or cirrhosis
    - Lymphoma
    - Hodgkin disease
    - Renal cell cancer
    - Ovarian cancer
    - Testicular cancer
    - Pancreatic cancer
    - Effectiveness of cancer treatment

NURSING IMPLICATIONS

- Assess
    - If the patient smokes. Smoking increases the AFP level in the blood.
    - If the patient is an insulin-dependent diabetic.
    - The patient's weight.
    - The gestational age of the fetus.
    - The patient's race.
    - If the patient has undergone a test, involving radioactive tracers in the previous 2 weeks since this might affect the test results.
- If high levels of AFP are detected, the healthcare provider will likely order an ultrasound and/or amniocentesis.
- Ease the patient's concerns. Tell the patient that the AFP test results provide a sign of a disease or medication condition, but are not used to make a definitive diagnosis. AFP levels can be high, but the fetus may be normal.

UNDERSTANDING THE RESULTS

- The result is available quickly. The laboratory determines normal values based on calibration of testing equipment with a control test. Test results are reported as high, normal, or low based on the laboratory's control test.
- Generally the normal range
    - Pregnant women 12 to 22 weeks' gestation: 19 to 75 IU/mL or 7 to 124 mcg/L or 7 to 124 ng/mL
    - Nonpregnant women and men: 0 to 6.4 IU/mL or 0 to 20 mcg/L or 0 to 20 ng/mL
- High levels may indicate
    - Neural tube defect
    - Omphalocele (abdominal organs are outside of the abdomen wall; repaired by surgery)
    - Multiple fetuses
    - Dead fetus
    - Incorrect gestational age

    - Alcohol abuse
    - Cirrhosis
    - Hepatitis B
    - Lymphoma, Hodgkin disease, ovarian cancer, testiclular cancer, renal cell cancer, and pancreatic cancer
- Low levels may indicate
    - Down syndrome (60% accuracy). Accuracy increases to 80% with the maternal serum triple screening test.
    - Incorrect gestational age.

HINT

An ultrasound confirms the gestational age. If the gestational age is different than the estimated gestational age, the AFP is adjusted using the multiple of median (MoM) factor. An AFP value that is 0.5 to 2.5 times the multiple of median factor is considered normal.

CAUTION

A normal AFP level does not rule out neural tube defects or Down syndrome. Likewise, an abnormal AFP level is not an indication of a definitive diagnosis. Additional testing such as an ultrasound or amniocentesis will likely be ordered for abnormal results. The patient may have an abnormal AFP level in their blood and a normal AFP level in the amniocentesis indicating low risk to the fetus.

TEACH THE PATIENT

- Explain
    - Why blood sample is taken.
    - That weight, race, gestational age is used to assess the test results.
    - That there is no special preparation required for the test.
    - That an abnormal test request does not mean there are developmental problems with the fetus. The test result is one of several signs that help the healthcare provider reach a diagnosis.
    - That a neural tube defect is not related to the mother's age. Neural tube defects occur without a family history of this problem.

Follicle-Stimulating Hormone 3

The follicle-stimulating hormone (FSH) is produced by the pituitary gland and controls sperm production by the testes and egg production in the ovaries. The FSH level is constant in men and changes with the menstrual cycle in women, with the highest level occurring during ovulation. The FSH test measures the level of the FSH in blood.

Hint

The healthcare provider may order the luteinizing hormone blood test, estrogen blood test, and the progesterone blood test in addition to the FSH test. The healthcare provider may order a sperm count or assessment of the patient's ovarian reserve.

WHAT IS BEING MEASURED?

- FSH level in blood

HOW IS THE TEST PERFORMED?

- A blood sample may be taken each day for a period ordered by the healthcare provider if the patient experiences menstrual cycle problems or is unable to become pregnant.
- See How to Collect Blood Specimen from a Vein in Chapter 1.

RATIONALE FOR THE TEST

- Assess
    - For the underlying cause of infertility
    - For abnormal menstrual periods
    - For precocious puberty
    - The function of the pituitary gland
    - Abnormal development of sexual organs

NURSING IMPLICATIONS

- Assess
    - The patient's age. Test results depend on the patient's age.
    - If the patient uses herbal or natural substances.

- The first day of the patient's last menstrual period. Test results depend on the patient's menstrual cycle.
- Which day the patient experiences the heaviest bleeding during the menstrual period.
- If the patient has undergone a radioactive tracer in the previous 7 days before the test.
- If the patient has taken digitalis, cimetidine, levodopa, clomiphene, birth control pills, estrogen, or progesterone 4 weeks before the test, since these medications may affect the test results.

UNDERSTANDING THE RESULTS

- FSH test results are available in 1 day. The laboratory determines normal values based on calibration of testing equipment with a control test. Test results are reported as high, normal, or low based on the laboratory's control test.
- Normal FSH range
  - Menstruating
    - Follicular/luteal phase: 5 to 20 IU/L
    - Midcycle peak: 30 to 50 IU/L
  - Postmenopause: Greater than 49 IU/L
  - Men: 5 to 15 IU/L
  - Children before puberty: Less than 7 IU/L
- High FSH may indicate
  - Women
    - Polycystic ovary syndrome (PCOS)
    - Ovarian failure before 40 years of age
    - Menopause
  - Men
    - Abnormal testicle functionality
    - Klinefelter syndrome
  - Children
  - Start of puberty
- Low FSH may indicate
  - Women
    - Loss of ovulation

- Men
    - Testicles not producing sperm
- Malnutrition
- Hypothalamus disorder
- Pituitary gland disorder
- Stress

TEACH THE PATIENT

- Explain
    - Why blood sample is taken.
    - That several blood samples may be necessary, one each for a period, ordered by the healthcare provider if the patient experiences menstrual cycle problems or is unable to become pregnant.
    - That the healthcare provider may ask the patient to stop taking digitalis, cimetidine, levodopa, clomiphene, birth control pills, estrogen, or progesterone for 4 weeks prior to the test.

Human Chorionic Gonadotropin 4

When a fertilized egg implants into the uterine wall, the placenta begins development. By the ninth day, the placenta produces Human Chorionic Gonadotropin (hCG) hormone, which is detectable in the patient's blood. The hCG test measures the level of hCG in blood. The level of the hCG hormone is used as a sign of pregnancy, which is confirmed by other tests. In a normal pregnancy, the level of the hCG hormone increases until 16 weeks of gestation and then gradually decreases until birth when no hCG hormone is detectable. A lower level of hCG hormone might indicate an ectopic pregnancy and a higher level may indicate multiple fetuses. The hCG hormone test is typically ordered as part of a maternal serum screening test that is administered after 15 weeks' gestation.

Hint

The level of the hCG hormone can be detected before the patient misses her menstrual period and as early as 6 days after attachment of the egg. The hCG hormone test is also available as a urine test (home pregnancy test), which determines if the hCG hormone is present but does not measure the hormone level. The hCG hormone level is high 4 weeks following an abortion.

CAUTION

The hCG hormone can also be produced by a molar pregnancy, choriocarcinoma (uterine cancer), ovarian cancer, or other tumors. Testicular cancer also produces the hCG hormone in men. A normal hCG hormone level does not rule out cancer.

WHAT IS BEING MEASURED?

- The hCG level in blood.

HOW IS THE TEST PERFORMED?

- See How to Collect Blood Specimen from a Vein in Chapter 1.

RATIONALE FOR THE TEST

- Assess
  - For pregnancy
  - For ectopic pregnancy
  - For molar pregnancy
  - Treatment of molar pregnancy
  - For testicular cancer
  - For choriocarcinoma

NURSING IMPLICATIONS

- Assess if the patient
  - Has received hCG hormone infertility treatment prior to administering the test
  - Has taken Anergan, Phenergan, Prorex, diuretics, heparin, Ambien, Mellaril, Serentil, Stelazine, or Compazine

UNDERSTANDING THE RESULTS

- The hCG hormone test results are available quickly. The laboratory determines normal values based on calibration of testing equipment with a control test. Test results are reported as high, normal, or low based on the laboratory's control test.

- Normal hCG hormone level may indicate
    - Men: Less than 5 IU/L
    - Nonpregnant women: Less than 5 IU/L
    - Pregnant women
        - 24 to 28 days' gestation: 5 to 100 IU/L
        - 4 to 5 weeks' gestation: 50 to 500 IU/L
        - 5 to 6 weeks' gestation: 100 to 10,000 IU/L
        - 14 to 16 weeks' gestation: 12,000 to 270,000 IU/L
        - After 16 weeks' gestation: Levels decrease
- High hCG hormone level may indicate
    - Men
        - Lung cancer
        - Pancreatic cancer
        - Colon cancer
        - Liver cancer
        - Testicular cancer
    - Nonpregnant women
        - Lung cancer
        - Pancreatic cancer
        - Colon cancer
        - Liver cancer
        - Ovarian cancer
        - Choriocarcinoma
    - Pregnant women
        - Multiple fetuses
        - Down syndrome
        - Molar pregnancy
        - Incorrect gestational estimate
- Low hCG hormone level may indicate
    - Pregnant women
        - Ectopic pregnancy
        - Incorrect gestational estimate

- Fetal death
- Risk for spontaneous abortion

TEACH THE PATIENT

- Explain
  - Why sample is taken.
  - That the hCG hormone blood test confirms a home urine pregnancy test; however, other tests are necessary to confirm the pregnancy.
  - That the healthcare provider may ask the patient to refrain from taking hCG hormone infertility treatment, Anergan, Phenergan, Prorex, diuretics, heparin, Ambien, Mellaril, Serentil, Stelazine, or Compazine for several weeks prior to the test.

Inhibin A Test 5

Inhibin A hormone is secreted by the placenta and the level of the hormone is measured by the hormone inhibin A test. The hormone inhibin A test is a component of the quadruple screen tests that is administered at the 20th week of gestation to determine if there is a risk of birth defect(s) in the fetus such as Down syndrome. The quadruple screen test also includes the AFP test, β-hCG test, and the uE3 test.

CAUTION

The hormone inhibin A test is not used to diagnose the potential birth defects. Further testing is required if the hormone inhibin A test is abnormal.

WHAT IS BEING MEASURED?

- Inhibin A hormone level in blood

HOW IS THE TEST PERFORMED?

- See How to Collect Blood Specimen from a Vein in Chapter 1.

RATIONALE FOR THE TEST

- Assess risk for Down syndrome and other birth defects.

NURSING IMPLICATIONS

- Assess if the patient
  - Smokes
  - Is obese
  - Is a diabetic

UNDERSTANDING THE RESULTS

- The inhibin A hormone test results are available quickly.
- Normal inhibin A hormone test results.
- High inhibin A hormone level may indicate
  - Birth defect. Additional testing is required for positive diagnosis.

TEACH THE PATIENT

- Explain
  - Why sample is taken.
  - Abnormal test results do not necessarily mean there is a birth defect. Further testing might be ordered by the healthcare provider.

Prolactin 6

Prolactin is a hormone produced by the pituitary gland that increases during pregnancy, causing an increase in milk production and enlargement of the mammary glands. A high level of progesterone that occurs during pregnancy prevents milk from ejecting. Progesterone levels fall after delivery. Sucking on the nipple by the newborn causes ejection of milk from the breast, which simulates release of prolactin thus causing lactogenesis, resulting in increased production of milk. Prolactin levels return to normal after delivery if the mother is not breast-feeding. The prolactin test measures the level of prolactin in the blood.

WHAT IS BEING MEASURED?

- Prolactin level in blood

HOW IS THE TEST PERFORMED?

- Avoid eating or drinking 12 hours before the test.
- Take the blood sample 3 hours after the patient awakens. Prolactin levels are normally higher in the morning.
- Rest for 30 minutes before taking the blood sample.
- Avoid stimulating the nipples for 1 day prior to the test.
- See How to Collect Blood Specimen from a Vein in Chapter 1.

RATIONALE FOR THE TEST

- Assess for
  - Prolactinoma (pituitary gland tumor)
  - Underlying cause of amenorrhea
  - Underlying cause of infertility
  - Underlying cause of nipple discharge
  - Erectile dysfunction

NURSING IMPLICATIONS

- Assess if the patient
  - Exercised 12 hours prior to the test
  - Is under emotional stress
  - Has difficulty sleeping
  - Has undergone a radioactive tracer test 1 week prior to the test
  - Is taking tricyclic antidepressants, birth control pills, phenothiazines, hypertension medication, or cocaine
  - Has stimulated her nipples within one day prior to the test

UNDERSTANDING THE RESULTS

- The prolactin test results are available quickly. The laboratory determines normal values based on calibration of testing equipment with a control test. Test results are reported as high, normal, or low based on the laboratory's control test.
- Normal prolactin test results
  - Pregnant women: 20 to 400 ng/mL

    - Nonpregnant women: Less than 25 ng/mL
    - Men: Less than 20 ng/mL
- High prolactin level may indicate
    - Prolactinoma
    - Idiopathic hyperprolactinemia
    - Hypothyroidism
    - Cirrhosis
    - Kidney disease

TEACH THE PATIENT

- Explain
    - Why blood sample is taken.
    - That the blood sample must be taken 3 hours after the patient awakens.
    - That the patient must avoid simulating the nipples for 1 day prior to the test.
    - That the patient will be asked to rest for 30 minutes before the blood sample is taken.
    - That the healthcare provider may ask the patient to refrain from taking tricyclic antidepressants, birth control pills, phenothiazines, hypertension medication, or cocaine 12 hours prior to the test.
- Avoid eating or drinking 12 hours before the test.

Phenylketonuria Test

Phenylalanine is an amino acid in breast milk, formula, dairy products, and meats. The body requires the phenylalanine hydroxylase enzyme to metabolize phenylalanine into tyrosine. Phenylketonuria (PKU) is a genetic disorder whereby the patient is missing the phenylalanine hydroxylase enzyme and is therefore unable to metabolize phenylalanine, causing a buildup of phenylalanine level in the blood thus resulting in mental retardation and seizures. The PKU test measures the level of the phenylalanine hydroxylase enzyme in blood. Newborns are administered the PKU test between 12 and 28 hours after birth and again a week after birth.

HINT

Infants older than 6 weeks of age may be administered a PKU urine test if a PKU test was not performed at birth. Newborns who are ill are retested 3 weeks after birth.

WHAT IS BEING MEASURED?

- Measures the level of the phenylalanine hydroxylase enzyme in blood

HOW IS THE TEST PERFORMED?

- The newborn should ingest formula or breast milk for 48 hours prior to the test.
- See How to Collect Blood Specimen from a Heel Stick in Chapter 1.

Hint

The healthcare provider may order a blood sample taken from a vein to confirm a positive PKU test result that used a sample from a heel stick.

RATIONALE FOR THE TEST

- Assess for PKU.

NURSING IMPLICATIONS

- Assess if the
  - Newborn ingested formula or breast milk for 48 hours prior to the test.
  - Newborn was born prematurely.
  - Patient has vomited prior to the test.
  - Patient is taking antibiotics.

UNDERSTANDING THE RESULTS

- The PKU test results are available quickly.
- Normal PKU test results: Less than 3 mg/dL.
- High PKU test results may indicate
  - PKU

TEACH THE PATIENT

- Explain
  - Why a blood sample is taken.

- To the parents that the newborn must ingest formula or breast milk for 48 hours prior to the test.
- That the test may be repeated a week after birth.

Tay-Sachs Test 8

Hexosaminidase A is an enzyme that metabolizes ganglioside, which is a fatty acid. If the hexosaminidase A is not present, ganglioside accumulates in the brain and nerve cells, resulting in neural damage. This is referred to as Tay-Sachs disease, which is an inherited disease. The Tay-Sachs test measures the amount of hexosaminidase A in the blood.

Hint

The healthcare provider may order an amniocentesis or the chorionic villus sampling of the placenta to determine if the fetus has the hexosaminidase A enzyme.

Caution

A positive result will be confirmed by genetic testing.

WHAT IS BEING MEASURED?

- Hexosaminidase A enzyme in blood

HOW IS THE TEST PERFORMED?

- See How to Collect Blood Specimen from a Vein in Chapter 1.

RATIONALE FOR THE TEST

- Assess
    - For Tay-Sachs disease
    - If the female patient is carrying the Tay-Sachs trait

NURSING IMPLICATIONS

- Assess if the female patient
  - Has had a blood transfusion 90 days prior to the test
  - Is taking birth control pills
  - Is pregnant

UNDERSTANDING THE RESULTS

- The creatinine blood level test and creatinine clearance test results are available quickly. The laboratory determines normal values based on calibration of testing equipment with a control test. Test results are reported as high, normal, or low based on the laboratory's control test.
- Normal test results:
  - Total hexosaminidase: 10.4 to 23.8 U/L
  - Percentage of blood level: 56% to 80%
- One-half of the normal level may indicate: Patient is a carrier of the Tay-Sachs trait.
- No hexosaminidase A enzyme may indicate
  - Tay-Sachs disease
  - Sandhoff disease (missing hexosaminidase A and hexosaminidase B)

TEACH THE PATIENT

- Explain
  - Why a blood sample is taken.
  - That the test results may indicate the patient carries the Tay-Sachs trait and genetic counseling is recommended.
  - That the healthcare provider may ask the patient to refrain from taking birth control pills prior to the test.

Sickle Cell Test 9

Normal red blood cells contain hemoglobin A. In sickle cell disease, red blood cells contain hemoglobin S, which causes the red blood cell to form a sickle shape, therefore called sickled blood cells. Sickle cell disease is an autosomal recessive disease in which the sickle cell gene must be inherited from both parents. Patients with sickled blood cells can experience a sickle cell crisis when sickled blood cells block blood vessels, resulting in decreased blood flow. Sickled blood cells are destroyed faster than normal red blood cells leading to sickle cell anemia. The sickle cell test determines if the patient has sickled blood cells.

HINT

The healthcare provider may order the high performance liquid chlomatography (HPLC) test to examine the patient's DNA for the sickle cell gene. Sickle cell disease is more prevalent in African Americans. Healthcare providers commonly test newborns for the sickle cell trait, although infants younger than 6 months can have false-negative test results since they have fetal hemoglobin in their blood and, therefore, the sickle cell test is repeated after 6 months of age. The healthcare provider may order the sickle cell test for a genus using chorionic villus sampling (CVS) or amniocentesis.

CAUTION

The patient should undergo genetic counseling if he/she has the sickle cell trait or sickle cell disease.

WHAT IS BEING MEASURED?

- Existence of sickled blood cells

HOW IS THE TEST PERFORMED?

- See How to Collect Blood Specimen from a Vein in Chapter 1.

RATIONALE FOR THE TEST

- Assess for
  - Sickle cell trait
  - Sickle cell disease

NURSING IMPLICATIONS

- Assess if the patient
  - Has received a blood transfusion within 3 months before the test is administered
  - Should undergo genetic counseling if the test result is positive for sickle cell trait or sickle cell disease.

UNDERSTANDING THE RESULTS

- The sickle cell test results are available quickly.
- Normal sickle cell test results: Normal hemoglobin presence.
- Abnormal sickle cell test results:
  - Sickle cell trait: One-half of red blood cells contain hemoglobin A and the other half contain hemoglobin S.
  - Sickle cell disease: Nearly all red blood cells contain hemoglobin S.

TEACH THE PATIENT

- Explain
  - Why sample is taken.
  - That the patient should receive genetic counseling if the sickle cell trait or sickle cell disease is present.

Hemochromatosis Gene Test

The hemochromatosis (HFE) gene increases the absorption of iron (HFE), which causes a buildup of iron in the liver, heart, blood, joints, skin, and pancreas, resulting in joint pain, weight loss, and decreased energy. This can lead to arrhythmia, cirrhosis, diabetes, heart failure, arthritis, and change in skin color. The HFE gene test determines if the patient has the HFE gene.

Hint

The healthcare provider will order the test if close family members have HFE. The healthcare provider might order the ferritin level test and transferring saturation test to measure the level of iron in the patient's blood.

Caution

Existence of the HFE gene means that the patient has an increased chance of having HFE, but does not mean that the patient has HFE. It is advised that the patient consult a genetic counselor before the test is administered to discuss the risk of developing HFE.

WHAT IS BEING MEASURED?

- Existence of the HFE gene in blood

HOW IS THE TEST PERFORMED?

- See How to Collect Blood Specimen from a Vein in Chapter 1.

RATIONALE FOR THE TEST

- Screen for presence of the HFE gene.
- Assess
  - Underlying cause of HFE. If positive, other tests for HFE are not necessary.
  - Treatment of HFE.

NURSING IMPLICATIONS

- Assess if the patient
  - Has received a blood transfusion within a week before the test is administered
  - Has undergone genetic counseling prior to the test
- The HFE gene test may not rule out the existence of the HFE gene. The test detects the most common HFE gene mutation, but not all mutations.

UNDERSTANDING THE RESULTS

- The HFE gene test results are available in 2 weeks.
- Normal HFE gene test result (negative): No mutation found.
- Abnormal HFE gene test result (positive): Mutation found. The patient has a risk of developing HFE.

TEACH THE PATIENT

- Explain
  - Why sample is taken.
  - That the patient should receive genetic counseling prior to the test.

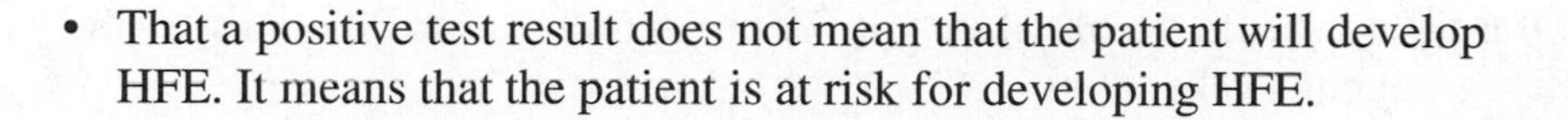

- That a positive test result does not mean that the patient will develop HFE. It means that the patient is at risk for developing HFE.

Summary

Enzymes, protein, hormones, and antibodies in a blood sample give healthcare providers clues to the underlying cause of infertility and to the existence or the risk for genetic disorders.

Infertility is sometimes caused by an immune system response to semen, where semen is recognized as an antigen, causing antibodies to attack and kill the sperm. Identifying the antisperm antibody in blood, vaginal fluids, or semen enables healthcare providers to find the cause of infertility.

The AFP level in blood is examined between 15 and 20 weeks of gestation for clues of genetic disease in the fetus. An unusually high or low level is a sign that further examination is necessary. Hormones in blood samples can indicate infertility, pregnancy, Down syndrome, and tumor.

Quiz

1. The lack of hexosaminidase A enzyme indicates that the patient
   a. May have multiple sclerosis (MS)
   b. May have Tay-Sachs disease
   c. Is pregnant
   d. Is infertile

2. The PKU test might indicate the risk of mental retardation and seizures.
   a. True
   b. False

3. The inhibin A test is used to diagnose
   a. Birth defects
   b. Tay-Sachs disease
   c. MS
   d. None of the above

4. Why should the blood sample for prolactin be taken 3 hours after the patient awakens?
   a. The test should always be taken immediately prior to lunch.
   b. The patient requires a good night sleep before the test.
   c. Prolactin levels are normally high when the patient first awakens.
   d. None of the above.

5. The presence of hCG in a blood sample may indicate
   a. Pregnancy
   b. Ectopic pregnancy
   c. Testicular cancer
   d. All of the above

6. What test is administered to determine if the patient may have precocious puberty?
   a. FSH test
   b. AFP test
   c. hCG test
   d. Inhibin A test

7. What might adversely affect the prolactin test?
   a. Exercising 12 hours before the test
   b. Natural birth
   c. Multiple births
   d. Sleeping the night before the test

8. What must the patient do before taking the prolactin test?
   a. Avoid the sun.
   b. Rest for 30 minutes.
   c. Exercise for 30 minutes.
   d. Avoid alcohol.

9. What will the healthcare provider do if the Tay-Sachs test result is positive?
   a. Order a CT scan.
   b. Order an MRI.
   c. Order genetic testing.
   d. None of the above.

10. What will the healthcare provider do if an ill newborn is administered the PKU test?
    a. Nothing.
    b. Test the newborn 3 months after birth.
    c. Test the newborn 3 weeks after birth.
    d. Order a genetic consult.

Answers

1. b. May have Tay-Sachs disease.
2. a. True.
3. a. Birth defects.
4. c. Prolactin levels are normally high when the patient first awakens.
5. d. All of the above.
6. a. FSH test.
7. a. Exercising 12 hours before the test.
8. b. Rest for 30 minutes.
9. c. Order genetic testing.
10. c. Test the newborn 3 weeks after birth.

CHAPTER 11

Tests for Infection

Signs and symptoms of an infection typically indicate that a microorganism has invaded the patient's body. There has been a tendency for healthcare providers to prescribe an antibiotic at the first sign of an infection. In doing so, the healthcare provider assumes that the patient is experiencing a bacterial infection and that the antibiotic will eliminate the bacteria. Typically, if the infection does not improve within 7 to 10 days, the patient will undergo an additional assessment.

Over time, this approach to treating infection has been one of many factors leading to antibiotic-resistant bacteria. The bacteria adapted to common antibiotics, making the antibiotic useless. Today, healthcare providers are encouraged to order tests that identify the microorganism and the best medication to kill them. These tests are called culture and sensitivity tests.

A sample of the patient's blood or infected tissue is sent to the laboratory where the microorganism is encouraged to replicate in a culture dish. Laboratory specialists then conduct tests to identify the microorganism. Once identified, the microorganism is exposed to medication known to kill it. The laboratory specialist determines the best medication and the minimum dose to administer to the patient that will kill the microorganism. In this way, the proper dose of the right medication can be prescribed, reducing the risk that the microorganism will become resistant to the medication.

In this chapter, you will learn about tests used to identify microorganisms and tests to identify medications to kill the microorganism.

Learning Objectives

1. Antibody Tests
2. Blood Culture
3. Mononucleosis Tests
4. *Helicobacter pylori* Tests
5. Herpes Simplex Virus Tests (HSV)
6. Lyme Disease Test
7. Rubella Test
8. Syphilis Tests

Key Words

Antibodies
Blood transfusion reaction
Congenital rubella syndrome (CRS)
Enzyme-linked immunosorbent assay (ELISA)
Epstein-Barr virus (EBV)
Heterophil antibodies
HSV (herpes simplex virus)
HSV-1
HSV-2
Indirect Fluorescent antibody (IFA)
Monospot test
POCkit tests
Rh antigen
Rh immune globulin (RhoGAM)
Varicella zoster

Antibody Tests 1

Antibodies are proteins made by the immune system that bind to bacteria, viruses, and other microorganisms to destroy the microorganism. Antibodies can also bind to red blood cells destroying them also. Antibody tests determine if antibodies are attacking red blood cells.

Antibodies are created as a result of

- Blood transfusion reaction: The transfused blood has different antigens on the surface of its red blood cells than the patient's red blood cells. This causes the immune system to produce antibodies that attack the transfused red blood cells.
- Rhesus factor sensitization: The Rh antigen is in the fetus's blood (Rh positive), but not in the pregnant woman's blood (Rh negative). During delivery, the fetus's blood mixes with the mother's blood causing the mother's immune system to create antibodies against the fetus's red blood cells. The mother becomes Rh sensitive. The antibodies can attack the fetus's red blood cells in future pregnancies if the fetus is Rh positive. Women with Rh-negative blood are given the Rh immune globulin (RhoGAM) that typically stops Rh sensitivity.
  - Rh antibody titer: The test is performed in early pregnancy to determine the mother's blood type and to determine if the mother is Rh negative.
- Autoimmune hemolytic anemia: The patient's immune system creates antibodies against the patient's red blood cells.
  - Direct Coombs test: Identifies antibodies attached to the patient's red blood cells. This test is performed
    - On a newborn whose mother is Rh negative to determine if the antibodies crossed the placenta into the newborn's blood
    - On a patient who received a blood transfusion to determine if there is a transfusion reaction
    - On a patient to determine if an autoimmune response is occurring
  - Indirect Coombs test: Identifies antibodies that have not but could attach to the patient's red blood cells if the patient's blood is mixed. This test is performed
    - On the blood transfusion recipient or donor before the transfusion to identify if antibodies exist in their blood.

WHAT IS BEING MEASURED?

- Antibodies that might attack the patient's red blood cells

HOW IS THE TEST PERFORMED?

- See How to Collect Blood Specimen from Vein in Chapter 1.

RATIONALE FOR THE TEST

- Assess
  - Blood transfusion reaction or potential reaction
  - Rh factor sensitization
  - Autoimmune hemolytic anemia

NURSING IMPLICATIONS

- Assess reason(s) patient should refrain from taking the test
  - History of blood transfusions
  - Recently administered contrast material IV or dextran injections
  - Was pregnant 3 months ago
  - Has taken cephalosporins, tetracyclines, tuberculosis medication, sulfa medication, or insulin

UNDERSTANDING THE RESULTS

- Negative test results
  - No antibodies found in the patient's blood (direct Coombs test).
  - The patient's blood is compatible with transfusion donor's blood (indirect Coombs test).
  - Rh-negative mother is not Rh sensitized (indirect Coombs test). The mother is administered the RhoGAM vaccine.
- Positive test results
  - The patient's blood contains antibodies for the patient's red blood cells (direct Coombs test)
  - The patient's blood is incompatible with transfusion donor's blood (indirect Coombs test). The blood is not transfused.
  - Rh-negative mother is Rh sensitized (indirect Coombs test). The healthcare provider monitors the pregnancy closely to prevent problems with the fetus if the fetus is Rh positive.

TEACH THE PATIENT

- Explain
  - That no special preparation is needed for the test.

- Why the test is ordered.
- Rh sensitivity and the administration of RhoGAM vaccine if the patient is Rh negative.

Blood Culture 2

Blood can be infected by bacteria or fungi. A blood culture identifies the bacteria or fungi by allowing the microorganism to grow in a controlled environment and then examining the microorganism under a microscope.

HINT

The healthcare provider typically orders a sensitivity test along with the blood culture. The sensitivity test identifies medication that kills the microorganism.

WHAT IS BEING MEASURED?

- Bacteria or fungi in the blood

HOW IS THE TEST PERFORMED?

- A blood sample is obtained from three different veins and at two different times to assure that bacteria on the skin does not contaminate the blood sample.
- See How to Collect Blood Specimen from a Vein in Chapter 1.

HINT

Patients who have catheters in their vein will have the blood sample taken from the catheter.

RATIONALE FOR THE TEST

- Assess
  - The existence of bacteria or fungi in blood
  - For endocarditis
  - Medication that will kill the microorganism
  - The cause of unexplained fever
  - The effect of treatment of a microorganism infection

NURSING IMPLICATIONS

- Kidney, lung, and throat infections can lead to bacterial infection of the blood.
- Assess if the patient has recently been administered antibiotics, since this can affect the test results.

UNDERSTANDING THE RESULTS

- Test results for bacterial infection are normally available in 3 days but can take longer than 10 days. Test results for a fungal infection can take up to 30 days.
- Normal is negative: No bacteria or fungi found in the blood sample.
- Normal false-negative: Improper processing or improper sampling results in bacteria or fungi not found in the blood sample but patient's blood is infected.
- Abnormal is positive: Bacteria or fungi found in the blood sample.
- Abnormal false-positive: Contaminated sample or improper processing results in bacteria or fungi found in the blood sample but not in the patient's blood.

TEACH THE PATIENT

- Explain
  - Why blood sample is taken.
  - Multiple samples may be taken from different sites and at different times.
  - That the results of the sensitivity test will tell the healthcare provider about the medication that will kill the microorganism.

Mononucleosis Tests 3

The Epstein-Barr virus (EBV) causes mononucleosis. The mononucleosis tests identify antibodies for the Epstein-Barr virus in the blood sample. There are two kinds of mononucleosis tests

- Monospot test: This test identifies heterophil antibodies that form between 2 and 9 weeks after the patient becomes infected.

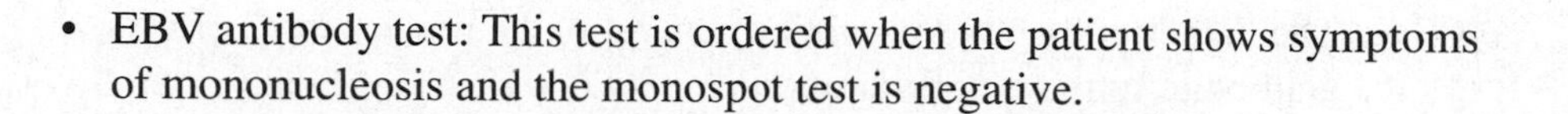

- EBV antibody test: This test is ordered when the patient shows symptoms of mononucleosis and the monospot test is negative.

WHAT IS BEING MEASURED?

- Antibodies for the Epstein-Barr virus in blood

HOW IS THE TEST PERFORMED?

- See How to Collect Blood Specimen from a Vein in Chapter 1.
- See How to Collect Blood Specimen from a Finger Stick in Chapter 1.

RATIONALE FOR THE TEST

- Assess for infectious mononucleosis

NURSING IMPLICATIONS

- A false-negative result can occur if the test is administered within 2 weeks of the patient being infected.
- Assess if the patient has lymphoma, hepatitis, rubella, cytomegalovirus, or lupus.

UNDERSTANDING THE RESULTS

- Monospot test results are available in 1 hour.
  - Normal (negative) heterophil antibody range is 0.
  - High (positive) heterophil antibody levels greater than 0 may indicate
    - Mononucleosis
    - Rheumatoid arthritis
    - Leukemia
    - Hepatitis
    - Lymphoma
- EBV antibody test results are available in 3 days.
  - Results are reported as a titer. A titer specifies how much of the sample of blood is diluted with saline before antibodies are no longer detected. The titer is reported as a ratio of parts of the blood sample and saline.

The higher the second number of the ratio, the greater number of antibodies in the blood sample.

- Normal (negative) EBV antibody titer range is below 1 to 40 (1:40).
- High (positive) EBV antibody titer may indicate
  - Mononucleosis

Hint

The presence of IgG antibody indicates that the patient was exposed to the Epstein-Barr virus in the past. Most adults have the IgG antibody.

TEACH THE PATIENT

- Explain
  - Why blood sample is taken.
  - That a false-negative result can occur if the test is administered within 2 weeks of the patient being infected.

Helicobacter pylori Tests 4

Helicobacter pylori is a bacterium that infects the stomach and duodenum and may result in a peptic ulcer. Many patients have *H pylori* but few develop peptic ulcer disease. There are four tests used to detect *H pylori*

- *Helicobacter pylori* blood antibody test: This test determines if the blood sample has *H pylori* antibodies.
- Urea breath test: This test determines the presence of *H pylori* in the stomach
- *Helicobacter pylori* stool antigen test: This test determines the presence of *H pylori* antigens in feces.
- Stomach biopsy: This is the endoscopic removal of the lining of the stomach and small intestine, which is examined for the presence of *H pylori*.

Caution

All tests may produce a false-negative result if the H pylori count is low and undetectable. The stomach biopsy may produce a false-negative result if the sample is not infected. The blood test for H pylori antibodies may give a false-positive result since antibodies are present years after the H pylori infection is resolved.

WHAT IS BEING MEASURED?

- *Helicobacter pylori* antibody levels in blood

HOW IS THE TEST PERFORMED?

- See How to Collect Blood Specimen from a Vein in Chapter 1.

RATIONALE FOR THE TEST

- Assess for the presence of *H pylori*.
- Assess if the treatment for *H pylori* is successful.

NURSING IMPLICATIONS

- Assess if the patient
  - Is pregnant or breast-feeding. The urea breath test is not performed.
  - Has taken antibiotics. This can reduce the *H pylori* count.
  - Has taken Prilosec, Carafate, Pepcid, AcipHex, Zantac, or Pepto-Bismol, since these can affect the test results.

UNDERSTANDING THE RESULTS

- Blood test results are available in 1 day.
  - Normal results: No *H pylori* antibodies
  - Abnormal blood test results: Presence of *H pylori* antibodies
- Urea breath test results are available in 3 hours.
  - Normal results: No tagged hydrocarbon
  - Abnormal blood test results: Tagged hydrocarbon present
- Stool antigen test results are available in 3 hours.
  - Normal results: No *H pylori* antigens
  - Abnormal blood test results: Presence of *H pylori* antigens
- Stomach biopsy test result is available between 2 and 10 days.
  - Normal results: No *H pylori*
  - Abnormal blood test results: Presence of *H pylori*

TEACH THE PATIENT

- Explain
  - Why blood sample is taken.
  - How to take a stool collection at home, if necessary.
  - That there is no long-term effect of swallowing a capsule or water containing radioactive material in the urea breath test.
  - That the healthcare provider may want the patient to refrain from taking Prilosec, Carafate, Pepcid, AcipHex, Zantac, or Pepto-Bismol 12 hours before the test.

Herpes Simplex Virus Tests 5

Herpes simplex virus (HSV) causes painful blisterlike sores on the skin and mucous membranes of the mouth, vagina, urethra, rectum, nose, and throat. There are two types of herpes simplex virus. These are

- Herpes simplex virus type 1 (HSV-1): Commonly called a fever blister or cold sore that appears on the lips and is spread by direct contact or indirectly through sharing eating utensils
- Herpes simplex virus type 2 (HSV-2): Commonly called genital herpes that appears on the penis or vagina and is spread by direct contact.

There are four common tests of herpes simplex virus

- HSV antibody test: This test identifies HSV antibodies in the blood but cannot differentiate between HSV-1 and HSV-2 and can produce a false-negative result, since the immune system takes several days to develop sufficient antibodies to be detected by the herpes simplex virus antibody test.
- Polymerase chain reaction (PCR) test: This test differentiates between HSV-1 and HSV-2 in cell scraping.
- Herpes virus antigen detection test: This test detects antigens on cells scraped from the herpes simplex virus sore using a microscope.
- Herpes viral culture: This test cultures cells or fluid from a herpes simplex virus sore to determine if the sore is from HSV-2.

Hint

HSV-1 can infect the genitals. HSV-2 can infect the newborn if the mother has HSV-2. It can also infect the mouth. There is no cure for herpes simplex virus; however, the herpes simplex virus can go into remission. Fatigue, stress, or sunlight can cause herpes simplex virus sore to recur. Varicella zoster is a type of herpes virus that is better known as shingles and chickenpox.

WHAT IS BEING MEASURED?

- Herpes simplex virus
  - Antibodies in blood
  - Antibodies in cell scraping
  - Antigens in cell scraping
  - Antigens in culture

HOW IS THE TEST PERFORMED?

- See How to Collect Blood Specimen from a Vein in Chapter 1.
- The POCkit tests for HSV-2 antibodies in blood in 10 minutes.

RATIONALE FOR THE TEST

- Assess for herpes simplex virus.
- Identify the types of herpes simplex virus.

NURSING IMPLICATIONS

- Assess if the patient
  - Has had sexual contact before the test results are known.
  - Urinated 2 hours before the test (sore is on the urethra).
  - Douched 24 hours before the test (sore is on the cervices).
  - Has taken ganciclovir, acyclovir, valacyclovir, or famciclovir.
  - If aculture was taken from a crusted sore.

UNDERSTANDING THE RESULTS

- Rapid herpes simplex virus culture test results take 3 days.
- The herpes simplex virus culture test results take 14 days.
- The herpes simplex virus antigen test results take 1 day.
- The polymerase chain reaction (PCR) test results take 3 days.
- The antibody blood tests take 2 days.
- Normal herpes simplex virus test results (negative): No HSV, no HSV antigen, no HSV DNA, no HSV antibodies.
- Abnormal herpes simplex virus test results (positive): HSV, HSV antigen, HSV DNA, HSV antibodies present.

CAUTION

A negative test does not rule out a herpes simplex virus infection. Additional testing is necessary if symptoms appear.

TEACH THE PATIENT

- Explain
  - Why blood sample is taken.
  - That the patient should not douche 24 hours before the test if the sore is on the crevices.
  - That the patient should not urinate 2 hours before the test if the sore is on the urethra.
  - That the patient should not have sexual contact before the test results are known.
  - That the healthcare provider may request that the patient may avoid taking ganciclovir, acyclovir, valacyclovir, or famciclovir for 2 weeks prior to the test.

Lyme Disease Test 6

The *Borrelia burgdorferi* bacterium is carried by ticks and transmitted by a tick bite. The Lyme disease test detects *B burgdorferi* bacteria antibodies in blood. The healthcare provider will order at least two of three Lyme disease tests.

- Indirect fluorescent antibody (IFA)
- Enzyme-linked immunosorbent assay (ELISA): quickest and most sensitive test for Lyme disease
- Western blot test: confirms positive IFA and ELISA test results

---

HINT

The healthcare provider orders the Lyme disease test if the patient has symptoms of Lyme disease or is known to have been exposed to ticks.

CAUTION

The Lyme disease test can produce a false-negative result if performed within 2 months of the tick bite because the patient's immune system may take 2 months to produce a detectable amount of B burgdorferi bacteria antibodies. Likewise, test can produce a false-positive result because B burgdorferi bacteria antibodies remain in the patient's blood years after the B burgdorferi bacteria was killed.

WHAT IS BEING MEASURED?

- *Borrelia burgdorferi* bacteria antibodies level in blood

HOW IS THE TEST PERFORMED?

- See How to Collect Blood Specimen from a Vein in Chapter 1.

RATIONALE FOR THE TEST

- Assess
  - For Lyme disease

NURSING IMPLICATIONS

- Assess if the patient
  - Has a viral or bacterial infection
  - Lipid level is high
  - Has ever been diagnosed with Lyme disease
  - Has ever been exposed to ticks

UNDERSTANDING THE RESULTS

- The Lyme disease test results are available in 2 weeks. The laboratory determines normal values based on calibration of testing equipment with a control test. Test results are reported as high, normal, or low based on the laboratory's control test.
- Results are reported as a titer. A titer specifies how much of the sample of blood is diluted with saline before antibodies are no longer detected. The titer is reported as a ratio of parts of the blood sample and saline. The higher the second number of the ratio, the greater number of antibodies in the blood sample.
- Normal Lyme disease test results (negative)
  - IFA: Less than 1:256
  - ELISA: No *B burgdorferi* bacteria antibodies found
  - Western blot test: No *B burgdorferi* bacteria antibodies found
- Abnormal Lyme disease (positive).
  - IFA: Greater than 1:256
  - ELISA: *B burgdorferi* bacteria antibodies found
  - Western blot test: *B burgdorferi* bacteria antibodies found

TEACH THE PATIENT

- Explain
  - Why blood sample is taken.
  - That a negative test results does not rule out Lyme disease since the immune system can take 2 months to develop detectable antibodies.
  - That a positive test results must be confirmed by another Lyme disease test, which is usually the Western blot test.
  - That a positive test result does not mean that the patient has Lyme disease since detectable antibodies remain in the blood for years after a *B burgdorferi* bacteria infection. The patient might have been infected in the past but is not currently infected.

Rubella Test 7

Rubella (German measles) is a virus that causes congenital rubella syndrome (CRS) if a pregnant woman is infected with the rubella virus and transmits the rubella virus to the fetus in the first trimester. Congenital rubella syndrome consists of birth defects and possibly a miscarriage or stillbirth. The rubella test detects rubella virus antibodies in the blood. There are two rubella virus antibodies that are detected:

- IgG antibody: The patient has immunity against the rubella virus. The rubella virus antibody developed from a previous rubella virus infection or from the rubella virus vaccination.
- IgM antibody: The patient currently has or recently had a rubella virus infection.

WHAT IS BEING MEASURED?

- Rubella virus antibodies in blood

HOW IS THE TEST PERFORMED?

- See How to Collect Blood Specimen from a Vein in Chapter 1.

RATIONALE FOR THE TEST

- Assess if the patient
  - Is immune to the rubella virus
  - Currently or recently has had a rubella virus infection

NURSING IMPLICATIONS

- A woman who wants to become pregnant must receive the rubella test to determine if she is immune to the rubella virus.
- A woman who is not immune to the rubella virus can receive the rubella vaccination but must wait 1 month after receiving the vaccination before becoming pregnant.
- A pregnant woman who is not immune to the rubella virus cannot receive the rubella vaccination during the pregnancy and must avoid anyone who is at risk of being infected with the rubella virus.

UNDERSTANDING THE RESULTS

- The rubella test results are available quickly.
- Results are reported as a titer. A titer specifies how much of the sample of blood is diluted with saline before antibodies are no longer detected. The titer is reported as a ratio of parts of the blood sample and saline. The higher the second number of the ratio, the greater number of antibodies in the blood sample.

- Normal test results (positive): 1:10 to 1:20.
- Lower levels (negative) may indicate: Risk for rubella virus infection.

TEACH THE PATIENT

- Explain
    - Why blood sample is taken.
    - That if the patient is pregnant, she should avoid others who might be at risk of rubella virus infection unless she is immune to the rubella virus.
    - That the patient should have the rubella test before becoming pregnant. The healthcare provider will likely order the rubella vaccination if she is not immune. She must wait 1 month after the vaccination before becoming pregnant.

Syphilis Tests 8

Treponema pallidum is the bacterium that causes syphilis. The syphilis tests identify *T pallidum* antibodies in a blood sample. There are seven types of syphilis tests:

- Veneral disease research laboratory (VDRL): This test identifies anti-cardiolipin antibodies that are produced by the patient who has syphilis. Diseases including syphilis cause the production of anticardiolipin antibodies; therefore, this test is used for screening of syphilis and not diagnosing syphilis.
- Rapid plasma reagin (RPR): This test is similar to VDRL except antibodies can be detected without the aid of a microscope.
- Enzyme-linked immunosorbent assay (ELISA): This test identifies *T pallidum* antibodies and is used for screening of syphilis. Additional test is necessary to diagnose syphilis.
- Fluorescent treponemal antibody absorption (FTA-ABS): This test identifies *T pallidum* antibodies after the fourth week of the initial infection and is used to confirm other positive test results.
- *Treponema pallidum* particle agglutination assay (TPPA): This test is similar to the FTA-ABS except it is not used to test spinal fluid.
- Darkfield microscopy: This test identifies the *T pallidum* bacterium under a darkfield microscope and is used to diagnose the early stage of syphilis.
- Microhemagglutination assay (MHA-TP): This test is similar to TPPA.

HINT

Many syphilis tests can use either blood sample or a sample of spinal fluid.

CAUTION

A positive VDRL test or RPR test does not mean that the patient has syphilis. Other conditions can cause a positive test result. FTA-ABS, MHA-TP, and TPPA tests remain positive even after the patient is successfully treated for syphilis. VDRL and RPR tests are negative when treatment for syphilis is successful. A negative result does not rule out syphilis, since detectable antibodies can take 4 weeks following the initial infection to develop.

WHAT IS BEING MEASURED?

- *Treponema pallidum* antibodies in blood or spinal fluid
- Anticardiolipin antibodies in blood or spinal fluid
- *Treponema pallidum* bacterium in blood or spinal fluid

HOW IS THE TEST PERFORMED?

- See How to Collect Blood Specimen from a Vein in Chapter 1.
- The healthcare provider will insert a needle into the spinal canal and remove a sample of spinal fluid.

RATIONALE FOR THE TEST

- Assess
  - For syphilis infection
  - Treatment for syphilis infection

NURSING IMPLICATIONS

- Assess if the patient has
  - Had a blood transfusion recently
  - Liver disease, yaws, lupus, or HIV
  - Taken antibiotics
- Consult the healthcare facility's policy for reporting diagnosed cases of syphilis to the local health department.

UNDERSTANDING THE RESULTS

- The syphilis test results are available in 10 days.
- Normal test results (nonreactive/negative): No antibodies found. No *T pallidum* bacterium seen under the microscope.
- Abnormal test results (reactive/positive): Antibody reagin found. *Treponema pallidum* bacterium seen under the microscope.
- Inconclusive test results (equivocal): unclear if antibody reagin found. Unclear if *T pallidum* bacterium is seen under the microscope.

TEACH THE PATIENT

- Explain
  - Why sample is taken.
  - That the sex partner must also be tested.
  - That the patient should refrain from sex until test shows that the patient and the sex partner are no longer infected.
  - That a diagnosis of syphilis may have to be reported to the health department.
  - That a negative test may not rule out syphilis since detectable antibodies can take 4 weeks following the initial infection to develop.

Summary

Prescribing medication for infection without first identifying the microorganism that is causing the infection has led to medication-resistant microorganisms, especially antibacterial-resistant bacteria.

Healthcare providers are encouraged to take samples of the infected blood or tissue and order laboratory tests that identify the microorganism before prescribing medication to the patient.

A culture and sensitivity test is ordered to identify the microorganism and to determine the medication and dose to kill the microorganism. A laboratory specialist grows the microorganism taken from the sample in a culture dish and then performs tests that definitively identify the microorganism.

Medication is then applied to the sample to determine if the medication kills the microorganism. Once the correct medication is found, the laboratory specialist

reports the findings to the healthcare provider, who then prescribes the proper medication at the proper dosage to the patient.

Quiz

1. Why is the Rh antibody titer administered early in pregnancy?
   a. To determine the Rh factor of the fetus
   b. To determine the Rh factor of both parents
   c. To determine the mother's blood type and if the mother is Rh negative
   d. None of the above

2. What is the purpose of the monospot test?
   a. This test identifies heterophil antibodies that form between 2 and 9 weeks after the patient becomes infected.
   b. This test identifies heterophil antibodies that form between 1 and 9 weeks after the patient becomes infected.
   c. This test identifies heterophil antibodies that form between 2 and 4 weeks after the patient becomes infected.
   d. This test identifies heterophil antibodies that form between 2 and 9 weeks before the patient becomes infected.

3. The herpes simplex virus type 1 is commonly known as
   a. Fever blister
   b. Genital herpes
   c. HSV-2
   d. HSV-3

4. What is the purpose of the tissue type test?
   a. To determine the compatibility between the patient and donor's tissues
   b. To assesse the antigen pattern associated with an autoimmune disease
   c. To identify the antigen pattern for the patient
   d. All of the above

5. What test is ordered to determine if the patient has a *B burgdorferi* bacteria infection?
   a. Indirect fluorescent antibody
   b. Enzyme-linked immunosorbent assay
   c. ELISA
   d. All of the above

6. What does it mean if the patient has the IgG antibody?
   a. The patient currently has a rubella virus infection.
   b. The patient recently had a rubella virus infection.
   c. The patient will be infected by the rubella virus.
   d. The patient has immunity against the rubella virus.

7. What is the purpose of the VDRL test?
   a. To identify *P treponema* antibodies
   b. To identify *T pallidum* antibodies
   c. To identify the anti-cardiolipin antibodies
   d. To identify the *T pallidum* bacterium

8. What is the purpose of the urea breath test?
   a. To determine the presence of *H pylori* in stool
   b. To determine the presence of *H pylori* in the blood
   c. To determine the presence of *H pylori* in the stomach
   d. To estimate the alcohol content of blood

9. Why is the direct Coombs test administered?
   a. To determine on an adult patient if an autoimmune response is occurring
   b. To determine on an adult patient who received a blood transfusion if there is a transfusion reaction
   c. To determine in a newborn whose mother is Rh-negative if the antibodies crossed the placenta into the newborn's blood
   d. All of the above

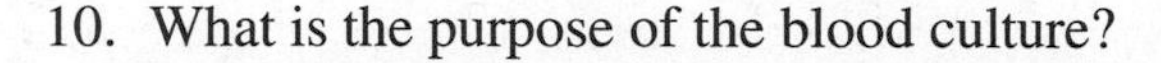

10. What is the purpose of the blood culture?
    a. To identify the bacteria or fungi
    b. To determine the Rh factor of a fetus
    c. To determine the Rh factor of the mother
    d. To identify the medication that will kill the microorganism

Answers

1. c. To determine the mother's blood type and if the mother is Rh negative.
2. a. This test identifies heterophil antibodies that form between 2 and 9 weeks after the patient becomes infected.
3. a. Fever blister.
4. d. All of the above.
5. d. All of the above.
6. d. The patient has immunity against the rubella virus.
7. c. To identify the anticardiolipin antibodies.
8. c. To determine the presence of *H pylori* in the stomach.
9. d. All of the above.
10. a. To identify the bacteria or fungi.

CHAPTER 12

Renal Function Tests

Metabolic waste is carried by the blood to the kidneys. The glomerulus in the kidneys acts as a filter to remove waste from the blood, which is collected in a renal tubule as urine. Metabolic waste such as sodium, potassium, and phosphorus can be reused by the body and is returned to the blood by the kidneys. The remaining waste is excreted as urine.

Renal function is measured in percentages. A person with two healthy kidneys has 100% renal function. Likewise, a person with one healthy kidney and one kidney with total renal failure is said to have 50% renal function. A person will experience health problems if he/she has 25% or less renal function. Dialysis is typically ordered for a patient with less than 15% renal function.

Renal failure occurs when the glomerulus no longer filters waste from the blood. This can occur suddenly (acute renal failure) in response to illness, medications, accidents, and poisons. It can also happen slowly (chronic kidney disease) from illnesses such as diabetes and high blood pressure. Chronic kidney disease can lead to end-stage renal disease, where all or nearly all the renal functions are permanently destroyed.

There are two tests used to determine renal function. These are blood urea nitrogen (BUN) and creatinine. You will learn about these tests in this chapter.

Learning Objectives

1. Blood Urea Nitrogen
2. Creatinine and Creatinine Clearance

Key Words

Ammonia
Blood creatinine level test
Creatine
Creatine phosphate
Nitrogen
Urea

Blood Urea Nitrogen 1

Ammonia is formed when bacteria in the intestines break down protein. Ammonia is then converted by the liver into the waste product urea, which is excreted by the kidneys in urine. Urea contains nitrogen. The blood urea nitrogen (BUN) test measures the level of nitrogen in the blood derived from urea.

Hint

The BUN test is typically performed with the creatinine test. The healthcare provider then uses the BUN/creatinine ratio to evaluate the patient's condition.

WHAT IS BEING MEASURED?

- BUN level in blood

HOW IS THE TEST PERFORMED?

- See How to Collect Blood Specimen from a Vein in Chapter 1.

RATIONALE FOR THE TEST

- Screen for
  - Kidney function
  - Dehydration
- Assess
  - The treatment for kidney disease
  - Need for kidney dialysis

NURSING IMPLICATIONS

- Assess
  - If the patient has eaten meat or protein prior to the administration of the test
  - If the patient has taken Fungizone, Garamycin, Nebcin, Chloromycetin, nafcillin, Kantrex, tetracycline, or diuretics
  - The patient's age, since BUN levels increase with age

UNDERSTANDING THE RESULTS

- Test results are available in 1 day. The laboratory determines normal values based on calibration of testing equipment with a control test. Test results are reported as high, normal, or low based on the laboratory's control test.
- Normal BUN range: 10 to 20 mg/dL
- Normal BUN/creatinine ratio: 10:1 to 20:1
- High BUN values may indicate
  - Kidney disease
  - Kidney stone
  - Addison disease
  - Kidney tumor
  - Burns
  - Heart failure

    - GI tract bleeding
    - Dehydration
    - High protein diet
- High BUN/creatinine ratio may indicate
    - Respiratory tract bleeding
    - GI tract bleeding
    - Shock
    - Kidney failure
    - Urinary tract blockage
    - Dehydration
- Low BUN values may indicate
    - Malnutrition
    - Liver damage
    - Overhydration
    - Low protein diet
- Low BUN/creatinine ratio may indicate
    - Rhabdomyolysis
    - Cirrhosis
    - Syndrome of inappropriate secretion of antidiuretic hormone (SIADH)
    - Muscle injury

HINT

BUN level is higher than the creatinine level in dehydration. Both BUN and creatinine levels are high when there is blockage of urine flow.

TEACH THE PATIENT

- Explain
    - Why blood sample is taken.
    - That the patient should avoid eating meat or protein for 24 hours before the test is administered.
    - That the healthcare provider may ask the patient to stop taking Fungizone, Garamycin, Nebcin, Chloromycetin, nafcillin, Kantrex, tetracycline, or diuretics for 2 weeks prior to administration of the test.

Creatinine and Creatinine Clearance 2

Creatine phosphate provides energy to skeletal muscles. After 7 seconds of intense effort, creatine phosphate converts to creatine. Creatine is metabolized into creatinine and is carried in blood to the kidneys for filtering and excretion in urine. If kidneys are malfunctioning, creatinine levels in the blood increase and creatinine levels in urine decrease. There are three types of creatinine tests

- Blood creatinine level test: This test measures the level of creatinine in blood.
- Creatinine clearance test: This test measures creatinine in a 24-hour urine sample and measures the level of creatinine in blood.
- Blood urea nitrogen/creatinine ratio (BUN/creatinine): This test compares the results of the blood urea test with the blood creatinine level test to assess for dehydration.

Hint

Urea is a by-product of protein metabolism in the liver that is excreted in urine. Fetal kidney function is assessed by testing the level of creatinine in amniotic fluid. The healthcare provider may order the glomerular filtration rate test to determine kidney function.

Caution

A normal blood creatinine level does not rule out kidney disease.

WHAT IS BEING MEASURED?

- Creatinine level in blood and urine

HOW IS THE TEST PERFORMED?

- Avoid
  - Strenuous exercise for 2 days prior to the test
  - Ingesting protein for 1 day prior to the test
  - Drinking coffee and tea prior to the creatinine clearance test, since these increase urine production
- See How to Collect Blood Specimen from a Vein in Chapter 1.

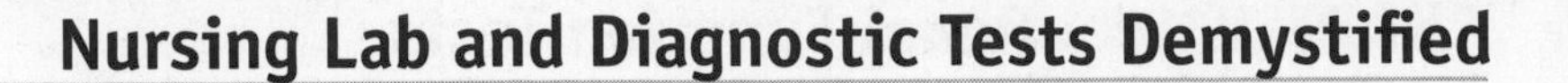

RATIONALE FOR THE TEST

- Screen for
  - Kidney function
  - Dehydration

NURSING IMPLICATIONS

- Assess if the patient
  - Has performed strenuous exercise for 2 days prior to the test
  - Has taken
    - Diuretics, Proloprim, Trimpex, Aldomet, Tagamet, ascorbic acid, cephalosporins, or Mefoxin, since these can affect the results of the blood creatine level test.
    - Dilantin, cephalosporins, Captopril, ascorbic acid, Garamycin, Tagamet, Proloprim, Trimpex, Cardioquin, Quinaglute, Quinine, Quinidex, amphotericin B, or procainamide, since these can affect the results of the creatinine clearance test.
    - Tagamet or tetracycline, since these can affect the results of the BUN/creatine ratio test.
  - Has ingested protein for 1 day prior to the test
  - Drank coffee or tea before the creatinine clearance test

UNDERSTANDING THE RESULTS

- The creatinine blood level test and creatinine clearance test results are available quickly. The laboratory determines normal values based on calibration of testing equipment with a control test. Test results are reported as high, normal, or low based on the laboratory's control test.
- Normal test results
  - Blood creatinine level
    - Men: 0.6 to 1.2 mg/dL
    - Women: 0.5 to 1.1 mg/dL
    - 13 to 19 years old: 0.5 to 1.0 mg/dL
    - 6 months to 12 years old: 0.3 to 0.7 mg/dL
    - Newborn: 0.3 to 1.2 mg/dL

- Creatinine clearance
  - Men: 90 to 140 mL/min
  - Women: 87 to 107 mL/min
- BUN/creatinine ratio
  - Over 11 months of age: 10:1 to 20:1
  - Under 12 months of age: 30:1 or less

- Higher levels may indicate
  - Blood creatinine level
    - Kidney disorder
    - Shock
    - Kidney stones
    - Dehydration
    - Shock
    - Rhabdomyolysis
    - Acromegaly
    - Gigantism
    - Polymyositis
    - Muscular dystrophy
    - Urinary tract blockage
    - Cancer
  - Creatinine clearance
    - Muscle injury
    - Pregnancy
    - Hypothyroidism
    - Burns
    - Strenuous exercise
    - Carbon monoxide poisoning
  - BUN/creatinine ratio
    - Dehydration
    - Kidney stones
    - Shock
    - Acute kidney failure

- Lower levels may indicate
  - Blood creatinine levels
    - Protein-deficient diet
    - Decreased muscle mass
    - Pregnancy
    - Liver disease
  - Creatinine clearance
    - Cirrhosis
    - Dehydration
    - Infection
    - Kidney disorder
    - Urinary tract blockage
    - Reduced blood to the kidneys
  - BUN/creatinine ratio
    - Protein-deficient diet
    - Cirrhosis
    - Rhabdomyolysis
    - SIADH
    - Cancer
    - Pregnancy

TEACH THE PATIENT

- Explain
- Why blood sample is taken.
  - That the healthcare provider may ask the patient to refrain from taking the following medications prior to the individual test
    - Blood creatinine level test: Diuretics, Proloprim, Trimpex, Aldomet, Tagamet, ascorbic acid, cephalosporin, or Mefoxin
    - Creatinine clearance test: Dilantin, cephalosporins, Captopril, ascorbic acid, Garamycin, Tagamet, Proloprim, Trimpex, Cardioquin, Quinaglute, Quinine, Quinidex, amphotericin B, or procainamide
    - BUN/creatinine ratio test: Tagamet or tetracycline

- To avoid
  - Strenuous exercise for 2 days prior to the test
  - Ingesting protein for 1 day prior to the test
  - Drinking coffee and tea for the creatinine clearance test, since these increase urine production

Summary

Glomerulus in the kidneys filters metabolic waste carried by blood, which is collected in a renal tubule as urine. The kidneys recycle some waste that can be utilized by the body while the remaining waste is excreted as urine.

Healthcare providers use the BUN and creatinine clearance tests to assess renal function. A patient experiences health problems if renal function falls to 25% or below. Waste that normally is excreted builds up in the blood causing a toxic result.

Renal function can decrease suddenly or slowly over time. An accident, illness medications, or poison can cause acute renal failure, which can be reversible with little or no ongoing kidney damage. Chronic kidney disease can cause a gradual decrease in renal function and can eventually lead to end-stage renal disease where nearly all renal function is permanently destroyed. The patient then undergoes dialysis or kidney transplantation.

Quiz

1. What is measured by the BUN test?
   a. Ammonia level in blood
   b. Urea level in blood
   c. Creatinine level in blood
   d. Nitrogen in the blood

2. What must be assessed before administering the BUN test?
   a. Age of the patient
   b. If the patient has taken diuretics
   c. If the patient has ingested meat
   d. All of the above

3. What does a low BUN level indicate?
    a. Liver damage
    b. Malnutrition
    c. Low protein diet
    d. All of the above

4. Aside from renal failure, what would a high BUN level indicate?
    a. Shock
    b. Urinary tract blockage
    c. Respiratory tract bleeding
    d. All of the above

5. What important fact should be known about renal failure?
    a. Creatinine blood levels increase and creatinine urine levels decrease.
    b. Creatinine blood levels decrease and creatinine urine levels increase.
    c. Creatinine blood levels decrease and creatinine urine levels decrease.
    d. Always use abbreviations to save time and space on the chart.

6. What test measures creatinine in a 24-hour urine sample?
    a. Creatinine clearance test
    b. Creatinine level test
    c. BUN/creatinine level test
    d. None of the above

7. Why should the patient avoid drinking coffee or tea before the creatinine clearance test?
    a. These effect how the sample is taken.
    b. These decrease urine production.
    c. These increase urine production.
    d. None of the above.

8. What is the purpose of the BUN/creatinine test?
   a. Compares the ratio of BUN and creatinine in blood to assess for pregnancy
   b. Compares the ratio of BUN and creatinine in blood to assess for testing errors
   c. Compares the ratio of BUN and creatinine in blood to assess for dehydration
   d. None of the above

9. What is creatinine?
   a. After intense effort by skeletal muscles, creatine phosphate converts to creatine, which is metabolized into creatinine and is filtered by the liver.
   b. After intense effort by skeletal muscles, creatine phosphate converts to creatine, which is metabolized into creatinine and is excreted by the kidney.
   c. After intense effort by the liver, creatine phosphate converts to creatine, which is metabolized into creatinine and is excreted by the kidney.
   d. After intense effort by the lungs, creatine phosphate converts to creatine, which is metabolized into creatinine and is excreted by the kidney.

10. What should the patient avoid before taking the creatinine clearance test?
    a. Strenuous exercise
    b. Eating meat
    c. Drinking coffee
    d. All of the above

Answers

1. d. Nitrogen in the blood.
2. d. All of the above.
3. d. All of the above.

4. d. All of the above.
5. a. Creatinine blood levels increase and creatinine urine levels decrease.
6. a. Creatinine clearance test.
7. c. These increase urine production.
8. c. Compares the ratio of BUN and creatinine in blood to assess for dehydration.
9. b. After intense effort by skeletal muscles, creatine phosphate converts to creatine, which is metabolized into creatinine and is excreted by the kidney.
10. d. All of the above.

CHAPTER 13

Pancreatic and Lipid Metabolism Tests

The pancreas produces insulin and glucagon along with digestive enzymes that are used by the small intestine to break down carbohydrates, protein, and fat. Health-providers administer pancreatic tests to determine pancreatic function.

Lipids are compounds used to store energy, to develop cell membranes, and are elements of vitamins and hormones. Lipids combine with proteins to form lipoproteins. Common lipoproteins in the body are cholesterol and triglycerides.

Cholesterol is released into the blood stream mostly by the liver and other organs, although some cholesterol is ingested in food. Two types of cholesterol are high-density lipoprotein (HDL) and low-density lipoprotein (LDL).

There is a balance between LDL and HDL. LDL is distributed to cells throughout the body through the blood stream. Excess LDL is removed by HDL from the blood and transported to the liver, where LDL is metabolized into bile acids and excreted from the body. An imbalance occurs when too much LDL is in the blood, leading to accumulation of LDL on the artery walls, which causes a narrowing of the arteries that leads to a blockage.

Healthcare providers order lipid metabolism tests to determine the level of lipids in the patient's blood stream. You will learn about these pancreatic and lipid tests in this chapter.

Learning Objectives

1. Amylase
2. Lipase
3. Cholesterol and Triglycerides Tests

Key Words

High-density lipoprotein (HDL)
Low-density lipoprotein (LDL)
Total cholesterol
Triglycerides
Very low-density lipoprotein (VLDL)

Amylase 1

Amylase is an enzyme that breaks down starch into sugar. Amylase is produced by salivary glands and the pancreas. The amylase test measures the amount of amylase in the blood. There is normally a low level of amylase in blood unless the salivary glands or pancreas is blocked or damaged.

WHAT IS BEING MEASURED?

- Amylase level in blood

HOW IS THE TEST PERFORMED?

- Refrain from eating or drinking, except water, for 2 hours before the test.
- Avoid drinking alcohol for 24 hours before the test.
- See How to Collect Blood Specimen from a Vein in Chapter 1.

RATIONALE FOR THE TEST

- Screen for
  - Pancreatic disease
  - Pancreatitis
  - Inflammation of the salivary glands
- Assess
  - Treatment for pancreatic and salivary gland disease

NURSING IMPLICATIONS

- Assess if the patient
  - Refrained from eating and drinking, except water, 2 hours before the test.
  - Drank alcohol 24 hours before the test.
  - Has taken codeine, morphine, diuretics, Coumadin, Indocin, aspirin, or birth control pills, since these can affect the test results.
  - Has chronic kidney disease, since this can cause high levels of amylase.
  - Liver or pancreatic damage, since this can affect the test results.

Hint

The healthcare provider is likely to order a lipase test along with the amylase test. Lipase is an enzyme that is produced only by the pancreas. The amylase test may also be performed with the creatinine test to help the healthcare provider diagnose pancreatitis.

UNDERSTANDING THE RESULTS

- Test results are available within 3 days. The laboratory determines normal values based on calibration of testing equipment with a control test. Test results are reported as high, normal, or low based on the laboratory's control test.
- Generally the normal range is 60 to 180 U/L.
- High levels may indicate
  - Pancreatitis
  - Pancreatic cancer
  - Mumps
  - Bowel infarction

  - Inflammation of the salivary glands
  - Cystic fibrosis
  - Diabetic ketoacidosis
  - Gallstones
  - Stomach ulcer
  - Macroamylasemia
  - Ectopic pregnancy
- Low levels may indicate
  - Preeclampsia
  - Advanced cystic fibrosis
  - Severe liver disease
  - Macroamylasemia

HINT

High values are normal for older adults and pregnant women. The patient may have chronic pancreatitis even if there is a low amylase level. Children have very little amylase until they reach their first birthday and then their amylase increases to the normal adult level.

TEACH THE PATIENT

- Explain
  - Why blood sample is taken.
  - The function of amylase and the meaning of abnormal values.
  - That the patient must refrain from eating or drinking, except water, for 2 hours before the test and avoid drinking alcohol for 24 hours before the test.
  - That the patient should tell the healthcare provider if they are taking codeine, morphine, diuretics, Coumadin, Indocin, aspirin, or birth control pills, since these can affect the test results.

Lipase 2

Lipase is an enzyme in the pancreas. Level of lipase in the blood increases when the pancreatic duct is blocked or there is damage to the pancreas. The lipase test measures the level of lipase in the blood.

Hint

The lipase test does not diagnose pancreatic disorder. A high level of lipase in the blood requires additional testing. The healthcare provider may order the amylase test at the same time as the lipase test.

WHAT IS BEING MEASURED?

- Lipase level in blood

HOW IS THE TEST PERFORMED?

- Avoid eating or drinking, except water, 12 hours before the test is administered.
- See How to Collect Blood Specimen from a Vein in Chapter 1.

RATIONALE FOR THE TEST

- Screen for
  - Pancreatic disease
  - Pancreatitis
  - Cystic fibrosis
- Assess treatment for
  - Pancreatitis
  - Cystic fibrosis

NURSING IMPLICATIONS

- Assess if the patient has
  - Eaten or drunk, except water, 12 hours before the test is administered.
  - High cholesterol, varicose veins, kidney disease, diabetes, or high blood pressure.
  - Taken narcotics, anticoagulants, or cholinergics.

UNDERSTANDING THE RESULTS

- The lipase test results are available in 12 hours. The laboratory determines normal values based on calibration of testing equipment with a control test. Test results are reported as high, normal, or low based on the laboratory's control test.

- Normal lipase test result: Less than 200 U/L
- High lipase levels may indicate
  - Pancreatic cancer
  - Pancreatitis
  - Pancreatic disease
  - Cholecystitis
  - Gallstones
  - Bowel obstruction
  - Infarction
  - Chronic kidney disease
  - Inflammation
  - Infection
  - Drug abuse
  - Peptic ulcer

TEACH THE PATIENT

- Explain
  - Why blood sample is taken.
  - That the patient must not eat or drink, except water, for 12 hours before the test is administered.
  - That the healthcare provider may ask the patient to refrain from taking narcotics, anticoagulants, or cholinergics 2 weeks before the test is administered.

Cholesterol and Triglycerides Tests 3

Cholesterol is produced in the liver and is used for cell growth and hormone production. Cholesterol attaches to proteins in the blood forming lipoproteins. Excess cholesterol in the blood forms plaque on the side of blood vessels that can lead to cardiovascular disorders. The cholesterol and triglycerides tests profile lipoproteins to measure components of cholesterol. These components are

- Total cholesterol: Total cholesterol is the total amount of LDL and HDL in the blood sample.

- Low-density lipoprotein (LDL): LDL transports lipids from the liver. High levels of LDL increase the risk of cardiovascular disorders.
- Very low-density lipoprotein (VLDL): VLDL distributes triglyceride that is produced in the liver. High levels of VLDL increase the risk of cardiovascular disorders.
- High-density lipoprotein (HDL): HDL binds with lipids in the blood returning lipids to the liver, where lipids are metabolized. High levels of HDL decrease the risk of cardiovascular disorders.
- Triglycerides: Triglycerides are stored lipids. High levels of both triglycerides and LDL increase the risk of cardiovascular disorders.

WHAT IS BEING MEASURED?

- Total cholesterol, low-density lipoprotein, very low-density lipoprotein, high-density lipoprotein, and triglycerides levels in blood.

HOW IS THE TEST PERFORMED?

- See How to Collect Blood Specimen from a Vein in Chapter 1.

RATIONALE FOR THE TEST

- Screen
  - For lipid disorder
- Assess
  - Treatment for lipid disorder

NURSING IMPLICATIONS

- Assess if the patient
  - Has eaten or drink 12 hours before the test is administered.
  - Has eaten a high fatty diet 24 hours before the test is administered.
  - Has exercised before the test is administered.
  - Has undergone a radioactive scan within a week before the test.
  - Has an infection, hyperthyroidism, hypothyroidism, cirrhosis, hepatitis, diabetes, kidney disease.
  - Has malnutrition.

- Abuse alcohol or medication.
- Is undergoing alcohol or medication withdrawal.
- Is pregnant.
- Has taken androgens, birth control pills, corticosteroids, estrogen, niacin, antibiotics, tranquilizers, or diuretics.

UNDERSTANDING THE RESULTS

- Test results are available in 1 day. The laboratory determines normal values based on calibration of testing equipment with a control test. Test results are reported as high, normal, or low based on the laboratory's control test.
- Normal total cholesterol range: Less than 240 mg/dL.
- Normal HDL range: Greater than 40 mg/dL.
- Normal total cholesterol: HDL ratio less than 5:1.
- Normal LDL range: Less than 160 mg/dL.
- Normal VLDL range: Less than 160 mg/dL.
- Normal triglycerides range: Less than 200 mg/dL.
- Lower HDL and higher LDL, VLDL, and triglycerides increase the risk of cardiovascular disorders including
  - Coronary artery disease
  - Acute coronary syndrome
  - Myocardial infarction

TEACH THE PATIENT

- Explain
  - Why blood sample is taken.
  - Not to eat or drink for 12 hours before the test.
  - Not to eat a fatty diet for 24 hours before the test.
  - Not to exercise before the test.
  - That the healthcare provider may request the patient to stop taking androgens, birth control pills, corticosteroids, estrogen, niacin, antibiotics, tranquilizers, or diuretics for 2 weeks prior to the test.

Summary

The pancreas produces digestive enzymes that are used in the small intestine to break down carbohydrates, protein, and fat. In addition, the pancreas excretes insulin and glucagon to ensure an adequate blood glucose level.

Lipids combine with proteins to form lipoproteins, the most commonly known are cholesterol and triglycerides. Types of cholesterol include HDL and LDL.

Excess LDL is removed by HDL from the blood and transported to the liver, where LDL is metabolized into bile acids and excreted from the body. An imbalance occurs when too much LDL is in the blood, leading to accumulation of LDL on the artery walls causing a narrowing of the arteries thus leading to a blockage.

Healthcare providers order pancreatic tests to determine if the pancreas is functioning properly and lipid metabolism tests to determine the level of lipids in the patient's blood.

Quiz

1. What should the patient refrain from before taking the amylase test?
   a. Eating 2 hours before the test
   b. Drinking anything except water 2 hours before the test
   c. Drinking alcohol 24 hours before the test
   d. All of the above

2. A high level of amylase might indicate
   a. Mumps
   b. Pancreatitis
   c. Inflammation of the salivary glands
   d. All of the above

3. The very low-density lipoprotein test examines
   a. LDL
   b. Triglycerides
   c. HDL
   d. None of the above

4. Before taking a lipid metabolism test, the patient should avoid
   a. Eating a high fatty diet 24 hours before the test
   b. Drinking 12 hours before the test
   c. Exercising
   d. All of the above

5. Which of the following test results indicates a high risk of cardiovascular disorders?
   a. High HDL, low LDL
   b. Low HDL, high LDL
   c. Low HDL, low LDL
   d. Low VLDL, low LDL

6. Before administering the lipase, assess
   a. If the patient has taken cholinergics
   b. The patient's age
   c. The patient's lifestyle
   d. The patient's glucose level

7. A high lipase level may indicate
   a. Gallstones
   b. Bowel obstruction
   c. Chronic kidney disease
   d. All of the above

8. HDL removes excess LDL from the blood and transports it to the small intestine where it is metabolized.
   a. True
   b. False

9. An increased amylase level in blood may indicate
   a. Measles
   b. Cardiac infarction
   c. A blocked salivary gland
   d. None of the above

10. What medication might affect the amylase test?
    a. Birth control pills
    b. Diuretics
    c. Aspirin
    d. All of the above

Answers

1. d. All of the above
2. d. All of the above
3. b. Triglycerides
4. d. All of the above
5. b. Low HDL, high LDL
6. a. If the patient has taken cholinergics
7. d. All of the above
8. b. False
9. c. A blocked salivary gland
10. d. All of the above

CHAPTER 14

Diagnostic Radiology Tests

Radiology tests using X-rays enable the healthcare provider to view inside the body without opening the skin. Although an X-ray provides a primitive view when compared with CT, or CAT scans, and MRIs, it remains a cost-effective way to identify many common disorders.

X-rays are based on the principle that the X-ray is absorbed by dense objects and will pass through lesser dense objects. Dense objects such as bone appear white on the X-ray file and lesser dense objects, such as air, appear black or a lighter shade of gray, such as fluid and fat.

There are many kinds of X-ray tests that a healthcare provider can order, each focusing on a particular area of the body. You will learn about each of these tests in this chapter.

Learning Objectives

1. How an X-ray Is Taken
2. Abdominal X-ray
3. Extremity X-ray
4. Spinal X-ray
5. Mammogram
6. Chest X-ray
7. Dental X-ray
8. Facial X-ray
9. Skull X-ray

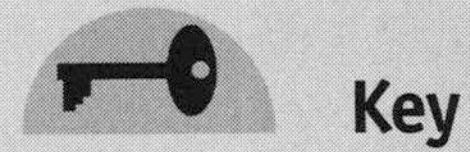

Key Words

Bitewing
Cartilage
Lead apron
Ligaments
Nonpalpable cysts
Occlusal
Orthopantogram
Panoramic
Periapical
Photographic film
Soft tissue
Tendons
X-ray beam

How an X-ray Is Taken

The patient is positioned between the X-ray gun and a piece of photographic film. The X-ray gun focuses the X-ray beam at the area of the patient's body that is being examined. A portion of the X-ray beam passes through the patient's body striking the photographic film, leaving a black area on the photographic film. Another portion of the X-ray beam is absorbed by bone and other dense tissue in the patient's body, which appears as shades of gray on the photographic film depending on the density of the tissue. Areas that are not being X-rayed are protected by a lead apron, where possible, to prevent radiation from reaching those areas.

HINT

An X-ray does not provide a good image of cartilage, ligaments, tendons, and other soft tissues. The healthcare provider will likely order a computed tomography (CT) scan or a magnetic resonance imaging (MRI) to examine soft tissue.

TEACH THE PATIENT

- Explain
  - X-ray may not be taken if the patient is pregnant.
  - The level of radiation will not have any long-term effects on the patient.
  - The healthcare provider may require specific preparation depending on the part of the body being X-rayed.
  - The any jewelry in the vicinity of the X-ray site must be removed.
  - A radiologist is usually the healthcare professional who reads the X-ray.
  - The patient may be required to remove clothing and wear a gown during the X-ray.
  - The patient will be asked to lie still and hold his/her breath while the X-ray is being taken.
  - The X-ray room is cool.
  - The patient may feel uncomfortable while positioning for the X-ray.
- Assess
  - If the patient has signed a consent form.

Abdominal X-ray 2

An abdominal X-ray shows the position, size, and shape of the stomach, diaphragm, liver, spleen, large and small intestines.

HINT

The healthcare provider may order a CT scan, ultrasound, or IV pyelograph also. A KUB X-ray is an abdominal X-ray that examines the kidneys, uterus, and bladder.

CAUTION

Ovaries are not protected by a lead apron during the X-ray because of their location at the site of the X-ray. Abdominal X-rays cannot detect ulcers or bleeding.

WHAT IS BEING EXAMINED?

- The position, size, and shape of the stomach, diaphragm, liver, spleen, large and small intestines
- Position of nasogastric tube, nephrostomy tube, V-P shunt, or dialysis catheter

HOW IS THE TEST PERFORMED?

- The patient may be asked to stand between the X-ray machine and the photographic film or to lie on his/her back or side on the X-ray table.
- When the patient lies on his/her back
    - A lead apron is placed over the lower pelvic area to protect the area from X-ray beam.
    - The X-ray machine is positioned over the abdomen.
    - The patient is asked to lie still and hold his/her breath while the X-ray is being taken.
    - An X-ray is taken.
- When the patient is asked to stand between the X-ray machine and the photographic film.
    - The patient is asked to stand still and hold his/her breath when the X-ray is being taken.
    - An X-ray is taken.
- See How an X-ray Is Taken as discussed earlier in the chapter.

RATIONALE FOR THE TEST

- Assess
    - Underlying cause of abdominal or flank pain.
    - Confirmation of position of nasogastric tube, nephrostomy tube, V-P shunt, or dialysis catheter
    - Location of an ingested foreign body
    - Underlying cause of vomiting and nausea
    - For intestine blockage
    - For perforation in the intestines or stomach

NURSING IMPLICATIONS

- Assess if the patient
  - Is pregnant. An ultrasound is ordered instead of an X-ray to protect the fetus from radiation exposure.
  - Has taken Pepto-Bismol (bismuth) or barium 4 days prior to the test. Bismuth and barium might obstruct the X-ray beam.
  - Has an empty bladder.

UNDERSTANDING THE RESULTS

- The test takes 10 minutes and the results are ready within 15 minutes or 2 days depending on the patient's condition.
- Normal test results indicate
  - Normal position, size, and shape of the stomach, diaphragm, liver, spleen, large and small intestines
  - No growths
  - Normal amounts of air, fluid, or stool
- Abnormal test results indicate
  - Abnormal position, size, and shape of the stomach, diaphragm, liver, spleen, large and small intestines
  - Blockage
  - Perforation in the intestine or stomach (Conn syndrome)
  - Abnormal amounts of air, fluid, or stool
  - Kidney stones
  - Gallstones
  - Foreign object located
  - Growths

TEACH THE PATIENT

- Explain
  - Why X-ray is being taken.
  - What the patient can expect in the X-ray room.

- That the patient should not take Pepto-Bismol (bismuth) or barium 4 days prior to the test.
- That the patient will need to empty his/her bladder before the test.

Extremity X-ray 3

An extremity X-ray shows damage to hands, wrists, arms, feet, ankles, knees, hip, or legs.

HINT

The healthcare provider may order a bone scan, CT scan, or MRI if the X-ray does not reveal a disorder.

CAUTION

The healthcare provider may not order an X-ray if the results of the X-ray would not alter treatment of the disorder.

WHAT IS BEING EXAMINED?

- Bones of the hands, wrists, arms, feet, ankles, knees, hip, or legs

HOW IS THE TEST PERFORMED?

- The patient sits, stands, or lies down depending on extremity being examined.
- The extremity is positioned in front of the X-ray machine an is held in place by an apparatus if necessary.
- An X-ray is taken.
- See How an X-ray Is Taken as discussed earlier in the chapter.

HINT

The healthcare provider may order opposing extremities X-rayed to make a comparison.

RATIONALE FOR THE TEST

- Assess for
  - Fracture
  - Dislocation
  - Tumors
  - Deformities/degeneration
  - Fluid around joints
  - Growth
  - The underlying cause of pain in the extremity
  - Alignment following treatment
  - Infection
  - Foreign objects

NURSING IMPLICATIONS

- Assess if the patient
  - Is pregnant. An ultrasound is ordered instead of an X-ray if the patient is pregnant to protect the fetus from radiation exposure.
  - Remained still during the X-ray.

UNDERSTANDING THE RESULTS

- The test takes 10 minutes and the results are ready within 15 minutes or 2 days depending on the patient's condition.
- Normal test results indicate
  - No fracture.
  - No dislocation.
  - No tumors.
  - No deformities.
  - No fluid around joints.
  - Normal growth.
  - Extremity is aligned.

- No infection.
- No foreign objects.
- Abnormal test results indicate
    - Signs of fracture
    - Signs of dislocation
    - Signs of tumors
    - Signs of deformities
    - Signs of fluid around joints
    - Abnormal growth
    - Extremity misaligned
    - Signs of infection
    - Signs of foreign objects

TEACH THE PATIENT

- Explain
    - Why X-ray is being taken.
    - What the patient can expect in the X-ray room.
    - That the patient will need to hold the extremity still while the X-ray is being taken.

Spinal X-ray 4

The spine consists of 33 vertebrae, nearly all separated by a disc that absorbs shock related to movement. There are four types of spinal X-rays:

- Cervical spine: 7 vertebrae in the cervical area of the spine
- Thoracic spine: 12 vertebrae in the thoracic area of the spine
- Lumbosacral spine: 5 vertebrae in the lumbar area of the spine and 5 vertebrae in the sacrum area
- Sacrum/coccyx: 5 vertebrae in the sacrum area and 4 vertebrae in the coccyx area

---

HINT

A disc is cartilage.

Caution

Strained back muscles or ligaments are not visible on an X-ray.

WHAT IS BEING EXAMINED?

- Structure and position of the spine

HOW IS THE TEST PERFORMED?

- The patient lies on an X-ray table.
- The X-ray machine is positioned over the site.
- The patient is asked to lie still and hold his/her breath while the X-ray is being taken.
- An X-ray is taken (up to five X-rays may be taken).
- See How an X-ray Is Taken discussed earlier in the chapter.

Hint

If the patient has a cervical collar, an X-ray will be taken and read to determine if the cervical collar can be removed without causing further injury.

RATIONALE FOR THE TEST

- Assess for
  - Fracture
  - Dislocation
  - Tumors
  - Deformities/degeneration
  - Curvature of the spine
  - Bone spurs
  - The underlying cause of weakness, pain, or numbness
  - Alignment following treatment

NURSING IMPLICATIONS

- Assess if the patient
  - Is pregnant. An ultrasound is ordered instead of an X-ray if the patient is pregnant to protect the fetus from radiation exposure.

- Is morbidly overweight, since this can blur details of the X-ray.
- Has taken Pepto-Bismol (bismuth), or barium, 4 days prior to the test. Bismuth and barium might obstruct the X-ray beam.
- Can remain still during the X-ray.

UNDERSTANDING THE RESULTS

- The test takes 15 minutes and the results are ready within 15 minutes or 2 days depending on the patient's condition.
- Normal test results indicate
  - No fracture.
  - No dislocation.
  - No tumors.
  - No deformities/degeneration.
  - Spine is aligned.
  - No bone spurs.
- Abnormal test results indicate
  - Signs of fracture.
  - Signs of dislocation.
  - Signs of tumors.
  - Signs of deformities/degeneration.
  - Spine is misaligned.
  - Signs of bone spurs.

TEACH THE PATIENT

- Explain
  - Why X-ray is being taken.
  - What the patient can expect in the X-ray room.
  - That the patient should not take Pepto-Bismol (bismuth) or barium 4 days prior to the test.
  - That the patient will need to lie still while the X-ray is being taken.

Mammogram 5

A mammogram is an X-ray that detects palpable and nonpalpable cysts or masses in the breast and is used to screen for signs of breast cancer. Suspicious masses are biopsied to determine if the mass is cancerous.

Hint

A mammogram can be performed on a patient who has had breast implants. A mammogram cannot be performed if the patient is breast-feeding. A breast ultrasound determines if a mass is a cyst or solid mass. A digital mammogram is considered to have the same accuracy as an X-ray mammogram. The healthcare provider may order the breast cancer (BRCA) gene test if there is a history of breast cancer in the patient's family.

Caution

A mammogram is not normally performed if the patient is pregnant. If the mammogram must be performed, a lead apron is placed over the patient's abdomen.

Warning

The healthcare provider must provide the patient with the original mammogram images if requested.

WHAT IS BEING EXAMINED?

- The breast for signs of cysts or solid masses

HOW IS THE TEST PERFORMED?

- Remove all powders, deodorant, ointments, or perfume from the breast since these may appear on the X-ray image.
- Schedule the mammogram for within 2 weeks after the end of the patient's menstrual period to reduce tenderness of her breasts.
- Remove the patient's clothes above the waist and provide her with a gown.
- The patient stands or sits.

- Each breast is placed on the X-ray plate.
- A second plate is placed on top of the breast.
- The breast is compressed.
- The patient is asked to hold her breath while the X-ray is taken.
- X-ray images are taken from the top and side of the breast.
- See How an X-ray Is Taken as discussed earlier in the chapter.

RATIONALE FOR THE TEST

- Assess
  - If the patient has cysts, solid masses, or calcification in the breast
  - Underlying cause of breast discomfort

NURSING IMPLICATIONS

- Assess if the patient
  - Is pregnant
  - Is breast-feeding
  - Has breast implants
  - Removed all powders, deodorant, ointments, or perfume from the breast since these may appear on the X-ray image
  - Assess if the mammogram was scheduled for 2 weeks after the end of the patient's menstrual period to reduce tenderness of her breasts
- Identify the location of any previous breast biopsies.

UNDERSTANDING THE RESULTS

- The test takes 15 minutes and the results are ready within 15 minutes or 10 days depending on the patient's condition.
- Normal test results indicate
  - No cyst
  - No solid mass
  - Normal tissue
  - Clear ducts
- Abnormal test results indicate
  - Signs of cyst

- Signs of solid mass
- Signs of calcification

Caution

A normal mammogram does not rule out breast cancer. Results may be difficult to interpret if the patient is obese.

TEACH THE PATIENT

- Explain
  - Why mammogram is being taken.
  - The procedure for taking the mammogram.
  - That the breast is compressed to obtain the best possible view of the breast.
  - That the test will be uncomfortable but not painful.
  - That the patient should not use powders, deodorant, ointments, or perfume on the breasts on the day of the mammogram.
  - That the patient should schedule the mammogram for within 2 weeks after the end of her menstrual period to reduce tenderness of the breasts.

Chest X-ray 6

A chest X-ray shows the position, size, and shape of the collarbone, breastbone, heart, airway, lungs, thoracic spine, ribs, lymph nodes, and blood vessels.

Hint

The healthcare provider may also order an echocardiogram (ECG), ultrasound, MRI, or a CT.

WHAT IS BEING EXAMINED?

- The position, size, and shape of the collarbone, breastbone, heart, airway, lungs, thoracic spine, ribs, lymph nodes, and blood vessels

HOW IS THE TEST PERFORMED?

- Remove the patient's clothes above the waist and provide her with a gown.
- The patient stands or sits or lies down on the X-ray plate.
- The patient is asked to hold his/her breath while the X-ray is being taken.
- X-ray images are taken from the front and side.
- See How an X-ray Is Taken as discussed earlier in the chapter.

RATIONALE FOR THE TEST

- Assess
  - For pulmonary disease or disorders
  - The underlying cause of chest pain or respiratory problems
  - The underlying cause of cardiac problems
  - A chest injury
  - Positioning of a medical device
- Identify foreign objects in the airway and esophagus.

NURSING IMPLICATIONS

- Assess if the patient
  - Is pregnant
  - Is able to hold his/her breath
  - Has removed all jewelry in the thoracic area since these may appear on the X-ray image
  - Is obese
- Assess scars in the thoracic area that might appear on the X-ray.

UNDERSTANDING THE RESULTS

- The test takes approximately 15 minutes and the results are ready within 15 minutes or 2 days depending on the patient's condition.
- Normal test results indicate
  - Normal position, size, and shape of the collarbone, breastbone, heart, airway, lungs, thoracic spine, ribs, lymph nodes, and blood vessels.

- No abnormal fluid seen.
- Medical device is in the proper position.
- Abnormal test results indicate
  - Abnormal position, size, and shape of the collarbone, breastbone, heart, airway, lungs, thoracic spine, ribs, lymph nodes, and blood vessels.
  - Abnormal fluid seen.
  - Medical device is not in the proper position.

TEACH THE PATIENT

- Explain
  - Why X-ray is being taken.
  - The procedure for taking the X-ray.
  - That the patient should remove all jewelry in the thoracic area.

Dental X-ray 7

Dental X-rays are used to assess the condition of the patient's jaw, mouth, and teeth. There are four types of dental X-rays:

- Panoramic (orthopantogram): This assesses the temporomandibular joints, the jaw, sinuses, teeth and nasal area for tumors, fractures, cysts, and impacted teeth, but not cavities.
- Occlusal: This assesses the palate and lower portions of the mouth for fractures, cleft palate, abscesses, tumors, cysts, and immature teeth.
- Bitewing: This is a single view of the upper and lower back teeth and is used to assess the formation of teeth, bone loss, infection, and tooth decay.
- Periapical: This is a view of a tooth and used to assess abscesses, tumors, cysts, and the overall status of the patient's teeth.

WHAT IS BEING EXAMINED?

- The temporomandibular joints, the jaw, sinuses, teeth, mouth, and nasal area

HOW IS THE TEST PERFORMED?

- The patient sits.
- The patient bites down on the X-ray film carrier that holds the X-ray film.
- X-ray image is taken.
- In a panoramic X-ray, the patient places his/her head in the arm of the X-ray machine and arm rotates while taking the X-ray image.
- See How an X-ray Is Taken as discussed earlier in the chapter.

RATIONALE FOR THE TEST

- Assess
  - For cysts, tumors, abscesses
  - The underlying cause of mouth and sinus pain
  - The health of teeth
  - The position of teeth
  - For abnormal structures in the mouth and jaw

NURSING IMPLICATIONS

- Assess if the patient
  - Is pregnant
  - Is able to open his/her mouth enough to insert the X-ray film carrier in the mouth
  - Has foreign objects in his/her mouth such as body piercing, retainers, dentures, braces, or bridges

UNDERSTANDING THE RESULTS

- The test takes a few minutes and the results are ready within 15 minutes.
- Normal test results indicate
  - Normal position, size, and shape of the temporomandibular joints, the jaw, sinuses, teeth, mouth, and nasal area
- Abnormal test results indicate
  - Tooth decay
  - Abnormal position, size, and shape of the temporomandibular joints, the jaw, sinuses, teeth, mouth, and nasal area

TEACH THE PATIENT

- Explain
  - Why X-ray is being taken.
  - The procedure for taking the X-ray.

Facial X-ray 8

A facial X-ray is used to assess facial bones, sinuses, and the orbital cavity.

HINT
The healthcare provider may also order a CT scan.

WHAT IS BEING EXAMINED?

- The facial bones, sinuses, and the orbital cavity

HOW IS THE TEST PERFORMED?

- The patient sits.
- The patient must remove eye glasses and body piercing in the area of the X-ray.
- The patient must hold his/her head still. A brace may be provided to stabilize the patient's head.
- X-ray image is taken.
- See How an X-ray Is Taken as discussed earlier in the chapter.

RATIONALE FOR THE TEST

- Assess for
  - Cysts, tumors, abscesses
  - The underlying cause of sinus pain
  - Fractures
  - Foreign objects

NURSING IMPLICATIONS

- Assess if the patient
  - Is pregnant
  - Is able to sit still for the X-ray
  - Has a prosthetic eye
  - Has a neck injury
  - Has body piercing in the area being X-rayed
  - Has removed eye glasses

UNDERSTANDING THE RESULTS

- The test takes 30 minutes and the results are ready within 15 minutes.
- Normal test results indicate
  - Normal position, size, and shape of bones in the face and the orbital cavity
- Abnormal test results indicate
  - Inflammation
  - Infection
  - Abnormal position, size, and shape of bones in the face and the orbital cavity
  - Cysts, tumors, abscesses
  - Foreign object

TEACH THE PATIENT

- Explain
  - Why X-ray is being taken.
  - Procedure for taking the X-ray.

Skull X-ray 9

A skull X-ray is used to assess the skull and sinuses.

Hint

The healthcare provider may also order a CT scan.

WHAT IS BEING EXAMINED?

- The skull and sinuses

HOW IS THE TEST PERFORMED?

- The patient sits still.
- The patient must remove eye glasses and body piercing in the area of the X-ray.
- The patient must hold his/her head still. A brace may be provided to stabilize the patient's head.
- X-ray image is taken. Several images are taken: front, back, top, and sides.
- See How an X-ray Is Taken as discussed earlier in the chapter.

RATIONALE FOR THE TEST

- Assess for
  - Cysts, tumors, abscesses
  - The underlying cause of sinus pain
  - Fractures
  - Foreign objects

NURSING IMPLICATIONS

- Assess if the patient
  - Is pregnant
  - Is able to sit still for the X-ray
  - Has a prosthetic eye
  - Has a neck injury
  - Has body piercing in the area being X-rayed
  - Has removed eye glasses

UNDERSTANDING THE RESULTS

- The test takes 30 minutes and the results are ready within 15 minutes.
- Normal test results indicate
  - Normal position, size, and shape of bones in the skull and the orbital cavity

- Abnormal test results indicate
  - Inflammation
  - Infection
  - Abnormal position, size, and shape of bones in the skull
  - Cysts, tumors, abscesses
  - Foreign object

TEACH THE PATIENT

- Explain
  - Why X-ray is being taken.
  - The procedure for taking the X-ray.

Summary

X-rays, although primitive to today's CT or CAT scans, and MRI, provide a cost-effective method of viewing the inside of the body without opening the skin. There are many types of X-ray tests, which you learned about in this chapter. Each works basically the same way.

An X-ray is absorbed by dense material such as bone and passes through less dense material such as soft tissue. The X-ray beam passes through the patient and onto photographic film located on the opposite side of the patient.

The photographic film turns black when struck by the X-ray beam. Those areas of the photographic film that are not struck by the X-ray beam appear gray. As a result, dense objects in the patient's body appear as shades of gray on the photographic film. Any unexpected gray area on the photographic film requires additional assessment by the healthcare provider.

Quiz

1. What should the patient avoid taking 4 days prior to an X-ray?
   a. Tea
   b. Aspirin
   c. Pepto-Bismol (bismuth)
   d. Alcohol

2. When might a healthcare provider not order an X-ray for a disorder of an extremity?
   a. If the results of the X-ray would not alter the treatment
   b. If the patient is less than 6 years of age
   c. If the patient is older than 80 years of age
   d. If the patient has been diagnosed with cancer

3. If the patient arrives in the emergency room wearing a cervical color, the healthcare provider will
   a. Remove the cervical collar so that the X-ray can be taken.
   b. Keep the cervical collar in place when taking the X-ray to determine if the cervical collar can be removed.
   c. Order that no X-rays be taken.
   d. None of the above.

4. A mammogram is not performed
   a. If the patient is breast-feeding
   b. If the patient is younger than 30 years old
   c. If the patient has had breast implants
   d. If the patient has no palpable lumps on the breast

5. Prior to a mammogram, the patient should not use
   a. Deodorant
   b. Ointments
   c. Powders
   d. All of the above

6. The orthopantogram is used to assess the
   a. Temporomandibular joints
   b. Sinuses
   c. Jaw
   d. All of the above

7. During what type of X-ray might the patient be placed in a brace for the X-ray?
   a. Facial X-ray
   b. Orthopantogram
   c. Periapical
   d. Occlusal

8. What X-ray will probably be ordered to examine cartilage, ligaments, and tendons?
   a. Occlusal
   b. Periapical
   c. Temporomandibular
   d. None of the above

9. What is the best way to avoid tenderness of the breasts in a mammogram?
   a. Schedule the mammogram during the second week of patient's menstrual period.
   b. Schedule the mammogram during the patient's menstrual period.
   c. Schedule the mammogram for within 2 weeks after the end of the patient's menstrual period.
   d. Schedule the mammogram for within 3 weeks after the end of the patient's menstrual period.

10. Why is it important to assess if the patient is morbidly overweight before taking a spinal X-ray?
    a. Details in the X-ray can be blurred by the additional weight.
    b. The patient may break the X-ray machine.
    c. The patient may not be able to withstand the test.
    d. None of the above.

Answers

1. c. Pepto-Bismol (bismuth).
2. a. If the results of the X-ray would not alter the treatment.
3. b. Keep the cervical collar in place when taking the X-ray to determine if the cervical collar can be removed.

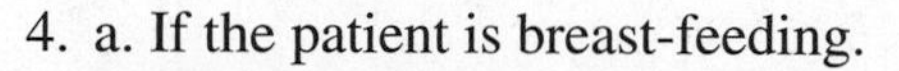

4. a. If the patient is breast-feeding.
5. d. All of the above.
6. d. All of the above.
7. a. Facial X-ray.
8. d. None of the above.
9. c. Schedule the mammogram within 2 weeks after the end of the patient's menstrual period.
10. a. Details in the X-ray can be blurred by the additional weight.

CHAPTER 15

Computed Tomography Scan

A computed tomography (CT) scan makes detailed images of structures within the body using a doughnut-shaped X-ray machine. While the patient lies within the scanner, an X-ray beam rotates around the patient, creating an image that represents a thin slice of the patient. Each rotation takes less than a second.

All slices are stored on a computer. The computer is used to reassemble slices of the patient enabling the healthcare provider to identify any abnormalities. Typically, the healthcare provider will print the image of any slices that indicate an abnormality, which is then saved with the patient's chart.

The patient may be administered contrast material such as iodine dye. The contrast material makes structures within the patient's body stand out on the computer by differentiating them with white, black, and shades of gray. Contrast material is administered intravenously or into joints or cavities of the body. The patient may also be asked to ingest other kinds of contrast material.

A CT scan may be used for staging cancer to assess if the cancer has spread to other sites in the body. CT scans are also used to identify masses or tumors, as well as fluid and the infection process. CT scans guide the healthcare provider when performing a procedure such as a biopsy.

You will learn about different types of CT scans in this chapter.

Learning Objectives

1. Full-Body CT Scan
2. CT Scan of the Head
3. CT Scan of the Spine

Key Words

Computed tomography
Contrast material
CT myelogram
Intrathecal space
Intrathecally
Intravenous pyelogram
Iodine dye
Kidneys, ureters, and bladder (KUB)
Positron emission tomograph (PET)
Shellfish allergy
Temporomandibular disorder

Full-Body CT Scan 1

A full-body CT scan creates an image of the patient's entire body. A healthcare provider orders a full-body CT scan if it is suspected that the patient may have disorders throughout the body and the healthcare provider is unable to narrow the disorder to specific areas of the body. This situation may occur if the patient is involved in a severe motor vehicle accident.

Hint

Typically, a healthcare provider orders a CT scan for a specific part of the body rather than ordering a full-body scan. A full CT scan is time consuming and usually provides more than enough information necessary for the healthcare provider to diagnose the patient's disorder. Some healthcare providers feel that a full-body scan identifies benign growths and other disorders that do not adversely affect the patient but could lead to additional tests and surgery that are unnecessary.

The result of a CT scan is commonly compared with the results of a positron emission tomograph (PET) to identify cancer.

Caution

Determine if the patient is allergic to shellfish or iodine. Contrast material may contain iodine and other substances that could cause the patient to have an allergic reaction. Also determine if the patient will be administered a sedative to relax her/him during the CT scan. If so, make sure that the patient does not drive any vehicle following the CT scan until the sedative has worn off.

WHAT IS BEING EXAMINED?

- Head
- Thorax
- Abdomen
- Kidneys, ureters, and bladder (KUB)
- Intravenous pyelogram for urinary tract blockage
- Liver
- Pancreas
- Bile ducts
- Gallbladder
- Adrenal glands
- Spleen
- Pelvis (ovaries, fallopian tubes, uterus, prostate gland)
- Extremities

HOW IS THE TEST PERFORMED?

- Depending on the nature of the CT scan, the patient may be administered an enema or asked not to eat after midnight prior to the CT scan.
- If contrast material is required for the test, then the patient is administered the contrast material before the test. The method for administering the contrast material depends on the nature of the CT scan.
    - Approximately 40 minutes before the test, the patient may be asked to ingest contrast material.

- Contrast material may be administered in a vein or in a cavity, such as the bladder or rectum, immediately before the test.
- The patient removes jewelry and clothes and is given a gown to wear during the test.
- The patient lies on the CT scanner table.
- The patient must lie still during the test.
- The patient will be in the CT room alone.
- The CT scan technician is in the next room observing through a window.
- The patient and the CT scan technician are able to converse during the test using an intercom.
- The CT scanner table moves into the opening of the CT scanner.
- The CT scanner moves around the patient when taking images of the patient.
- The patient hears a clicking sound as the CT scanner moves.
- The CT scan can take up to 2 hours.
- A radiologist, who is a medical doctor, interprets the results of the CT scan and writes a report that is given to the patient's healthcare provider.
- The patient is asked to drink large amounts of water and other fluids for 24 hours following the CT scan to flush the contrast material from the patient's body.

RATIONALE FOR THE TEST

- Assess for
  - Growths
  - Obstructions
  - Inflammation or infection
  - Foreign objects
  - Bleeding
  - Fluid collection
  - Pulmonary embolism

NURSING IMPLICATIONS

- Assess if the patient
  - Has allergies (shellfish, iodine)

- Is breast-feeding since contrast material can pass to the baby in breast milk, the patient should prepare to use formula instead of breast milk for 2 days following the CT scan if contrast material is administered
- Has heart disorder, asthma, thyroid or kidney disorders, or diabetes
- Takes Glucophage
- Has taken Pepto-Bismol 4 days prior to the CT scan

- Determine if the patient
  - Is claustrophobic
  - Can lie still during the test

UNDERSTANDING THE RESULTS

- The results are available within 2 days.
- Normal test results indicate
  - Normal size of organs and blood vessels
  - No blockages
  - No bleeding
  - No abnormal fluid collection
  - No growths
  - No inflammation
- Abnormal test results indicate
  - Abnormal size of organs and blood vessels
  - Blockages
  - Bleeding
  - Abnormal fluid collection
  - Growths
  - Inflammation

TEACH THE PATIENT

- Explain
  - Why the CT scan is being administered.
  - The CT scan test.

- Why contrast material may be administered.
- That the contrast material may leave a metallic taste and that the patient may feel flushed when the contrast material is administered IV.
- That the healthcare provider may ask the patient to stop taking Glucophage several days before the CT scan, since there might be a reaction with contrast material.
- That some patients may be allergic to the contrast material and that the healthcare team is ready to take measures to reverse any adverse reaction to the contrast material.
- That the healthcare provider may ask the patient to stop breast-feeding for 2 days following the CT scan if contrast material is administered.
- That the patient will be asked to drink a large amount of water for 24 hours following the CT scan if contrast material is administered.
- That the healthcare provider may administer a sedative prior to the test if the patient is unable to relax during the test. If a sedative is administered, the patient should arrange to be driven home following the test.

CT Scan of the Head 2

The patient's head is placed into the CT scanner as the CT scanner takes sliced images of the patient's skull, brain, and other parts of the patient's head. The healthcare provider may order a perfusion CT. A perfusion CT is used to determine the blood supply to areas of the brain. Contrast material is administered IV. Areas of the brain that receive blood are highlighted on the computer image by the contrast material. Areas without blood flow are not highlighted.

WHAT IS BEING EXAMINED?

- Head
- Brain
- Eyes
- Ears
- Nose

- Mouth
- Sinuses

HOW IS THE TEST PERFORMED?

- See Full-Body CT Scan discussed earlier in this chapter.
- The CT scan of the head takes approximately 30 minutes.

RATIONALE FOR THE TEST

- Assess for
  - Growths
  - Obstructions
  - Inflammation or infection
  - Foreign objects
  - Bleeding
  - Fluid collection
  - Headache
  - Vertigo
  - Vision problem
  - Broken bones
  - The result of facial surgery
  - Temporomandibular disorder
  - Paget disease
  - Stroke
  - Reasons for change in the level of consciousness
- Provide baseline images before surgery

NURSING IMPLICATIONS

- See Full-Body CT Scan discussed earlier in this chapter.
- Assess if the patient
  - Has removed glasses, contact lenses, and hearing aids

UNDERSTANDING THE RESULTS

- The results are available within 2 days.
- Normal test results indicate
  - Normal-sized skull, brain, ventricles, blood vessels, eyes, ears, sinuses
  - No blockages
  - No bleeding
  - No abnormal fluid collection
  - No foreign objects
  - No ischemia
- Abnormal test results indicate
  - Abnormal-sized skull, brain, ventricles, blood vessels, eyes, ears, sinuses
  - Blockages
  - Inflammation
  - Growth
  - Fluid collection
  - Foreign objects

TEACH THE PATIENT

- See Full-Body CT Scan discussed earlier in this chapter.
- Explain
  - That contrast material may be administered intravenously and flows throughout blood vessels in the brain, showing areas of the brain that are receiving blood and are not receiving blood.

CT Scan of the Spine 3

The CT scan of the spine creates images of the cervical, thoracic, and lumbosacral spine. All 33 vertebrae and discs are pictured along with the cerebrospinal fluid (CSF). During the scan, the CT scanner can be tilted to follow the curvature of the spine. Depending on the purpose of the scan, the healthcare provider may require that contrast material be administered intrathecally into the spinal canal.

WHAT IS BEING EXAMINED?

- Vertebrae
- Discs
- Cerebrospinal fluid

HOW IS THE TEST PERFORMED?

- See Full-Body CT Scan discussed earlier in this chapter.
- The healthcare provider may order a CT myelogram. A CT myelogram requires that a sample of cerebrospinal fluid be removed for microscopic examination before contrast material is administered. Contrast material is then administered in the intrathecal space around the spinal cord.
- If a CT myelogram is performed, the patient lies on his/her stomach. An area of the lumbar spine is anesthetized and contrast material is injected. The CT table is tilted to help distribute the contrast material throughout the spine.
- After a CT myelogram, the patient is asked to keep his/her head raised to prevent seizures and headaches.

RATIONALE FOR THE TEST

- Assess for
  - Growths
  - Obstructions
  - Narrowing of the spinal canal
  - Deformities
  - Fractures
  - Inflammation and infection
  - Bone compression
  - Osteoporosis
  - Congenital defects

NURSING IMPLICATIONS

- See Full-Body CT scan discussed earlier in this chapter.

UNDERSTANDING THE RESULTS

- The results are available within 2 days.
- Normal test results indicate
  - Normal size of vertebrae, discs, and spinal canal
  - No blockages
  - No inflammation or infection
- Abnormal test results indicate
  - Abnormal size of vertebrae, discs, and spinal canal
  - Blockages
  - Inflammation or infection
  - Osteoporosis, arthritis
  - Spinal stenosis
  - Growths

TEACH THE PATIENT

- See Full-Body CT Scan discussed earlier in this chapter.
- Tell the patient to contact his/her healthcare provider if he/she experiences
  - A headache that lasts more than a day or a severe headache following the CT scan
  - A temperature of 101.1°F or greater
  - Numbness, pain, or weakness in extremities
  - Irritablility
  - Difficult bowel movements or trouble urinating
- Explain that the patient
  - Should keep his/her head raised following the CT scan to avoid headaches.
  - May experience nausea and vomiting following the CT scan.
  - In rare situations may experience a seizure following the CT scan and if this occurs, should seek emergency medical help immediately.

Summary

A computed tomography (CT) scanner uses X-rays to create detailed images of the inside of a patient's body on a computer screen enabling a radiologist to identify abnormalities inside the patient.

A patient lies on a table that is slowly inserted into the doughnut-shaped CT scanner. The X-ray beam rotates around the patient creating a thin image of the cross-section of the patient's body in less than a second.

Each slice is stored on a computer and is reassembled, enabling the radiologist to examine slices in sequence for any abnormality. If an abnormality is found, the relative slices can be printed and saved in the patient's chart.

Depending on the nature of the CT scan, the patient may be administered contrast material. Contrast material such as an iodine dye causes areas of the body affected by this material to be highlighted on the computer image. Areas of the body unaffected by the contrast material are not highlighted, possibly indicating a blockage.

Quiz

1. Why is the patient asked to drink a large amount of water for 24 hours following a CT scan?
   a. To disperse X-rays
   b. To flush the contrast material
   c. To avoid cramps
   d. None of the above

2. What will a healthcare provider do if the patient is anxious about being placed in a CT scanner?
   a. Administer a sedative.
   b. Cancel the test.
   c. Wait for a calmer moment to administer the test.
   d. Tell the patient to behave like an adult.

3. What would you do if the patient tells you he/she is allergic to shellfish?
   a. Be prepared to respond to an allergic reaction.
   b. Schedule the CT scan.
   c. Notify the healthcare provider and radiologist.
   d. None of the above.

4. If the patient scheduled for a CT scan is breast-feeding
   a. Ask the patient to use formula for 2 days following the CT scan if contrast material is administered.
   b. Tell the patient that she cannot be administered the CT scan.
   c. Cancel the CT scan.
   d. None of the above.

5. Why would a patient be asked to stop taking Glucophage before being administered a CT scan?
   a. Glucophage may react with contrast material.
   b. Glucophage reacts to X-rays.
   c. Glucophage reacts to MRI.
   d. There is no need to stop taking Glucophage if contrast material is administered before the CT scan.

6. What is the purpose of a perfusion CT?
   a. To determine osteoporosis
   b. To determine results of the KUB
   c. To determine blood supply to the brain
   d. None of the above

7. Where is contrast material administered in a CT myelogram?
   a. Muscle
   b. In the intrathecal space
   c. Nerve
   d. Legs

8. After a CT scan of the spine with contrast material, what should the patient do?
   a. Lie on his/her back
   b. Lie on his/her stomach
   c. Bend over
   d. Keep his/her head elevated

9. What should the patient do if he/she experiences a seizure following the CT scan?
   a. Call his/her healthcare provider immediately.
   b. Call for emergency medical care immediately.
   c. Lie down on his/her bed.
   d. Keep his/her head elevated.

10. What should the patient do if he/she experiences weakness in the extremities following a CT scan?
    a. Call his/her healthcare provider.
    b. Call the radiologist.
    c. Lie down for 15 minutes.
    d. Call for emergency medical help immediately.

Answers

1. b. To flush the contrast material.
2. a. Administer a sedative.
3. c. Notify the healthcare provider and radiologist.
4. a. Ask the patient to use formula for 2 days following the CT scan if contrast material is administered.
5. a. Glucophage may react with contrast material.
6. c. To determine blood supply to the brain.
7. b. In the intrathecal space.
8. d. Keep his/her head elevated.
9. b. Call for emergency medical care immediately.
10. a. Call his/her healthcare provider.

CHAPTER 16

Ultrasound Scan

An ultrasound scan creates an image of organs and structures inside the body using sound waves similar in concept to the way in which ship crews are able to identify underwater objects while on the surface of the water.

High-frequency sound waves are transmitted by a transducer that is placed on the patient's skin. Sound waves penetrate the skin, bounce off organs and structures in the patient's body, and are detected by the transducer.

Sound waves detected by the transducer are translated into an image that appears on the ultrasound screen. The healthcare provider can then measure organs and structures that appear on the image to determine any abnormality.

Images can either be printed and included in the patient's chart or can be stored on a computer. An ultrasound can detect a growth but cannot differentiate between one that is malignant or benign, which is determined by a biopsy. An ultrasound can differentiate between a solid growth and a fluid-filled cyst.

An ultrasound scan is commonly ordered instead of a CT scan or MRI because it is less expensive and in many situations provides the healthcare provider with sufficient information to assist in diagnosis.

In this chapter you will learn about different kinds of ultrasound scans.

Learning Objectives

1. Benign Prostatic Hyperplasia Ultrasound
2. Transvaginal Ultrasound and Hysterosonogram
3. Testicular Ultrasound
4. Abdominal Ultrasound
5. Breast Ultrasound
6. Cranial Ultrasound
7. Doppler Ultrasound
8. Fetal Ultrasound
9. Pelvic Ultrasound
10. Thyroid and Parathyroid Ultrasound

Key Words

Color Doppler
Conductive gel
Continuous-wave Doppler
Duplex Doppler
Latex allergy
Power Doppler
Sonohysterogram
Transabdominal
Transrectal
Transvaginal
Valsalva maneuver

Benign Prostatic Hyperplasia Ultrasound 1

Middle-aged men might experience the urgency to void, hesitancy waiting for the urinary stream, or a weak urinary stream. These may be signs of an enlarged prostate that places pressure on the bladder and blocks the urinary stream. Noncancerous, enlarged prostate is referred to as benign prostatic hyperplasia (BPH) or hypertrophy. An ultrasound is used to assist the healthcare provider diagnose the condition.

The benign prostatic hyperplasia ultrasound is also used to help guide the healthcare provider when taking a biopsy of the prostate, which is commonly performed if the patient's prostate-specific antigen (PSA) level is elevated.

Hint

The healthcare provider may also evaluate the bladder and the kidneys while performing the benign prostatic hyperplasia ultrasound to determine urinary retention and kidney stones that may block urinary flow.

Caution

The benign prostatic hyperplasia ultrasound cannot determine if urinary flow is blocked by the prostate.

WHAT IS BEING EXAMINED?

- The prostate gland

HOW IS THE TEST PERFORMED?

- The ultrasound transducer is either inserted into the rectum (transrectal ultrasound [TRUS]) or placed on the patient's abdomen (transabdominal ultrasound).
- An image of the prostate, kidneys, and bladder appears on the screen.
- The healthcare provider estimates the size of the prostate, determines if there is any urinary retention or blockage.

RATIONALE FOR THE TEST

- Assess
  - The size of the prostate
  - Urinary retention
  - A urinary blockage
- Guide the healthcare provider when taking a biopsy of the prostate.

NURSING IMPLICATIONS

- Assess if the patient can lie still during the test.

UNDERSTANDING THE RESULTS

- The test takes 30 minutes and the results are ready immediately if the test is performed by the healthcare provider or within a few hours if a technician performs the test.

- Normal test results indicate
  - Prostate is an acceptable size
  - Kidneys show no obstruction
  - No urine or an acceptable volume of urine in the bladder
- Abnormal test results indicate
  - Enlarged prostate
  - Enlarged kidneys
  - Kidney obstruction
  - Enlarged bladder
  - Significant volume of urine in bladder

TEACH THE PATIENT

- Explain
  - Why ultrasound is being taken.
  - That clear gel will be placed on the ultrasound site to increase the conduction of sound.
  - That the patient will not feel any pain.
  - That the patient must lie still during the test.

Transvaginal Ultrasound and Hysterosonogram 2

When a woman has difficulty conceiving, the healthcare provider may perform a transvaginal ultrasound or a hysterosonogram, which is also known as a sonohysterogram. These scans enable the healthcare provider to assess ovarian follicle development and the endometrium. This assessment may help the healthcare provider determine when to perform intrauterine insemination.

HINT

The transvaginal ultrasound is the preferred method rather than the transabdominal ultrasound for assessing the uterine lining and follicle growth.

CAUTION

The transvaginal ultrasound may not display scars or small tumors.

WHAT IS BEING EXAMINED?

- The uterus
- Uterine lining
- The endometrial cavity
- Ovarian follicle development

HOW IS THE TEST PERFORMED?

- The patient may be required to have an empty bladder prior to the scan or a full bladder depending on what is being examined.
- The ultrasound transducer is inserted into the vagina.
- The healthcare provider performs the ultrasound scan from within the vagina.
- When preparing for a hysterosonogram, the uterus is filled with fluid prior to insertion of the transducer into the vagina.

RATIONALE FOR THE TEST

- Assess
  - The uterus
  - The fallopian tubes
  - Ovaries
  - The endometrial cavity
  - Uterine lining
  - Ovarian follicle development
- To schedule intrauterine insemination
- To guide the healthcare provider when removing follicles

NURSING IMPLICATIONS

- Assess if the patient
  - Can lie still during the test
  - Has an empty bladder prior to the test

UNDERSTANDING THE RESULTS

- The test takes 30 minutes and the results are ready immediately if the test is performed by the healthcare provider or within a few hours if a technician performs the test.
- Normal test results indicate
    - No growths.
    - No scar tissue.
    - The uterus is a normal shape.
    - The fallopian tubes are not blocked.
    - Ovaries are normal size.
    - The endometrial cavity is normal shape.
    - Uterine lining is normal size.
- Abnormal test results indicate
    - Few follicles
    - Uterine fibroids
    - Ovarian cysts
    - Hydrosalpinx
    - Thick uterine lining
    - Deformed uterine lining
    - Enlarged uterus
    - Abnormal fallopian tubes

TEACH THE PATIENT

- Explain
    - Why ultrasound is being taken.
    - That the patient must have an empty bladder before the scan.
    - That the patient will not feel any pain.
    - That the patient must lie still during the test.
    - That the transducer is inserted into the vagina.
    - That the uterus will be filled with fluid if the healthcare provider orders a hysterosonogram.

Testicular Ultrasound

A healthcare provider may order a testicular ultrasound if the patient shows signs and symptoms of testicular abnormalities or infertility. The testicular ultrasound displays an image of the patient's testicles and scrotum including the epididymis, which is the coiled tube behind the testicle that collects sperm. The testicular ultrasound also displays an image of the vas deferens, which is the tube that connects the prostate gland to the testicles.

WHAT IS BEING EXAMINED?

- The testicles
- The scrotum
- The epididymis
- The vas deferens

HOW IS THE TEST PERFORMED?

- A conductive gel is placed on the transducer or on the scrotum.
- The transducer is placed on the scrotum and moved around to capture the image.
- The patient may be asked to hold his breath for a few seconds while the image is being taken.

RATIONALE FOR THE TEST

- Assess
  - The testicles
  - The epididymis
  - The vas deferens
  - The scrotum
  - The spermatic cord
  - For hydroceles and spermatoceles
- Guide the healthcare provider when performing a testicular biopsy.

NURSING IMPLICATIONS

- Assess
  - If the patient can lie still during the test.
  - If the patient has open skin in the scrotum.
- A consent form must be signed if the patient is undergoing a testicular biopsy.
- The patient must remove all clothing from the waist down.
- Towels are used to lift the scrotum.
- Wipe the conductive gel from the patient's skin when the test is completed.

UNDERSTANDING THE RESULTS

- The test takes 30 minutes and the results are ready immediately if the test is performed by the healthcare provider or within a few hours if a technician performs the test.
- Normal test results indicate
  - No growths.
  - The testicles are a normal shape.
  - The epididymitis does not show inflammation or blockage.
  - The spermatic cord is not twisted.
  - There is no fluid in the scrotum.
- Abnormal test results indicate
  - There is a growth on one or both testicles.
  - There is inflammation, infection, or blockage in the epididymis.
  - The spermatic cord is twisted.
  - There is fluid in the scrotum.

TEACH THE PATIENT

- Explain
  - Why ultrasound is being taken.
  - That the patient will not feel any pain.
  - That the patient must lie still during the test.
  - That the patient will have to remove all clothing from the waist down.

- That towels are used to lift the scrotum.
- That conductive gel is placed on the scrotum and it will be wiped off after the test is completed.
- That the transducer is placed on and moved around the scrotum.
- That a consent form must be signed if the patient is undergoing a testicular biopsy.

Abdominal Ultrasound

4

An abdominal ultrasound is ordered to view upper abdominal organs and structures. These include the liver, gallbladder, spleen, pancreas, and kidneys. It is also ordered to assist the healthcare provider in assessing the abdominal aorta, which is the artery located at the back of the chest and abdomen that supplies blood to the legs, abdomen, and organs in the lower portion of the body.

HINT

The healthcare provider may order an ultrasound of specific organs or structures within the abdomen if the abdominal ultrasound shows an abnormal condition in the abdomen.

WHAT IS BEING EXAMINED?

- Liver
- Gallbladder
- Bile ducts
- Spleen
- Pancreas
- Kidneys
- Abdominal aorta

HOW IS THE TEST PERFORMED?

- The patient must remove jewelry from the abdominal area.
- The patient should eat a fat-free dinner the night before the test.
- The patient should avoid eating 12 hours before the test to prevent gas from building up in the intestines.

- The patient may be asked to drink six glasses of water an hour before the test.
- The patient must remove all clothing from the waist down.
- A conductive gel is placed on the abdomen.
- The transducer is placed on the abdomen and moved around to capture the image.
- The healthcare provider may push on the abdomen to move organs within the abdominal cavity to get a clearer view of other organs.
- The patient may be asked to change position during the test.
- The patient may be asked to hold his/her breath for a few seconds.

RATIONALE FOR THE TEST

- Assess
  - Liver
  - Gallbladder
  - Bile ducts
  - Spleen
  - Pancreas
  - Kidneys
  - Abdominal aorta
- Guide the healthcare provider when taking a biopsy or when performing a paracentesis.

NURSING IMPLICATIONS

- Assess if the patient
  - Can lie still during the test
  - Has had a barium enema for an upper GI test 2 days prior to the ultrasound scan
  - Has had a barium enema for a lower GI test 2 days prior to the ultrasound scan
  - Has removed jewelry from the abdominal area
  - Has eaten a fat-free dinner the night before the test
  - Has not eaten 12 hours before the ultrasound scan

- Has drunk six glasses of water an hour before the ultrasound scan
- Has an open wound in the abdominal area
- Is obese
- Has signed a consent form when undergoing a biopsy.
- Has removed all clothing from the waist down.

- Wipe the conductive gel from the patient's skin when the test is completed.

UNDERSTANDING THE RESULTS

- The test takes 60 minutes and the results are ready immediately if the test is performed by the healthcare provider or within a few hours if a technician performs the test.
- Normal test results indicate
  - No growths
  - All organs are normal shape
  - No inflammation or blockage of the bile ducts
- Abnormal test results indicate
  - There is a growth on one or multiple organs.
  - There is inflammation, infection, or blockage of the bile ducts.
  - The aorta is enlarged.
  - Fluid appears in the abdomen.

TEACH THE PATIENT

- Explain
  - Why ultrasound is being taken.
  - That the patient will not feel any pain.
  - That the patient must lie still during the test.
  - That the patient will have to remove all clothing from the waist down.
  - That the patient will need to remove jewelry from the abdominal area.
  - That the patient should eat a fat-free dinner the night before the test.
  - That the patient should avoid eating for 12 hours before the test.
  - That the patient may be asked to drink six glasses of water an hour before the test.

- That the healthcare provider may push on the abdomen to move organs within the abdominal cavity to get a clearer view of other organs.
- That the patient may be asked to change position during the test.
- That the patient may be asked to hold his/her breath for a few seconds while the ultrasound is being taken.
- That conductive gel is placed on the abdomen and it will be wiped off after the test is completed.
- That the transducer is placed on and moved around the abdomen.
- That a consent form must be signed if the patient is undergoing a biopsy.

Breast Ultrasound 5

The breast ultrasound creates an image of all areas of the breast, including portions of the breast that are near the chest. A mammogram typically doesn't show images of the breast that are near the chest. A breast ultrasound is sometimes performed on younger women whose breast tissues are dense, making it difficult to see breast abnormalities on a mammogram.

A healthcare provider frequently orders a breast ultrasound to assess a lump identified by either palpation or from a mammogram. The breast ultrasound can distinguish between a solid mass and a cyst. If it is a solid mass, then the healthcare provider performs a biopsy to determine if the mass is benign or malignant.

Caution

A breast ultrasound does not replace an annual mammogram.

WHAT IS BEING EXAMINED?

- The breast

HOW IS THE TEST PERFORMED?

- The patient must remove jewelry from the breast area.
- The patient must remove clothing above the waist.
- A conductive gel is placed on the transducer before placing it on the breast.
- The transducer is placed on the breast and moved around to capture the image.

RATIONALE FOR THE TEST

- Assess
  - A mass found on palpation or by a mammogram
  - The breast of a younger woman whose breast tissues are dense
  - Silicone breast implants
  - Breast pain
- Guide the healthcare provider when taking a biopsy or when draining a cyst.

NURSING IMPLICATIONS

- Assess if the patient
  - Can lie still during the test
  - Has removed jewelry from the breast area
  - Has an open wound in the breast area
- A consent form must be signed if the patient is undergoing a biopsy or a cyst drainage.
- The patient must remove all clothing above the waist.
- Wipe the conductive gel from the patient's skin when the test is completed.

UNDERSTANDING THE RESULTS

- The test takes 30 minutes and the results are ready immediately if the test is performed by the healthcare provider or within a few hours if a technician performs the test.
- Normal test results indicate
  - No mass found
  - Normal-shaped breasts
  - No inflammation
- Abnormal test results indicate
  - There is a cyst.
  - There is a solid mass.
  - There is inflammation or infection.

TEACH THE PATIENT

- Explain
    - Why ultrasound is being taken.
    - That the patient will not feel any pain.
    - That the patient must lie still during the test.
    - That the patient will have to remove all clothing from the waist up.
    - The patient will need to remove jewelry from the breast area.
    - That conductive gel is placed on the breast and it will be wiped off after the test is completed.
    - That the transducer is placed on and moved around the breast.
    - That a consent form must be signed if the patient is undergoing a biopsy or drainage of a cyst.

Cranial Ultrasound 6

A cranial ultrasound creates images of the brain and ventricles. Since ultrasound cannot penetrate bone, a cranial ultrasound is performed on babies up to 18 months old, whose cranium has yet to form. A cranial ultrasound is commonly used in premature newborns to assess complications of the premature birth.

A cranial ultrasound is also performed on adults during brain surgery to visualize any masses in the brain. The ultrasound is able to capture the image because a portion of the patient's cranium is removed during surgery, enabling the ultrasound to penetrate the brain tissue.

WHAT IS BEING EXAMINED?

- Brain
- Ventricles

HOW IS THE TEST PERFORMED?

- On babies
    - Feed the baby near the time of the ultrasound scan to keep the baby still and comfortable during the scan.
    - Place the baby on his/her back or have the parent hold the baby.
    - The transducer is placed on and moved around the fontanelle.

- On adults
  - A portion of the cranium is removed in surgery.
  - The transducer is placed on and moved around the exposed brain.

RATIONALE FOR THE TEST

- Assess for
  - Babies
    - Why the baby has an abnormally large head
    - For encephalitis
    - For meningitis
    - For hydrocephalus
    - For periventricular leukomalacia (PVL)
    - For intraventricular hemorrhage (IVH)
  - Adults
    - Brain mass

NURSING IMPLICATIONS

- Assess if
  - Baby can lie still during the test.
  - Parent can hold the baby still during the test.
  - Baby's cranium has closed.

UNDERSTANDING THE RESULTS

- The test takes 30 minutes and the results are ready immediately.
- Normal test results indicate
  - No mass or cyst found
  - Normal-shaped brain
  - No inflammation
  - No bleeding
  - There is no excessive cerebrospinal fluid (CSF)
- Abnormal test results indicate
  - There is a mass or cyst
  - There is inflammation or infection

- There is bleeding
- There is excessive cerebrospinal fluid (CSF)

TEACH THE PATIENT

- Explain
  - Why ultrasound is being taken.
  - That the patient will not feel any pain.
  - That the patient must lie still during the test and the parent can hold the baby during the test.
  - That conductive gel is placed on the fontanelle and it will be wiped off after the test is completed.
  - That the transducer is placed on and moved around the fontanelle.
  - That the test will be conducted during surgery, if the patient is an adult.

Doppler Ultrasound 7

A Doppler ultrasound is used to assess blood flow through the blood vessels. There are four types of Doppler ultrasound

- Continuous-wave: Produces a pulsating, audible sound reflecting pulsating blood through a blood vessel.
- Duplex: Produces an image of the blood vessel along with a computer-generated graph, which indicates the speed and direction of blood flow.
- Color: Produces an image of the blood vessel with the speed and direction of blood flow represented by colors on the image.
- Power: Similar to the color Doppler; however, the power Doppler is five times as sensitive in detecting blood flow.

WHAT IS BEING EXAMINED?

- Blood flow in blood vessels

HOW IS THE TEST PERFORMED?

- The patient should remove all jewelry around the site of the blood vessel.
- The patient will remove all clothing around the site of the blood vessel.

- A conductive gel is placed on the skin that covers the blood vessel.
- The transducer is placed over and moved around the blood vessel.
- If the extremities are being tested, the healthcare provider may test both extremities to compare the results. Furthermore, the test may be performed with the patient both lying and sitting. The healthcare provider may also place the blood pressure cuff on the extremities during the test.
- The patient may be asked to perform the Valsalva maneuver by pinching his/her nose, closing his/her mouth, and then exhaling.
- If arteries in the neck are being tested, the patient will place his/her head on a pillow.

RATIONALE FOR THE TEST

- Assess for
  - Narrow blood vessels
  - Blood clots (deep vein thrombosis)
  - Atherosclerosis
  - Stroke
- Use of mapping in vein grafts

NURSING IMPLICATIONS

- Assess if the
  - Patient can lie still during the test.
  - Patient has smoked cigarettes or chewed tobacco or otherwise ingested nicotine 2 hours before the test. Nicotine constricts blood vessels resulting in a false test result.
  - Patient can perform the Valsalva maneuver.
  - Patient is obese, since this can interfere with the test.
  - Patient has arrhythmias, which may alter blood flow through unobstructed blood vessels.

UNDERSTANDING THE RESULTS

- The test takes 60 minutes and the results are ready immediately.

- Normal test results indicate
  - No blood clots
  - Normal blood flow and pulse
- Abnormal test results indicate
  - There is a blockage in blood flow.
  - There is an abnormal pulse.
  - There is abnormal shape of blood vessels.

TEACH THE PATIENT

- Explain
  - Why ultrasound is being taken.
  - That the patient will not feel any pain.
  - That the patient must lie still during the test.
  - That conductive gel is placed on the skin over the blood vessel and it will be wiped off after the test is completed.
  - That the transducer is placed on and moved around the skin over the blood vessel.
  - That patient should avoid smoking, chewing tobacco, or otherwise ingesting nicotine 2 hours before the test.
  - That the healthcare provider may ask the patient to perform the Valsalva maneuver.
  - That arrhythmias may provide a false test result.

Fetal Ultrasound 8

A fetal ultrasound produces an image (sonogram) of the fetus, the placenta, and amniotic fluid during pregnancy. The healthcare provider orders a fetal ultrasound to determine the size, position, and sex of the fetus and to identify any abnormalities prior to birth.

There are two types of fetal ultrasound tests:

- Transabdominal: The ultrasound transducer is placed on the patient's abdomen.
- Transvaginal: The ultrasound transducer is covered in a latex sheath and inserted into the vagina.

Caution

Verify that the patient is not allergic to latex if the healthcare provider is performing a transvaginal ultrasound. A normal fetal ultrasound does not rule out fetal abnormalities or problems with the placenta and amniotic fluid.

WHAT IS BEING EXAMINED?

- The fetus
- The placenta
- Amniotic fluid

HOW IS THE TEST PERFORMED?

- Transabdominal ultrasound
  - The patient should remove all jewelry around the abdomen.
  - The patient will remove all clothing around the abdomen.
  - Early pregnancy: The patient needs a full bladder and must drink a large volume of water and avoid urinating until the test is completed. The water moves the intestines away from the uterus. If the patient is unable to maintain a full bladder, the healthcare provider may order the insertion of a urinary catheter through her urethra and fill the bladder with sterile water.
  - Late pregnancy: The patient does not need a full bladder since the intestines are repositioned by fetal growth.
  - The patient lies on her back.
  - A conductive gel is placed on the abdomen.
  - The transducer is placed over and moved around the abdomen.
  - If the patient becomes short of breath, the healthcare provider may raise the head of the bed or place the patient on her side.
- Transvaginal ultrasound
  - The patient does not need a full bladder for this test.
  - The patient lies on her back with hips raised.
  - Verify before the test that the patient is not allergic to latex.
  - The healthcare provider places a transducer covered with a latex sheath into the vagina to capture the image.

RATIONALE FOR THE TEST

- Assess
  - Progress of the pregnancy
  - The gestational age of the fetus
  - For fetal defects
  - The number of fetuses
  - The placenta
  - The amniotic fluid
  - The fetal position
  - The cervix
- Detection of ectopic pregnancy

NURSING IMPLICATIONS

- Assess if the patient
  - Can lie still during the test
  - Has a full bladder if a transabdominal ultrasound is being performed
  - Is allergic to latex if a transvaginal ultrasound is being performed
  - Is obese

UNDERSTANDING THE RESULTS

- The test takes 60 minutes or less and the results are ready immediately.
- Normal test results indicate
  - Expected fetal gestation
  - Normal placenta
  - Expected volume of amniotic fluid
  - No birth defects
- Abnormal test results indicate
  - Underdeveloped or abnormally developed fetus
  - Unexpected fetal position
  - Placenta previa
  - Molar pregnancy

- Suspected birth defect discovered
- Unexpected volume of amniotic fluid

TEACH THE PATIENT

- Explain
  - Why ultrasound is being taken.
  - That the patient will not feel any pain.
  - That the patient must lie still during the test.
  - That the patient will require a full bladder. If the patient is unable to drink a sufficient volume of water, then a urinary catheter will be inserted and sterile water will be placed into her bladder (transabdominal).
  - That the patient will not urinate until after the test is concluded (transabdominal).
  - That conductive gel is placed on the skin over the abdomen and it will be wiped off after the test is completed (transabdominal).
  - That the transducer is placed on and moved around the abdomen (transabdominal).
  - That the ultrasound transducer may be covered in a latex sheath and inserted into the patient's vagina (transvaginal).
  - That if the patient feels short of breath during the test, the head of the bed will be raised or she will be repositioned on her side (transvaginal).
  - That a normal fetal ultrasound does not rule out abnormalities with the fetus, placenta, or amniotic fluid.

Pelvic Ultrasound 9

A pelvic ultrasound creates images of the bladder, ovaries, uterus, cervix, fallopian tubes, prostate gland, and seminal vesicles. There are three types of pelvic ultrasound tests.

- Transabdominal: The transducer is moved along the abdomen.
- Transrectal: The transducer is inserted into the rectum.
- Transvaginal: The transducer is covered with a latex sheath and inserted in the vagina.

CAUTION

Verify that the patient is not allergic to latex before a transrectal or transvaginal ultrasound is performed.

WHAT IS BEING EXAMINED?

- Bladder
- Ovaries
- Uterus
- Cervix
- Fallopian tubes
- Prostate gland
- Seminal vesicles

HOW IS THE TEST PERFORMED?

- Transabdominal ultrasound
  - The patient removes all jewelry from below the waist.
  - The patient will remove all clothing from below the waist.
  - The patient will be required to fill his/her bladder to push the intestines away from the pelvic organs. If the patient is unable to drink water, the healthcare provider will insert a catheter into the patient's bladder and infuse sterile water.
  - The patient will be required to sign a consent form if the ultrasound is used to guide the healthcare provider to take a biopsy.
  - The patient lies on his/her back.
  - A conductive gel is placed on the abdomen.
  - The transducer is placed over and moved around the abdomen.
- Transrectal ultrasound
  - The patient is given an enema an hour before the test.
  - Verify that the patient is not allergic to latex.
  - A latex sheath is placed over the ultrasound transducer.
  - The patient lies on his/her left side with knees bent.
  - The ultrasound transducer is lubricated and then inserted into the patient's rectum to capture the image.

- Transvaginal ultrasound
  - The patient should avoid drinking fluids 4 hours before the test.
  - The patient lies on her back with hips raised.
  - Verify before the test that the patient is not allergic to latex.
  - The healthcare provider places a transducer covered with a latex sheath into the vagina to capture the image.

RATIONALE FOR THE TEST

- Assess
  - The cause of urinary disorders
  - The bladder
  - For growths
  - For pelvic inflammatory disease (PID)
  - For placement of intrauterine device (IUD)
  - The size of pelvic organs and structures
  - The fetal position
  - For the cause of infertility
- Guide the healthcare provider when performing a biopsy.

NURSING IMPLICATIONS

- Assess if the patient
  - Has ingested contrast material 2 days before the ultrasound
  - Can lie still during the test
  - Has a full bladder (transabdominal)
  - Is allergic to latex (transrectal or transvaginal)
  - Has removed jewelry and clothing from below the waist
  - Signed a consent if the ultrasound is used for a biopsy
  - Has been given an enema an hour before the test (transrectal)
  - Can lie on left side and bend knees (transrectal)
  - Has not drunk fluids 4 hours before the test (transvaginal)
  - Is obese

UNDERSTANDING THE RESULTS

- The test takes 30 minutes or less and the results are ready immediately.
- Normal test results indicate
  - Pelvic organs are normal size
  - No growth
  - No blockages
  - Intrauterine device is in the expected position
- Abnormal test results indicate
  - Unexpected size of one or more pelvic organs.
  - Growths are seen.
  - Inflammation and infection noticed.
  - Unexpected fluid.
  - Blockage found.
  - Intrauterine device is not in the expected position.

TEACH THE PATIENT

- Explain
  - Why ultrasound is being taken.
  - That the patient will not feel any pain.
  - That the patient must lie still during the test.
  - That the patient cannot ingest contrast material 2 days before the test.
  - That the patient will require a full bladder. If the patient is unable to drink a sufficient volume of water, then a urinary catheter will be inserted and sterile water will be placed into the bladder (transabdominal).
  - That the patient will not urinate until after the test is concluded (transabdominal).
  - That the patient will be given an enema an hour before the test (transrectal).
  - That the patient must avoid drinking 4 hours before the test (transvaginal).
  - That conductive gel is placed on the skin over the abdomen and it will be wiped off after the test is completed (transabdominal).

- That the transducer is placed on and moved around the abdomen (transabdominal).
- That the ultrasound transducer may be covered in a latex sheath and inserted into the rectum or vagina (transrectal or transvaginal).
- That if she feels short of breath during the test, the head of the bed will be raised or that she will be repositioned on her side (transvaginal).

Thyroid and Parathyroid Ultrasound 10

The thyroid and parathyroid ultrasound is used to create an image of the thyroid and the parathyroid glands. The thyroid gland produces thyroxine that controls body's metabolism. The parathyroid gland produces the parathyroid hormone (PTH) that controls body's calcium and phosphorus balance in the blood.

Caution

The thyroid and parathyroid ultrasound enables the healthcare provider to assess the size of these glands but not the production of hormones.

WHAT IS BEING EXAMINED?

- Thyroid gland
- Parathyroid gland

HOW IS THE TEST PERFORMED?

- The patient removes all jewelry from the head and neck area.
- The patient may be asked to remove all clothing from above the waist.
- The patient lies on his/her back and a pillow is placed under the patient's shoulders.
- A conductive gel is placed on the neck.
- The healthcare provider may place a bag filled with water or a sponge of gel over the patient's throat to increase conductivity.
- The transducer is placed over and moved around the neck. The patient may be asked to turn his/her head.

RATIONALE FOR THE TEST

- Assess
  - The size of the thyroid gland
  - The size of the parathyroid gland
  - For growths
- Guide the healthcare provider when performing a biopsy.

NURSING IMPLICATIONS

- Assess if the patient
  - Can lie still during the test
  - Has removed jewelry and clothing from above the waist
  - Signed a consent if the ultrasound is used for a biopsy
  - Can move his head
  - Is obese

UNDERSTANDING THE RESULTS

- The test takes 30 minutes and the results are ready immediately.
- Normal test results indicate
  - Thyroid is normal size
  - Parathyroid is normal size
  - No growth
- Abnormal test results indicate
  - Unexpected size of thyroid or parathyroid
  - Growths are seen

TEACH THE PATIENT

- Explain
  - Why ultrasound is being taken.
  - That the patient will not feel any pain.
  - That the patient must lie still during the test.
  - That conductive gel is placed on the skin over the neck, and it will be wiped off after the test is completed.

Summary

High-frequency sound waves penetrate the patient's body. Some waves are reflected and others pass through the body. The reflected waves are received by a transducer and passed through to a computer that generates an image on the screen.

The healthcare provider is able to view the image and determine if there are any abnormal organs or structures in the patient's body. An ultrasound can differentiate between a fluid-filled cyst and a solid mass but cannot tell if the solid mass is benign or malignant. Only a biopsy study of the mass can make this determination.

An ultrasound is less expensive than alternative tests such as a CT scan or MRI. Although these other alternative tests provide greater detail, the ultrasound typically provides sufficient information to help the healthcare provider reach a diagnosis.

Quiz

1. A gel is placed on the ultrasound transducer and on the patient's skin to
   a. Increase conductivity of sound waves
   b. Increase conductivity of radio waves
   c. Increase conductivity of light waves
   d. Decrease ambient noise

2. The patient's bladder is filled during an abdominal ultrasound because
   a. Water increases conductivity of sound.
   b. The full bladder increases the surface area of the patient's abdomen.
   c. Full bladder repositions the intestines.
   d. None of the above.

3. What test is used to indicate blood flow?
   a. Doppler ultrasound
   b. Continuous-wave Doppler
   c. Color Doppler
   d. All of the above

4. A fetal ultrasound produces
   a. A sonogram
   b. An X-ray of the fetus
   c. An X-ray of the mother
   d. None of the above

5. Before a transvaginal ultrasound is administered, you must
   a. Inform the woman that conductive gel will be placed on her abdomen.
   b. Verify the woman's age.
   c. Verify that the patient is not allergic to latex.
   d. Inform the woman that the transducer will be pressed down on her abdomen.

6. What happens if the patient is unable to drink a large volume of fluid before a transabdominal ultrasound?
   a. The test is cancelled.
   b. An MRI is ordered.
   c. Sterile water is inserted into the bladder using a urinary catheter.
   d. A CT scan is ordered.

7. A thyroid ultrasound can detect if the thyroid gland is producing excess thyroxine.
   a. True
   b. False

8. The patient who is 8 months pregnant and is scheduled to receive a transabdominal ultrasound
   a. Does not have to have a full bladder for the test
   b. Must have a full bladder for the test
   c. Cannot have her bladder filled using a urinary catheter
   d. Must have her bladder filled using a urinary catheter

9. If the patient smoked several cigarettes a half hour before a Doppler ultrasound
   a. The test should be postponed because nicotine constricts blood vessels and could result in a false test result.
   b. The test should be continued as scheduled.
   c. The test should be continued as scheduled; however, the healthcare provider should be told of the situation.
   d. None of the above.

10. A cranial ultrasound can be performed on a 2-year-old child.
    a. True
    b. False

Answers

1. a. Increase conductivity of sound waves.
2. c. The full bladder repositions the intestines.
3. d. All of the above.
4. a. A sonogram.
5. c. Verify that the patient is not allergic to latex.
6. c. Sterile water is inserted into the bladder using a urinary catheter.
7. b. False.
8. a. Does not have to have a full bladder for the test.
9. a. The test should be postponed because nicotine constricts blood vessels and could result in a false test result.
10. b. False.

CHAPTER 17

Magnetic Resonance Imaging

Magnetic resonance imaging (MRI) uses pulsating radio waves in a magnetic field to produce an image of inside the patient's body. The patient lies on his/her back on a table. A coil is placed around the area of the patient that is being scanned and a belt is placed around the patient to detect breathing. The table moves into the magnetic field and the belt triggers the MRI scan so that breathing does not interfere with capturing the image.

A clanking/tapping noise is heard while the MRI scans the patient. The patient may listen to music through headphones to block out the noise.

There are two main types of MRI machines; one is closed the other is open. In the closed machine, the patient's body is entirely enclosed while only a portion of the patient's body is enclosed in an open machine.

The healthcare provider may order that contrast material be administered to the patient prior to the MRI. The contrast material highlights areas of the body that are being studied and may be ingested or administered intravenously.

The MRI produces digital images that are displayed on a computer screen and can be stored for further review by the patient's healthcare team. The MRI creates images that are more detailed than images produced by a CT scan, X-ray, or ultrasound.

No metal objects should be on or inside the patient during an MRI, including credit cards. Information on the credit card might be erased by the MRI's magnetic field. An X-ray may be ordered to determine if there is any metal inside the patient before the MRI is administered, especially if the patient was in an accident where metal fragments might be embedded throughout the body.

However, dental fillings are usually permitted although the patient is likely to feel tingling in the mouth during the MRI. The patient may experience skin irritation if he/she has iron pigment tattoos.

In this chapter you'll learn about different kinds of MRI tests.

Learning Objectives

1. Abdominal MRI
2. Breast MRI
3. Head MRI
4. Knee MRI
5. Shoulder MRI
6. Spinal MRI

Key Words

BRCA1
BRCA2
Cervical spine
Claustrophobic
Closed MRI
Contrast material
Diffusion-perfusion imaging
Gadolinium
Glucagon
Iodine allergy
Lumbosacral spine
Magnetic resonance angiogram (MRA)
Magnetic resonance spectroscopy
Open MRI
Sciatica
Shellfish allergy
Spinal disc disorders
Spinal stenosis
Thoracic spine

Abdominal MRI

An abdominal MRI produces detailed images of organs, structures, and tissues contained within the abdomen. The healthcare provider may order that the patient be administered contrast material prior to the MRI to highlight parts of the abdomen on the MRI image. This enables the healthcare provider to identify any subtle abnormalities that may exist in the abdomen.

WHAT IS BEING EXAMINED?

- Abdominal organs, structures, and tissues

HOW IS THE TEST PERFORMED?

- The patient is assessed for any metal that might be on or inside his/her body. The healthcare provider determines if the presence of any metal may require cancellation of the MRI.
- An assessment is made to determine if the patient can be administered contrast material over a 2-minute period, if required for the MRI. Contrast material is either ingested or administered IV, causing a flushing feeling for the patient. Contrast material may or may not be used if the patient is allergic to shellfish or iodine or if the patient has kidney abnormalities or sickle cell anemia, depending on the type of contrast material that the healthcare provider plans to use for the test.
- An assessment is made to determine if the patient is claustrophobic. If so, then the healthcare provider may administer a sedative to the patient or schedule the test to be performed using an open MRI machine.
- The patient removes all clothing and wears a gown during the MRI.
- The patient may be given headphones to avoid hearing the clanking/tapping noise created by the MRI.
- The patient may be administered glucagon to reduce intestinal movement during the test.
- The patient lies on his/her back on the MRI table.
- A coil is placed on top of the patient's abdomen.
- A belt is cinched around the abdomen to detect the patient's breathing patterns.
- The table is moved into the MRI machine as images are taken of the abdomen.
- Images are viewed during and after the MRI is completed to assist the healthcare provider reach a diagnosis.

RATIONALE FOR THE TEST

- Assess
  - The size of abdominal organs and structures
  - The existence of a growth
  - Identify a blockage
  - The existence of fluid within the abdomen
  - For inflammation
  - Blood flow

NURSING IMPLICATIONS

- Determine if the patient has any metal on or inside his/her body.
- Assess if the patient
  - Has eaten or drunk before the MRI is administered. Some MRI studies require that the patient refrain from eating or drinking 12 hours before the test is administered.
  - Is allergic to shellfish, iodine, or contrast material.
  - Is pregnant.
  - Is claustrophobic.
  - Has kidney disease.
  - Is wearing any medication patches.
  - Can lie still during the test.

UNDERSTANDING THE RESULTS

- The test takes 60 minutes and the results are ready immediately if the test is performed by the healthcare provider or within a few hours if a technician performs the test.
- Normal test results indicate
  - Normal size of abdominal organs and structures
  - No growth(s)
  - No blockage
  - No fluid within the abdomen
  - No inflammation
  - Normal blood flow

- Abnormal test results indicate
  - Unusual size of abdominal organs and structures.
  - The existence of a growth.
  - A blockage is identified.
  - Fluid exists within the abdomen.
  - Inflammation or infection is present.
  - Unusual blood flow.

TEACH THE PATIENT

- Explain
  - Why MRI is being taken.
  - What the patient will experience during the MRI.
  - That no metal can be on or inside the patient during the MRI.
  - That the patient will not feel any pain during the procedure, although the patient may feel a tingling sensation if the patient has metal fillings in his/her teeth.
  - That the patient may be administered a sedative if he/she is claustrophobic.
  - That some patients who are allergic to shellfish and iodine may also be allergic to contrast material. If the patient is allergic to contrast material, the healthcare provider will discuss the risk and benefit of administering the contrast material. If the patient agrees that the benefits outweigh the risk, then the healthcare provider may administer medication that counteracts the allergic reaction to the contrast material.
  - That the patient may be asked to drink contrast material or that the contrast material may be administered by IV, causing a flushing feeling.
  - That the patient may be asked to refrain from eating or drinking 12 hours before the MIR.

Breast MRI 2

A breast MRI produces detailed images of the breast that provide more information to the healthcare provider than a breast ultrasound or traditional mammography. Healthcare providers order breast MRIs typically when other tests such as a mammography indicate an abnormality. If the abnormality is inflammation, a growth, or blood flow to breast tissues, the healthcare provider may administer contrast material to enhance the image of those areas of the breast.

Women who are positive for the BRCA1 or BRCA2 gene or whose family members developed breast cancer before the age of 50 are considered high risk for developing breast cancer and may be recommended for annual breast MRIs to detect early signs of breast cancer.

The healthcare provider may also order annual breast MRIs for women who normally have dense breast tissue. An MRI is better suited to examine dense breast tissue than an ultrasound test.

WHAT IS BEING EXAMINED?

- Breasts

HOW IS THE TEST PERFORMED?

- The patient is assessed for any metal that might be on or inside her body. The healthcare provider determines if the presence of any metal may require cancellation of the MRI.
- An assessment is made to determine if the patient can be administered contrast material over a 2-minute period, if required for the MRI. Contrast material is either ingested or administered IV, causing a flushing feeling for the patient. Contrast material may not be used if the patient is allergic to shellfish and iodine or if the patient has kidney abnormalities or sickle cell anemia.
- An assessment is made to determine if the patient is claustrophobic. If so, then the healthcare provider may administer a sedative to the patient or schedule the test to be performed using an open MRI machine.
- The patient removes all clothing and wears a gown during the MRI.
- The patient may be given headphones to block out the clanking/tapping noise created by the MRI.
- The patient lies on her back on the MRI table.
- A coil is placed on top of the patient's chest area.
- A belt is cinched around the patient to detect the patient's breathing patterns.
- The table is moved into the MRI machine as images are taken of the breasts.
- Images are viewed during and after the MRI is completed to assist the healthcare provider in reaching a diagnosis.

RATIONALE FOR THE TEST

- Assess
    - For infection
    - The existence of a growth
    - For inflammation
    - Blood flow
    - Women who are at a high risk for breast cancer
    - Women who normally have dense breast tissue
    - Breast cancer treatment
    - Breast implants

NURSING IMPLICATIONS

- Determine if the patient has any metal on or inside her body.
- Assess if the patient
    - Is allergic to shellfish, iodine, or contrast material
    - Is pregnant
    - Is claustrophobic
    - Has kidney disease
    - Is wearing any medication patches
    - Can lie still during the test

UNDERSTANDING THE RESULTS

- The test takes 60 minutes and the results are ready immediately if the test is performed by the healthcare provider or within a few hours if a technician performs the test.
- Normal test results indicate
    - No growth(s)
    - No blockage
    - No infection
    - No inflammation
    - Normal blood flow
    - Breast implant correctly positioned

- Abnormal test results indicate
  - The existence of a growth
  - Breast implants improperly positioned
  - Inflammation or infection is present
  - Unusual blood flow or blockage

TEACH THE PATIENT

- Explain
  - Why MRI is being taken.
  - What the patient will experience during the MRI.
  - That no metal can be on or inside the patient during the MRI.
  - That the patient will not feel any pain during the procedure, although the patient may feel a tingling sensation if she has metal fillings in her teeth.
  - That the patient may be administered a sedative if she is claustrophobic.
  - That some patients who are allergic to shellfish and iodine may also be allergic to contrast material. If the patient is allergic to contrast material, the healthcare provider will discuss with the patient the risk and benefits of administering the contrast material. If the patient agrees that the benefits outweigh the risk, then the healthcare provider may administer medication that counteracts the allergic reaction to the contrast material.
  - That the contrast material may be administered IV causing a flushing feeling.

Head MRI 3

An MRI of the head is ordered to produce images of the brain and blood vessels that supply blood to the brain to determine the underlying cause of headache, assess for head injury, or determine if the patient has abnormal blood flow or a disorder that affects the brain. Unlike an ultrasound, the head MRI is a closed procedure and does not require that the patient's skull be opened.

There are three types of MRIs used to assess the brain:

- Magnetic resonance spectroscopy: Assesses changes in brain chemistry caused by disease

- Magnetic resonance angiogram (MRA): Assesses speed, direction, and flow of blood in the brain
- Diffusion-perfusion imaging: Assesses inflammation, tumors, and stroke and evaluates the fluid content of the brain

---

Caution

Healthcare providers may order an MRI with gadolinium containing contrast material. Gadolinium can cause nephrogenic fibrosing dermopathy in patients who have kidney failure.

WHAT IS BEING EXAMINED?

- Brain

HOW IS THE TEST PERFORMED?

- The patient is assessed for any metal that might be on or inside his/her body. The healthcare provider determines if the presence of any metal may require cancellation of the MRI.
- An assessment is made to determine if the patient can be administered contrast material, if required for the MRI. Contrast material is administered IV, causing a flushing feeling for the patient. Contrast material may not be used if the patient is allergic to shellfish and iodine or if the patient has kidney abnormalities or sickle cell anemia.
- An assessment is made to determine if the patient is claustrophobic. If so, then the healthcare provider may administer a sedative to the patient or schedule the test to be performed using an open MRI machine.
- The patient removes all clothing and wears a gown during the MRI.
- The patient may be given headphones to block out the clanking/tapping noise created by the MRI.
- The patient lies on his/her back on the MRI table.
- A belt is placed on the patient to detect his/her breathing patterns.
- The table is moved into the MRI machine as images are taken of the brain.
- Images are viewed during and after the MRI is completed to assist the healthcare provider reach a diagnosis.

RATIONALE FOR THE TEST

- Assess
  - For infection
  - The existence of a growth
  - For inflammation
  - Blood flow
  - For stroke
  - For suspected head injury
  - For hydrocephaly
  - For multiple sclerosis (MS)
  - For Alzheimer disease
  - For Parkinson disease
  - For Huntington disease

NURSING IMPLICATIONS

- Determine if the patient has any metal on or inside his body.
- Assess if the patient
  - Is allergic to shellfish, iodine, or contrast material
  - Is pregnant
  - Is claustrophobic
  - Has kidney disease
  - Is wearing any medication patches
  - Can lie still during the test

UNDERSTANDING THE RESULTS

- The test takes 60 minutes and the results are ready immediately if the test is performed by the healthcare provider or within a few hours if a technician performs the test.
- Normal test results indicate
  - No growth(s)
  - No blockage
  - No infection

- No inflammation
- Normal blood flow
- Abnormal test results indicate
  - The existence of a growth.
  - Inflammation or infection is present.
  - Unusual blood flow or blockage.

TEACH THE PATIENT

- Explain
  - Why MRI is being taken.
  - What the patient will experience during the MRI.
  - That no metal can be on or inside the patient during the MRI.
  - That the patient will not feel any pain during the procedure, although the patient may feel a tingling sensation, if he/she has metal fillings in his/her teeth.
  - That the patient may be administered a sedative if he/she is claustrophobic.
  - That some patients who are allergic to shellfish or iodine may also be allergic to contrast material. If the patient is allergic to contrast material, the healthcare provider will discuss with the patient the risk and benefit of administering the contrast material. If the patient agrees that the benefits outweigh the risk, then the healthcare provider may administer medication that counteracts the allergic reaction to the contrast material.
  - That the contrast material may be administered IV, causing a flushing feeling.

Knee MRI 4

A knee MRI produces detailed images of structures and tissues contained within the knee. These enables the healthcare provider to identify any abnormalities that may exist in the knee including damage to tendons, ligaments, cartilage, and fluid. The healthcare provider may order a knee MRI to assess if the patient requires arthroscopy of the knee.

WHAT IS BEING EXAMINED?

- Knee structures and tissues

HOW IS THE TEST PERFORMED?

- The patient is assessed for any metal that might be on or inside his/her body. The healthcare provider determines if the presence of any metal may require cancellation of the MRI.
- An assessment is made to determine if the patient can be administered contrast material, if required, for the MRI. Contrast material is administered IV causing a flushing feeling for the patient. Contrast material may not be used if the patient is allergic to shellfish or iodine or if the patient has kidney abnormalities or sickle cell anemia.
- An assessment is made to determine if the patient is claustrophobic. If so, the healthcare provider may administer a sedative to the patient or schedule the test to be performed using an open MRI machine.
- The patient removes all clothing and wears a gown during the MRI.
- The patient may be given headphones to blockout the clanking/tapping noise created by the MRI.
- The patient lies on his/her back on the MRI table.
- A coil is placed on top of the patient's knee.
- A belt is cinched around the patient to detect his/her breathing patterns.
- The table is moved into the MRI machine as images are taken of the abdomen.
- Images are viewed during and after the MRI is completed to assist the healthcare provider reach a diagnosis.

RATIONALE FOR THE TEST

- Assess
  - The knee structures and tissues
  - The existence of a growth
  - For arthritis
  - Tendons, ligaments, cartilage, meniscus
  - If arthroscopy is required

NURSING IMPLICATIONS

- Determine if the patient has any metal on or inside his/her body.

- Assess if the patient
  - Is allergic to shellfish, iodine, or contrast material
  - Is pregnant
  - Is claustrophobic
  - Has kidney disease
  - Is wearing any medication patches
  - Can lie still during the test
  - Has had prior knee surgeries requiring metal devices to be implanted in the knee

UNDERSTANDING THE RESULTS

- The test takes 60 minutes and the results are ready immediately if the test is performed by the healthcare provider or within a few hours if a technician performs the test.
- Normal test results indicate
  - Normal structures and tissues in the knee
  - No growth(s)
  - No blockage
  - No fluid within the knee
  - No inflammation
  - Normal blood flow
- Abnormal test results indicate
  - Unusual structures and tissues in the knee.
  - The existence of a growth.
  - A blockage is identified.
  - Fluid exists within the knee.
  - Inflammation or infection is present.
  - Unusual blood flow.

TEACH THE PATIENT

- Explain
  - Why MRI is being taken.
  - What the patient will experience during the MRI.

- That no metal can be on or inside the patient during the MRI.
- That the patient will not feel any pain during the procedure, although the patient may feel a tingling sensation if he/she has metal fillings in his/her teeth.
- That the patient may be administered a sedative if he/she is claustrophobic.
- That some patients who are allergic to shellfish and iodine may also be allergic to contrast material. If the patient is allergic to contrast material, the healthcare provider will discuss with the patient the risk and benefits of administering the contrast material. If the patient agrees that the benefits outweigh, the risk, then the healthcare provider may administer medication that counteracts the allergic reaction to the contrast material.
- That the patient may be administered contrast material IV, causing a flushing feeling.

Shoulder MRI 5

The shoulder MRI is ordered to show the healthcare provider detailed images of inside the shoulder including ligaments, cartilages, muscles, bone structure within the shoulder, and fluid. These images are more detailed than can be achieved using ultrasound and CT scans and are commonly ordered to assess shoulder pain that is unexplained by other signs or symptoms.

WHAT IS BEING EXAMINED?

- Shoulder structures and tissues

HOW IS THE TEST PERFORMED?

- The patient is assessed for any metal that might be on or inside his/her body. The healthcare provider determines if the presence of any metal may require cancellation of the MRI.
- An assessment is made to determine if the patient can be administered contrast material, if required, for the MRI. Contrast material is administered IV, causing a flushing feeling for the patient. Contrast material may not be used if the patient is allergic to shellfish or iodine or if has kidney abnormalities or sickle cell anemia.
- An assessment is made to determine if the patient is claustrophobic. If so, then the healthcare provider may administer a sedative to the patient or schedule the test to be performed using an open MRI machine.

- The patient removes all clothing and wears a gown during the MRI.
- The patient may be given headphones to block out the clanking/tapping noise created by the MRI.
- The patient lies on his/her back on the MRI table.
- A coil is placed on the patient's shoulder.
- A belt is cinched around the patient to detect his/her breathing patterns.
- The table is moved into the MRI machine as images are taken of the shoulder.
- Images are viewed during and after the MRI is completed to assist the healthcare provider reach a diagnosis.

RATIONALE FOR THE TEST

- Assess
  - The shoulder structures and tissues
  - The existence of a growth
  - For arthritis
  - Tendons, ligaments, cartilage, bones, muscles
  - Assess rotator cuff disorders
  - If arthroscopy is required

NURSING IMPLICATIONS

- Determine if the patient has any metal on or inside his/her body
- Assess if the patient
  - Is allergic to shellfish, iodine, or contrast material
  - Is pregnant
  - Is claustrophobic
  - Has kidney disease
  - Is wearing any medication patches
  - Can lie still during the test
  - Has had prior knee surgeries requiring metal devices to be implanted in the shoulder or other parts of the body

UNDERSTANDING THE RESULTS

- The test takes 60 minutes and the results are ready immediately if the test is performed by the healthcare provider or within a few hours if a technician performs the test.
- Normal test results indicate
  - Normal structures and tissues in the shoulder
  - No growth(s)
  - No blockage
  - No fluid within the shoulder
  - No inflammation
  - Normal blood flow
- Abnormal test results indicate
  - Unusual structures and tissues in the shoulder.
  - The existence of a growth.
  - A blockage is identified.
  - Fluid exists within the shoulder.
  - Inflammation or infection is present.
  - Unusual blood flow.
  - Rotator cuff disorders.
  - Ligament and tendon tear or injury.

TEACH THE PATIENT

- Explain
  - Why MRI is being taken.
  - What the patient will experience during the MRI.
  - That no metal can be on or inside the patient during the MRI.
  - That the patient will not feel any pain during the procedure, although the patient may feel a tingling sensation if he/she has metal fillings in his/her teeth.
  - That the patient may be administered a sedative if he/she is claustrophobic.
  - That some patients who are allergic to shellfish and iodine may also be allergic to contrast material. If the patient is allergic to contrast material, the healthcare provider will discuss with the patient the risk

and benefits of administering the contrast material. If the patient agrees that the benefits outweigh the risk, then the healthcare provider may administer medication that counteracts the allergic reaction to the contrast material.

- That the patient may be administered contrast material IV, causing a flushing feeling.

Spinal MRI 6

The spinal MRI shows detailed images of the patient's spine. This includes the cervical spine, thoracic spine, and lumbosacral spine. The spinal MRI helps the healthcare provider assess if the patient has spinal disc disorders, spinal stenosis, as well as tumors or arthritis. The healthcare provider also orders a spinal MRI to assess unexplained spinal pain.

WHAT IS BEING EXAMINED?

- Spine

HOW IS THE TEST PERFORMED?

- The patient is assessed for any metal that might be on or inside his/her body. The healthcare provider determines if the presence of any metal may require cancellation of the MRI.
- An assessment is made to determine if the patient can be administered contrast material, if required, for the MRI. Contrast material is administered IV, causing a flushing feeling for the patient. Contrast material may not be used if the patient is allergic to shellfish or iodine or if has kidney abnormalities or sickle cell anemia.
- An assessment is made to determine if the patient is claustrophobic. If so, then the healthcare provider may administer a sedative to the patient or schedule the test to be performed using an open MRI machine.
- The patient removes all clothing and wears a gown during the MRI.
- The patient may be given headphones to block out the clanking/tapping noise created by the MRI.
- The patient lies on his/her back on the MRI table.
- A belt is cinched around the patient to detect his/her breathing patterns.

- The table is moved into the MRI machine as images are taken of the shoulder.
- Images are viewed during and after the MRI is completed to assist the healthcare provider reach a diagnosis.

RATIONALE FOR THE TEST

- Assess
  - The spinal structures and tissues
  - For ruptured disc
  - For sciatica
  - For spinal stenosis
  - Growths
  - For arthritis
  - For damaged nerves

NURSING IMPLICATIONS

- Determine if the patient has any metal on or inside his body.
- Assess if the patient
  - Is allergic to shellfish, iodine, or contrast material
  - Is pregnant
  - Is claustrophobic
  - Has kidney disease
  - Is wearing any medication patches
  - Can lie still during the test
  - Has had prior surgeries where metal devices might have been implanted in the shoulder or other parts of the body

UNDERSTANDING THE RESULTS

- The test takes 60 minutes and the results are ready immediately if the test is performed by the healthcare provider or within a few hours if a technician performs the test.

- Normal test results indicate
  - Normal structures and tissues in the spine
  - No growth(s)
  - No blockage
  - No inflammation
  - Normal blood flow
  - No damaged nerves
- Abnormal test results indicate
  - Unusual structures and tissues in the spine.
  - The existence of a growth.
  - A blockage is identified.
  - Inflammation or infection is present.
  - Ruptured disc.
  - Sciatica.
  - Spinal stenosis.
  - Arthritis.

TEACH THE PATIENT

- Explain
  - Why MRI is being taken.
  - What the patient will experience during the MRI.
  - That no metal can be on or inside the patient during the MRI.
  - That the patient will not feel any pain during the procedure, although he/she may feel a tingling sensation if the patient has metal fillings in his teeth.
  - That the patient may be administered a sedative if he/she is claustrophobic.
  - That some patients who are allergic to shellfish or iodine may also be allergic to contrast material. If the patient is allergic to contrast material, the healthcare provider will discuss with the patient the risk and benefits of administering the contrast material. If the patient agrees that the benefits outweigh, the risk, then the healthcare provider may administer medication that counteracts the allergic reaction to the contrast material.
  - That the patient may be administered contrast material IV, causing a flushing feeling.

Summary

There are times when healthcare providers require seeing inside the patient's body at a higher resolution than is provided by the CT scan and ultrasound scan. In these situations, the healthcare provider may order an MRI.

An MRI uses pulsating radio waves in a magnetic field to produce an image of inside the patient's body. A coil and a belt are placed on the patient. The coil is placed over the area being scanned and the belt detects the patient's breathing, assuring that breaths do not interfere with taking the MRI image.

The patient lies on a table that is moved into the MRI machine. The MRI machine creates a clanking/tapping noise as images are taken. Some patients may be disturbed by this noise; therefore, headphones are provided to block out the noise. Contrast material may be administered to the patient prior to the MRI to highlight areas of the body that are being studied.

Some patients feel claustrophobic and become anxious when placed inside the MRI machine. The healthcare provider may administer a sedative to relax the patient prior to the scan or may schedule an open MRI, which is less enclosed than a traditional MRI machine.

The MRI generates digital images of the inside of the patient's body on a computer screen. Those images are stored for review by the patient's healthcare team.

No metal objects can be on or inside the patient during an MRI. However, dental fillings are permitted, although the patient is likely to feel tingling in his/her mouth during the MRI. The patient may also experience skin irritation if he/she has iron pigment tattoos.

Quiz

1. What might happen if the patient has dental fillings and undergoes an MRI?
    a. The MRI metal alert alarm sounds.
    b. The patient may experience a tingling sensation in his/her mouth.
    c. The MRI is cancelled.
    d. None of the above.

2. What happens if the patient is allergic to shellfish and is scheduled for an MRI?
    a. The MRI is cancelled.
    b. The healthcare provider and patient evaluate the benefit and risk of continuing with the MRI.
    c. The MRI continues as scheduled.
    d. The healthcare provider decides whether or not to continue with the MRI.

3. Why would a woman who is positive for BRCA1 receive annual MRI scans?
    a. The patient is at high risk for breast cancer and the MRI provides highly detailed views of the patient's breasts.
    b. BRCA1 interferes with a normal mammogram.
    c. Only radio waves can penetrate the BRCA1.
    d. None of the above.

4. What happens if the patient is claustrophobic?
    a. The patient can be given a sedative.
    b. The patient can be scheduled for an open MRI.
    c. The MRI might be cancelled.
    d. All of the above.

5. Why is a belt cinched around the patient to detect breathing patterns?
    a. To detect any metal on the patient's body
    b. To detect any metal inside the patient's body
    c. To ensure that images taken coordinate with the patient's breathing patterns
    d. To detect metal both on and inside the patient's body

6. What is the function of an MRA?
    a. To assess the speed, direction, and flow of blood
    b. To assess fluid content of the brain
    c. To assess changes in the brain chemistry
    d. All of the above

7. Why is a knee MRI ordered?
   a. To identify any broken bones in the knee
   b. To assess if arthroscopy is required
   c. To assess the meniscus
   d. All of the above

8. A patient is permitted to wear a nonmetallic medication patch during an MRI.
   a. True
   b. False

9. Why should a patient with kidney disease avoid being administered contrast material prior to an MRI?
   a. The patient might have difficulty excreting the contrast material.
   b. The contrast material contains metallic elements.
   c. Patients with kidney disease should never receive an MRI.
   d. None of the above.

10. Metal objects can be taken into the MRI room as long as the object is not on the patient during the MRI scan.
    a. True
    b. False

Answers

1. b. The patient may experience a tingling sensation in his/her mouth.
2. b. The healthcare provider and patient evaluate the benefit and risk of continuing with the MRI.
3. a. The patient is at high risk for breast cancer and the MRI provides highly detailed views of the patient's breasts.
4. d. All of the above.
5. c. To ensure that images taken coordinate with the patient's breathing patterns.

6. a. To assess the speed, direction, and flow of blood.
7. d. All of the above.
8. b. False.
9. a. The patient might have difficulty excreting the contrast material.
10. b. False.

CHAPTER 18

Positron Emission Tomography Scan

A positron emission tomography (PET) scan is a nuclear medicine test that creates a roadmap of blood flow in the patient's body, enabling the healthcare provider to visualize abnormal blood flow to the patient's tissues and organs.

A radioactive chemical called a tracer and a special camera that detects the tracer inside the patient's body are the keys to a PET scan. The healthcare provider administers the tracer into the patient's veins prior to the scan. The tracer gives off positrons, which are very small charged particles that can be detected by the PET scan camera. The PET scan camera takes a series of images, each capturing the position of positrons in the body. These images are stored and replayed on a computer screen.

These images show the tracer containing blood as the blood makes its way into organs and tissues, giving the healthcare provider a clear picture of blood flow within the body.

In this chapter you will learn about the PET scan.

Learning Objective

1 PET Scan

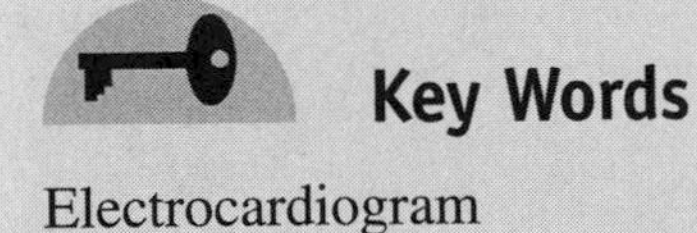

Key Words

Electrocardiogram
Photon
Positron
Radioactive
Single photon emission computed tomography (SPECT)
Tracer

PET Scan 1

The PET scan is ordered to study blood flow and metabolic activity within a patient's body. Healthcare providers frequently combine results from the PET scan with the CT scan results to obtain a thorough understanding of how well tissues and organs are being infused with blood.

Hint

Sometimes a CT scan is performed along with a PET scan. The tracer contains low-level radiation that will rarely lead to tissue damage. The tracer is flushed from the patient's body within 24 hours following the scan. It is rare that a patient will have an allergic reaction to the tracer. The healthcare provider may order a single photon emission computed tomography (SPECT) to determine if a patient with chest pain is at risk for cardiac arrest.

WHAT IS BEING EXAMINED?

- Blood flow and metabolic activity of tissues and organs

HOW IS THE TEST PERFORMED?

- The patient is administered the tracer intravenously.

- The patient lies on a table.
- The patient may be given a blindfold or earplugs to wear during the scan.
- Electrocardiogram electrodes are placed on the patient if the patient's heart is being studied.
- The PET scan camera moves around the patient.
- The patient might be asked to tell a story or read during the scan if his/her brain is being studied.
- The healthcare provider or PET scan technician is outside the PET scan room and is able to speak to the patient through an intercom and see the patient through a window.
- Patients must drink lots of fluid for a full day to flush the tracer from their bodies.

RATIONALE FOR THE TEST

- Assess
  - Blood flow to organs and tissues
  - Metabolic activity or organs
  - For stroke and transient ischemic attack (TIA)
  - For multiple sclerosis
  - For Parkinson disease and Alzheimer disease
  - For epilepsy
  - For coronary artery disease
  - For the presence of cancer and if the cancer has metastasized

NURSING IMPLICATIONS

- The patient will likely be asked to withhold medications for 24 hours before the PET scan.
- The healthcare provider may ask the patient to decrease the dose of insulin if the patient is a diabetic.
- Assess if the patient
  - Has ingested caffeine or alcohol 24 hours before the PET scan
  - Eaten 8 hours before the PET scan

- Is pregnant
- Is breast-feeding
- If the patient signed a consent form prior to the PET scan
- Can lie still during the test

UNDERSTANDING THE RESULTS

- The test takes 3 hours and the results are ready immediately if the test is performed by the healthcare provider or within a few hours if a technician performs the test.
- Normal test results indicate
  - Normal blood flow
  - No growth
  - No blockage
  - Normal metabolic activity
- Abnormal test results indicate
  - Unexpected blood flow.
  - The existence of a growth.
  - A blockage is identified.
  - Unusual metabolic activity.

TEACH THE PATIENT

- Explain
  - Why PET scan is being taken.
  - What the patient will experience during the PET scan.
  - That the patient will not feel any pain during the procedure.
  - That the patient may be administered a sedative if he/she is claustrophobic.
  - The patient must refrain from ingesting alcohol, caffeine, and tobacco for 24 hours prior to the PET scan.
  - That the patient may be asked to stop taking or to reduce the dose of medication prior to the PET scan.
  - That the PET scan may not be administered if the patient is pregnant or breast-feeding.

- That the healthcare provider may administer a sedative prior to the PET scan if the patient is anxious and unable to lie still for the test.
- That the healthcare provider may ask the patient to tell a story or read during the PET scan, and ECG electrodes may be attached to the patient's body during the test.

Summary

A PET scan creates an image of a radioactive tracer as the tracer travels throughout the patient's blood stream creating a roadmap of the blood stream for the healthcare provider.

Prior to the scan, the tracer is usually administered into the patient's vein. As the tracer moves along with blood, it gives off tiny charged particles called positrons that are detected by the PET scan camera.

These images are then stored and replayed on a computer, enabling the healthcare provider to study blood flow to organs and tissues and to assess the patient's metabolic activities.

Quiz

1. What might a diabetic be asked to do prior to the PET scan?
   a. Remove his diabetic medical alert necklace.
   b. Reduce the insulin dose.
   c. Increase the insulin dose.
   d. None of the above.

2. After the PET scan, why is the patient asked to drink lots of fluids?
   a. To protect the patient's heart from the tracer
   b. To increase the effectiveness of the tracer
   c. To flush the tracer from his/her body
   d. To protect the patient's organs from the tracer

3. The tracer may negatively affect the fetus.
   a. True
   b. False
4. What should the nurse do if the patient drank a large cup of coffee the morning of the PET scan?
   a. Tell the patient this is unacceptable and continue with the PET scan.
   b. Notify the healthcare provider.
   c. Ask the patient to drink five cups of water prior to the PET scan.
   d. Take a sample of the patient's blood for testing.
5. What will the healthcare provider likely do if the patient is too anxious and cannot lie still for the PET scan?
   a. Cancel the PET scan.
   b. Warn the patient that he/she could be injured if he/she moves during the PET scan.
   c. Ask the patient's significant other to remain in the PET scan room during the test.
   d. Administer a sedative per order.
6. Why would the healthcare provider ask the patient to read during the PET scan?
   a. The healthcare provider is studying brain activity.
   b. The healthcare provider is monitoring the patient's eye movements.
   c. The healthcare provider needs to be ensured that the patient is awake during the scan.
   d. The healthcare provider is probing the brain during the scan.
7. How would you respond if the patient is fearful that the tracer will destroy his/her tissues?
   a. Explain that the tracer contains a very low dose of radiation that remains in the body for 24 days after the test and causes minor tissue damage.
   b. Explain that the tracer contains a very low dose of radiation that is flushed from the body within 24 hours of the test and rarely causes any tissue damage.
   c. Explain that the tracer contains the same amount of radiation as the sun and causes no more than a minor sunburn.
   d. None of the above.

8. Why should a breast-feeding mother avoid undergoing a PET scan?
   a. The mother's breasts will be tender for 24 hours following the PET scan.
   b. The tracer is likely to pass to the baby in breast milk.
   c. The PET scan places the mother under extreme stress that decreases the volume of breast milk.
   d. None of the above.

9. A PET scan helps the healthcare provider in diagnosing
   a. Transient ischemic attack
   b. Multiple sclerosis
   c. Cancer
   d. All of the above

10. A patient with chest pain may be
    a. Administered a SPECT
    b. Unable to undergo any kind of PET scan
    c. At a high risk of complications from receiving the tracer
    d. None of the above

Answers

1. b. Reduce the insulin dose
2. c. To flush the tracer from his/her body
3. a. True
4. b. Notify the healthcare provider
5. d. Administer a sedative per order
6. a. The healthcare provider is studying brain activity
7. b. Explain that the tracer contains a very low dose of radiation that is flushed from the body within 24 hours of the test and rarely causes any tissue damage
8. b. The tracer is likely to pass to the baby in breast milk
9. d. All of the above
10. a. Administered a SPECT

CHAPTER 19

Cardiovascular Tests and Procedures

Cardiovascular tests are performed to assess the patient's heart and vascular system to determine if the blood is adequately being pumped and flowing throughout the patient's body. These tests measure cardiac contraction, the risk for coronary artery disease, and are used to identify blockage to coronary arteries and blood vessels of the extremities.

When a blockage is identified, the healthcare provider can perform one of several procedures to restore blood flow. The blockage might be surgically removed or pressed against the wall of the blood vessel and held in place by a stent. Alternatively, the healthcare provider may surgically bypass the blocked blood vessels using a vein from the patient's leg or by using an artificial blood vessel.

In this chapter you will learn about these tests and procedures.

Learning Objectives

1. Cardiac Blood Pool Scan
2. Cardiac Calcium Scoring
3. Electrocardiogram
4. Cardiac Perfusion Scan
5. Ankle-Brachial Index
6. Echocardiogram
7. Pericardiocentesis
8. Venogram

Key Words

Computed tomography (CT)
Doppler echocardiogram
Ejection fraction
First-pass scan
Gamma camera
Holter monitoring
Multigated acquisition (MUGA)
Peripheral arterial disease (PAD)
Plaque
Stress echocardiogram
Transesophageal echocardiogram (TEE)
Transthoracic echocardiogram (TTE)

Cardiac Blood Pool Scan 1

The cardiac blood pool scan is a test that measures the percentage of the patient's blood that is pumped in a cardiac contraction and therefore indicates how well the heart is contracting. This is referred to as the ejection fraction.

The healthcare provider administers radioactive material called a tracer into the patient vein. A gamma camera is used to monitor the tracer as the blood flows throughout the patient's heart. The healthcare provider can use one of the following two types of cardiac blood pool scans to estimate the ejection fraction.

- Multigated acquisition (MUGA): This is also known as gated scan. Each contraction of the heart triggers the gamma camera to take a picture of the heart, which is stored in a computer. These images are then placed back on the computer screen showing cardiac contractions. The healthcare provider may perform the MUGA scan twice—first without medication and the second after giving the patient nitroglycerin. Nitroglycerin dilates arteries and veins, reducing the amount of blood that must be pumped by the heart, which decreases the amount of oxygen requirements of the heart because the heart works less. The impact of the decreased cardiac workload should be reflected in the MUGA scan.
- First-pass: This captures images of the blood going through the heart and lungs for the first time.

HINT

The MUGA scan is not used on children and does not provide information about heart valves and thickness of cardiac walls. The first-pass scan is used on children to assess the existence of congenital heart disease.

CAUTION

The cardiac blood pool scan is not performed if the patient is pregnant.

WHAT IS BEING EXAMINED?

- Cardiac contractions

HOW IS THE TEST PERFORMED?

- The patient is administered the radioactive tracer into a vein.
- MUGA scan
  - A sample of blood is taken.
  - The radioactive tracer is mixed with the blood sample.
  - The blood sample is then reinjected into the patient's vein.
- It takes 4 hours for the patient's red blood cells to absorb the entire tracer. The patient may be able to leave the healthcare facility during this period.
- The patient is asked to remove dentures and jewelry before the scan is performed.

- The patient removes clothing.
- The patient lies on a table.
- An electrocardiogram is attached to the patient.
- The patient must lie still each time a picture is taken, which can take 5 minutes.
- The gamma camera is positioned close to the patient's chest and is repositioned during the scan.
- The patient may be asked to change position or to leave the table to perform exercises and then return to the table to continue the scan. The patient may be administered nitroglycerin.
- The patient will drink lots of fluid for 2 days following the scan to flush the tracer from his/her body.
- The MUGA scan takes 3 hours to complete and the first-pass scan takes 1 hour to complete.

RATIONALE FOR THE TEST

- To assess ventricle contractions
- To assess cardiac blood flow
- To diagnose cardiac abnormalities

NURSING IMPLICATIONS

- Determine if the patient
  - Has any allergies.
  - Has recently undergone radioactive tracer scans.
  - Has an implant in the chest such as a pacemaker.
  - Is wearing comfortable clothes if the patient is asked to exercise.
  - Can lie still on his back.
  - Has recently had a barium enema.
  - Is pregnant. The cardiac blood pool scan is not performed if the patient is pregnant.
- Assess if the patient
  - Has eaten 4 hours before the scan.
  - Has ingested caffeine 6 hours before the scan.

- Has smoked for 6 hours before the scan.
- Has provided the healthcare provider with all medications that are being taken. Digoxin and nitrate medication can affect the scan results.

UNDERSTANDING THE RESULTS

- The MUGA scan takes 3 hours and the first-pass scan takes 1 hour. Results are ready within a week of the scan.
- Normal test results indicate
  - Ejection fraction is 55% to 65%
  - Normal blood flow through the heart
  - Cardiac structures are normal and contracting normally
- Abnormal test results indicate
  - Ejection fraction is less than 55%
  - Inadequate cardiac contractions
  - Abnormal cardiac structures
  - Improper blood flow through the heart

TEACH THE PATIENT

- Explain
  - Why scan is being performed.
  - What the patient will experience during the scan.
  - That the patient will not feel any pain during the procedure except for a pinch from the needle used to administer the tracer.
  - That patient must tell the healthcare provider of any allergies, has recently undergone radioactive tracer scans, has an implant in the chest such as a pacemaker, or has recently had a barium enema.
  - That the patient cannot eat for 4 hour before the scan.
  - That the patient cannot ingest caffeine for 6 hours before the scan.
  - That the patient cannot smoke for 6 hours before the scan.
  - The healthcare provider must be informed of all medications that the patient takes.

Cardiac Calcium Scoring 2

Coronary arteries supply blood to cardiac muscles. Restricted coronary arteries decrease blood and oxygen supply to cardiac muscles. Blood flow through the coronary arteries is restricted by a plaque buildup on the coronary artery wall. Plaque contains calcium. Cardiac calcium scoring determines the level of calcium containing plaque on the coronary artery walls using a computed tomography (CT). Images of thin sections of the heart are taken using the CT scan and are stored in a computer and reviewed by the healthcare provider to assess signs of coronary artery disease.

HINT

Healthcare providers use a variety of assessment methods to determine if the patient has coronary artery disease.

CAUTION

A patient can have a high cardiac calcium score without having coronary artery disease.

WHAT IS BEING EXAMINED?

- Assessing the buildup of plaque containing calcium on the walls of the coronary arteries.

HOW IS THE TEST PERFORMED?

- The patient removes all jewelry.
- The patient removes clothing above the waist.
- The patient lies on a table.
- An electrocardiogram is attached to the patient.
- The patient may be administered medication to decrease cardiac contractions if the patient's heart rate is greater than 90 beats/minute.
- The patient may be administered a sedative if he/she feels claustrophobic being inside the CT machine.
- The patient will be alone in the CT room but can communicate with the CT technician using an intercom.

- The patient must lie still on a table.
- The table is slid into the CT scan (see Chapter 15).
- The patient is asked to hold breath and lie still for 30 seconds while images are taken by the CT machine.

RATIONALE FOR THE TEST

- To assess for coronary artery disease

NURSING IMPLICATIONS

- Assess if the patient
  - Is claustrophobic
  - Can lie still on the back
  - Has removed jewelry
  - Has ingested caffeine 12 hours before the test
  - Has smoked 12 hours before the test

UNDERSTANDING THE RESULTS

- The cardiac calcium scoring test takes 30 minutes. Results are ready within 1 week of the scan.
- Normal test results indicate
  - Score 0 = No plaque. There is a 5% chance of developing coronary artery disease.
  - Score < 11 = 10% chance of developing coronary artery disease.
- Abnormal test results indicate
  - Score 11 to 100 = Plaque was found. Patient has mild coronary artery disease.
  - Score 101 to 400 = Plaque was found. Patient has coronary artery disease and possibly a blockage of one or more coronary arteries.
  - Score > 400 = Plaque was found. Patient has coronary artery disease and there is a 90% chance that there is a blockage of one or more coronary arteries.

TEACH THE PATIENT

- Explain
  - Why scan is being performed.
  - What the patient will experience during the scan.
  - That the patient will not feel any pain during the procedure except for a pinch from the needle used to administer a sedative, if necessary.
  - That the patient will be administered a sedative if he/she feels claustrophobic being in the CT machine.
  - That the patient should not ingest caffeine 12 hours before the test.
  - That the patient should not smoke 12 hours before the test.

Electrocardiogram 3

An electrocardiogram (ECG) records electrical activity of the heart on paper. Electrical activity causes contractions of the heart, resulting in blood being pumped throughout the body. Any disruption of the electrical activity might cause the heart to perform less than normally, which appears on the electrocardiogram paper tracing.

Telemetry is a type of electrocardiogram that is used in a healthcare facility to constantly monitor the patient's cardiac activity. A monitoring device worn by the patient transmits cardiac electrical activity to a telemetry monitor at the nurse's station. An alarm is sounded whenever there is abnormal cardiac activity. A telemetry nurse interprets the tracing on the telemetry monitor and prints the tracing while taking appropriate action.

The Holter monitoring electrocardiogram, sometimes referred to as an ambulatory electrocardiogram, is a type of portable electrocardiogram that is attached to the patient for 24 hours. This device is worn on the patient's waist or around his shoulder, enabling the healthcare provider to detect abnormal electrical activity of the patient's heart during activities of daily living such as when the patient exercises and when sleeping. Holter monitoring can be continuously recorded or intermittently recorded. The two types of intermittently recorded Holter monitoring.

- Event monitor: The patient presses a button whenever a cardiac symptom occurs, causing the Holter monitor to record cardiac activity.
- Loop recorder: The Holter monitor constantly records cardiac electrical activity. The patient presses a button whenever a cardiac symptom occurs.

WHAT IS BEING EXAMINED?

- Electrical activity of the heart

HOW IS THE TEST PERFORMED?

- The patient removes all jewelry.
- The patient removes clothing covering the area where the electrocardiogram leads are attached to the patient.
- The patient lies on a table.
- An electrocardiogram is attached to the patient.
- The patient lies still.

RATIONALE FOR THE TEST

- To assess the cause of chest pain, palpitations, and other symptoms.
- To assess the effects of medication on the heart.
- To assess signs and symptoms of heart disease.
- To assess the performance of a pacemaker.

NURSING IMPLICATIONS

- Assess if the patient
  - Can lie still on the back
  - Has removed jewelry

UNDERSTANDING THE RESULTS

- The cardiac catheterization test takes 1 hour. Results are ready immediately.
- Normal test results indicate
  - Expected electrical activity on the trace paper
- Abnormal test results indicate
  - Unexpected electrical activity on the trace paper

TEACH THE PATIENT

- Explain
    - Why electrocardiogram is being performed.
    - What the patient will experience during the procedure.
    - That the patient will not feel any pain during the procedure.
    - That the patient will be required to remove jewelry.

Cardiac Perfusion Scan 4

The cardiac perfusion scan is commonly performed to determine the volume of blood in cardiac muscle during stress and at rest. Before the test, the patient is administered a radioactive tracer into his/her vein. A camera scans the patient's heart while at rest and stores the image on a computer. The patient's heart is then placed under stress by using medication or asking the patient to physically exercise. The patient's heart is then scanned by the camera and images are stored on the computer. The healthcare provider then compares the images to determine if the heart is receiving sufficient blood supply.

WHAT IS BEING EXAMINED?

- The volume of blood in heart muscle

HOW IS THE TEST PERFORMED?

- The patient removes all jewelry.
- The patient removes clothing above the waist.
- The patient lies on a table.
- An electrocardiogram is attached to the patient.
- Radioactive tracer is injected into the patient's vein.
- The camera is placed close to the patient's chest.
- The patient must remain still while the image is taken.
- The camera is moved and the patient is asked to exercise using a treadmill or stationary bike. During exercising the patient's blood pressure is monitored by an electrocardiogram.

- If the patient is unable to exercise, then medication is administered to simulate the stress of exercise on the patient's heart.
- The patient lies on the table.
- Radioactive tracer is injected into the patient's vein.
- The camera is placed close to the patient's chest.
- The patient remains still while the image is taken.
- The patient can resume normal activity following the test.

RATIONALE FOR THE TEST

- To assess the underlying cause of chest pain
- To assess the results of angioplasty and bypass surgery

NURSING IMPLICATIONS

- Assess if the patient
  - Can lie still on the back.
  - Has allergies.
  - Has taken Viagra, Cialis, or Levitra within 2 days of the procedure. These medications can cause the patient's blood pressure to fall if the patient is administered nitroglycerin during the procedure.
  - Is taking anticoagulants such as Coumadin, aspirin, or heparin.
  - Is taking Trental or Persantine.
  - Is pregnant.
  - Is breast-feeding. If so, the patient must discard breast milk collected 2 days following the test, since the breast milk will contact the radioactive tracer.
  - Has eaten or drunk 3 hours before the test.
  - Has drunk caffeine, alcohol, or smoked tobacco 1 day before the test.
  - Has removed jewelry.
  - Can exercise.
  - Recently had a heart attack or myocarditis.
  - Has a pacemaker.
  - Has an electrolyte imbalance.

UNDERSTANDING THE RESULTS

- The test takes 1 hour. Results are ready within 1 week.
- Normal test results indicate
    - Normal blood volume in the cardiac muscle
- Abnormal test results indicate
    - Decreased blood volume in the cardiac muscle

TEACH THE PATIENT

- Explain
    - Why the scan is being performed.
    - What the patient will experience during the procedure.
    - That the patient might feel flushed, nauseous, dizzy, or have a headache if medication is administered to place the patient's heart under stress.
    - That the patient will not feel any pain during the procedure except for a pinch from the needle used to administer the tracer material.
    - That the patient should not drink or eat for 3 hours before the procedure.
    - That the patient should tell the healthcare provider if he/she takes Viagra, Cialis, Levitra, Coumadin, aspirin, heparin, Trental, or Persantine.
    - That the patient should tell the healthcare provider during the test if the patient experiences fatigue, chest pain, shortness of breath, aching in the lower extremities, or lightheadedness, while the test is in progress.
    - That the patient will be required to remove jewelry.
    - That the patient may resume normal activities following the test.

Caution

The patient should call emergency medical care if he/she experiences difficulty in breathing or chest pains after leaving the healthcare facility following the test.

Ankle-Brachial Index 5

The ankle-brachial index (ABI) evaluates blood circulation in the lower extremity for peripheral arterial disease (PAD). The test may also be performed to determine how well blood is flowing to the extremities after procedures or surgery. The ankle-brachial index is determined by measuring the patient's blood pressure in both ankles and arms while the patient is resting.

Hint

The patient may be asked to exercise. If so, the same blood pressures are taken and compared to the resting blood pressure.

WHAT IS BEING EXAMINED?

- The patient is being screened for peripheral arterial disease (PAD).

HOW IS THE TEST PERFORMED?

- The patient lies on a table.
- Blood pressure is taken in both arms and legs.
- The ankle-brachial index is calculated as: ABI = ankle blood pressure + brachial blood pressure.

RATIONALE FOR THE TEST

- To assess for peripheral arterial disease
- To assess for risk for stroke

NURSING IMPLICATIONS

- Determine if the patient
  - Is able to have blood pressure taken in both arms and legs
  - Can lie on the back
  - Can walk on a treadmill
  - Is wearing comfortable clothes
- Assess if the patient has been diagnosed with arterial disease.

UNDERSTANDING THE RESULTS

- The test takes less than 1 hour. Results are ready immediately.
- Normal
  - An index of 1 or 1.1 indicating that blood pressure in the arms and ankles is approximately the same; peripheral blood flow is unobstructed.

- Abnormal
    - 0.98: narrowing of peripheral blood flow
    - 0.80: intermittent claudication
    - 0.25: high risk for peripheral arterial disease (PAD)

TEACH THE PATIENT

- Explain
    - Why test is being performed.
    - What the patient will experience during the test.
    - That the patient will not feel any pain during the test.
    - That the patient should wear comfortable clothes, since he/she will be walking on the treadmill.

Echocardiogram 6

An echocardiogram is an ultrasound scan of the patient's heart. Sound waves are transmitted by a transducer through the patient's chest to the heart. Sound waves echo off the heart and are detected by the transducer creating an image of the heart. Images are stored on a computer enabling the healthcare provider to play back images showing the beating heart. The three types of echocardiograms are

- Stress: Two sets of images of the heart are captured. The first set of images taken are of the resting heart. The second set of images are taken when the heart is under stress either following exercise or from medication.
- Transthoracic: This creates a set of views that illustrate the beating heart, which is the most common type of echocardiogram.
- Transesophageal: Under local anesthesia, the transducer is inserted down the patient's esophagus, enabling the generation of a clear image of the heart, since these images are not obscured by bones. This type of scan is commonly used to assess for vegetation on the heart valves.

WHAT IS BEING EXAMINED?

- Cardiac function and structures

HOW IS THE TEST PERFORMED?

- The patient removes clothing above the waist.
- The patient removes jewelry worn above the waist.
- The patient lies on a table.
- The patient is attached to an electrocardiogram.
- Stress echocardiogram
  - The patient is given a baseline ECG.
  - The patient is asked to pedal a stationary bicycle or walk on a treadmill while another ECG is taken.
  - If the patient is unable to exercise, the healthcare provider will administer dobutamine to chemically stress the patient's heart.
  - The patient is asked to lie on the table.
  - The patient may be administered saline through a vein to help the healthcare provider assess the patient's cardiac function.
  - If the healthcare provider is unable to obtain a clear view of the heart, the patient may be administered contrast material into the patient's vein. If this is done, then the patient will be placed on a mechanical ventilator to assist the patient's respiration.
  - Conductive gel is placed on the left side of the patient's chest.
  - A transducer is moved along the patient's chest as it generates sound waves through the body and receives reflected sound waves.
  - The healthcare provider asks the patient to change position, breathe slowly, or hold breath at different times during the test.
  - Reflected sound waves create an image of the heart on a computer screen.
- Transesophageal echocardiogram
  - A saline or heparin lock is inserted into the patient's arm.
  - The patient is administered medication that decreases saliva and stomach secretions.
  - The patient is administered a sedative. The patient remains conscious during the test.
  - The patient's blood oxygen level is measured by the pulse oximeter during the test.
  - The healthcare provider sprays an anesthetic over the back of the patient's throat to decrease the patient's gag reflex.

- The patient's head is tilted forward.
- A mouth guard is inserted into the patient's mouth to protect his/her teeth.
- A transducer probe is inserted down into the patient's esophagus as the patient swallows. The transducer probe remains in place for 20 minutes.
- The healthcare provider will suction excess saliva.
- The patient is asked not to swallow during the test.
- The transducer sends and receives sound waves and generates an image of the patient's heart.
- The transducer probe is removed once all images of the heart are captured.
- The patient is not permitted to ingest anything until the gag reflex is restored and the anesthetic wears off.

RATIONALE FOR THE TEST

- Assess for
  - Cardiac function
  - Cardiac abnormalities
  - Pericardial effusion
  - Valve abnormalities and vegetations
  - Ejection fraction

NURSING IMPLICATIONS

- Determine if the patient
  - Can lie on the back
  - Can swallow (transesophageal echocardiogram)
  - Can walk on a treadmill or pedal a stationary bicycle (stress echocardiogram)
  - Is wearing comfortable clothes
  - Has lung disease
  - Has large breasts

- Assess if the patient
  - Has eaten or drunk for 6 hours before the test
  - Has dentures
  - Has transportation home if the patient is to receive a sedative
  - Has signed a consent form
  - Has difficulty breathing
  - Has removed jewelry
  - Has removed clothing above the waist

UNDERSTANDING THE RESULTS

- The transthoracic, stress, and Doppler echocardiograms each take 1 hour to perform. The transesophageal echocardiogram takes 2 hours to perform. Results are available in a week.
- Normal test results indicate
  - Normal cardiac function
  - No cardiac growth
  - No fluid buildup around the heart
  - Normal ejection fraction
- Abnormal test results indicate
  - Unexpected cardiac function
  - Cardiac growth
  - Fluid buildup around the heart
  - Heart valve vegetations or other valve abnormalities
  - Abnormal ejection fraction

TEACH THE PATIENT

- Explain
  - Why the test is being performed.
  - How the test is performed.
  - What the patient will experience during the test.
  - That the patient will not feel any pain during the test.
  - That the patient should wear comfortable clothes, since he will be walking on the treadmill or pedaling a stationary bicycle (stress echocardiogram).

- That the patient will need to remove jewelry and clothing from the waist up.
- That the patient should refrain from eating and drinking 6 hours before the test.
- That the patient should remove all dentures prior to the test.
- That the patient will not be able to drive for 12 hours following the test if a sedative is administered.
- That conductive gel is wiped off the patient following the test.
- That the patient's heart rate may increase for 15 minutes and then return to normal in 3 minutes following administration of dobutamine, if to a patient undergoing a stress echocardiogram who is unable to exercise.
- That the patient can request to take a break during the test if he/she feels uncomfortable.
- That the patient may experience an unusual feeling in the throat if he/she is receiving a transesophageal echocardiogram.
- That the patient may feel sleepy or experience blurred vision, have trouble speaking, or have a dry mouth following the test. These should resolve shortly following the test.
- Although the patient is awake during the test, he/she probably will not remember the test itself.
- That the patient may experience hoarseness or a sore throat following a transesophageal echocardiogram. Lozenges and gargling with warm saltwater will provide relief.
- That the patient should avoid drinking alcohol for a day following the transesophageal echocardiogram.
- That the patient may experience nausea.

Caution

Prior to the test, the patient and the healthcare provider should arrange for a signal that the patient can give to the healthcare provider to indicate that the patient is uncomfortable during a transesophageal echocardiogram. The patient should call the healthcare provider if experiencing chest pains, difficulty swallowing, difficulty speaking, fast heartbeat, or difficulty breathing following the test.

Pericardiocentesis 7

There are disorders that cause pericardial effusion, which is the result of excess fluid build up in the pericardium (sac around the heart). As a result, the patient may experience cardiac tamponade, which inhibits cardiac contraction and is a life-threatening emergency. The healthcare provider may perform a pericardiocentesis to remove the excess fluid, thereby restoring normal cardiac contraction. Pericardial effusion can be caused by inflammation of the pericardium because of viral, bacterial, or fungal infection, blood from an injury, or disorders such as kidney failure, hypothyroidism, or rheumatoid arthritis.

HINT

Sometimes pericardiocentesis is performed in the emergency department.

WHAT IS BEING EXAMINED?

- Removal of excess fluid from the pericardium

HOW IS THE TEST PERFORMED?

- A blood test is taken prior to the procedure to determine if the patient has anticoagulation problems.
- The patient removes clothing above the waist.
- The patient removes jewelry worn above the waist.
- The patient angles his/her back on a table.
- A saline or heparin lock is inserted into the patient's arm.
- The patient is attached to an electrocardiogram.
- The patient is administered a sedative.
- The insertion site is cleaned with an antiseptic.
- The patient is administered a local anesthetic at the insertion site.
- A needle is inserted either between the patient's left ribs or below the sternum and into the pericardium guided by an echocardiogram.
- The patient may be asked to remain still or hold the breath while the needle is inserted.
- A catheter is slid over the needle into the pericardium and the needle is removed.

- The fluid is drained by the catheter.
- The catheter is removed once fluid is drained.
- Pressure is applied to the site for 5 minutes to stop bleeding.
- A chest X-ray is taken of the patient to ensure there is no collateral damage such as a collapsed lung.
- The patient may remain in the healthcare facility for observation.

RATIONALE FOR THE PROCEDURE

- Restore cardiac contraction
- Assess the source of pericardial effusion

NURSING IMPLICATIONS

- Determine if the patient
    - Can lie on the back
    - Can respond to direction during the procedure
    - Has taken anticoagulants (Coumadin, heparin) or antiplatelets (aspirin, clopidogrel)
    - Has allergies
    - Has recently been administered antibiotics
- Assess if the patient
    - Has eaten or drunk for 12 hours before the test
    - Has signed a consent form
    - Has removed jewelry
    - Has removed clothing above the waist

UNDERSTANDING THE RESULTS

- The procedure takes 30 minutes; however, the fluid may drain for several hours following the procedure. Results from analysis of the fluid removed from the pericardium is available within a week.
- Normal test results indicate
    - Less than or equal to 50 mL of fluid drained
    - Normal clear/pale yellow fluid

- No bacteria
- No abnormal cells
- Less than 500 white blood cells present in the fluid
- Abnormal test results indicate
  - Greater than 50 mL of fluid drained
  - Fluid appears cloudy
  - Bacteria cells found in the fluid
  - Abnormal cells found in the fluid
  - More than 500 white blood cells present in the fluid

TEACH THE PATIENT

- Explain
  - Why the procedure is being performed.
  - How the procedure is performed.
  - What the patient will experience during the procedure.
  - That the patient will not feel any pain during the procedure.
  - That the patient might experience an irregular heartbeat during the procedure.
  - That the patient will need to remove jewelry and clothing from the waist up.
  - That the patient should refrain from eating and drinking 12 hours before the test.
  - That the patient may remain in the healthcare facility for observation following the procedure.
  - That the patient may feel pressure as the needle is inserted.

CAUTION

The patient should tell the healthcare provider if experiencing chest pain or shortness of breath during the procedure. After returning home, the patient should call emergency medical help if experiencing chest pain, trouble breathing, sweating, lightheadedness, and signs of shock. The patient should call the healthcare provider if vomiting blood, is short of breath, has a fever, or feels dizzy.

Venogram 8

A venogram produces an X-ray image of blood flowing through the patient's veins. It is used by healthcare providers to assess the function of valves in the veins and if there is deep vein thrombosis (DVT). Venograms are used to evaluate

- Extremities
- Pelvis
- Kidneys

WHAT IS BEING EXAMINED?

- Blood flow through veins

HOW IS THE TEST PERFORMED?

- The patient removes all jewelry.
- The patient removes clothing.
- The patient must not eat 4 hours before the test.
- The patient can drink clear fluids up to 4 hours before the test but must not drink after 4 hours.
- The patient is asked to empty his bladder.
- The patient lies on a table that may be tilted during the test.
- For extremities, an elastic band is placed around the extremity that is being examined to fill the vein with blood.
- The patient may be administered a local anesthetic at the insertion site.
- Contrast material is inserted into the patient's vein.
- For kidneys
    - A catheter is inserted into the femoral vein in the groin and guided into the vein in the patient's kidney.
    - The contrast material is administered through the catheter. The catheter is removed once the test is completed.
    - The contrast material may also be administered into the inferior vena cava.
- X-ray images are taken of the area that is being studied.
- After the test is completed, saline is inserted into the vein to help flush the contrast material.

- The patient may be administered heparin to prevent a blood clot following the test.
- A dressing is placed on the insertion site.
- The patient is asked to drink lots of water to flush the contrast material from the body.

RATIONALE FOR THE TEST

- To identify thrombosis
- To identify sites for filters
- To assess valves in veins

NURSING IMPLICATIONS

- Assess if the patient
  - Has signed a consent
  - Can lie still on the back
  - Has allergies
  - Is pregnant
  - Is taking Coumadin, heparin, or aspirin
  - Has avoided eating or drinking for 4 hours before the test
  - Has emptied his/her bladder before the test
  - Has removed jewelry
  - Has kidney disorder
  - Has asthma
  - Has diabetes

UNDERSTANDING THE RESULTS

- Venogram takes 1 hour or less to perform.
- Normal test results indicate
  - Normal blood flow
  - No growth(s) in the kidney
- Abnormal test results indicate
  - Blockage of blood flow

- Varicose veins
- Growth(s) in the kidney

TEACH THE PATIENT

- Explain
  - Why the test is being performed.
  - What the patient will experience during the test.
  - That the patient will not feel any pain during the test except for a pinch when the contrast material is inserted into the vein.
  - That the patient may feel pins and needles in the arm or leg that is being studied, but this dissipates quickly following the test.
  - That the patient will have to sign a consent form.
  - That the patient must empty his/her bladder before the test.
  - That the patient will be required to rest in bed for 3 hours if a kidney venogram is performed.
  - That the patient should drink lots of water for a day following the procedure to flush the contrast material from the body.

Caution

The patient should call the healthcare provider if after the test he/she feels pain or experiences swelling in the area that was studied, or if experiences a fever.

Summary

In order to keep tissues oxygenated and fluid balanced within the body, the cardiovascular system must function adequately. In this chapter, you learned about several cardiovascular tests that are performed to assess cardiac function and blood flow throughout the body. These tests assess cardiac contraction, the risk for coronary artery disease, and are used to identify blockage to coronary arteries and blood vessels to the extremities.

Cardiovascular disorders can result in the blockage of blood flow to the heart or other vessels in the patient's body. In this chapter you also learned about several surgical procedures that remove or press plaque against the wall of the blood vessel, which is then held in place by a stent allowing blood to flow through the vessel.

Alternatively, the healthcare provider may surgically bypass the blocked blood vessels using a vein from the patient's leg or by using an artificial blood vessel.

Quiz

1. What is the purpose of a cardiac blood pool scan?
   a. To measure the ejection fraction of the heart
   b. To measure the amount of blood that pools in the heart
   c. To measure the amount of blood that pools on the extremities
   d. To measure the amount of blood that pools in the coronary arteries

2. What is the purpose of a gated scan?
   a. To assess cardiac contractions
   b. To stimulate cardiac contractions
   c. To assess cardiac oxygen requirements
   d. To assess the workload of the heart

3. What is Holter monitoring?
   a. An ambulatory electrocardiogram
   b. Monitors the patient's blood flow at rest
   c. Monitors the patient's blood flow at sleep
   d. Monitors the patient's blood flow after exercising

4. What occurs during a MUGA scan?
   a. A sample of blood is taken.
   b. The radioactive tracer is mixed with the blood sample.
   c. The blood sample is then reinjected into the patient's vein.
   d. All of the above.

5. What might be a problem if a MUGA scan detects a 45% ejection fraction?
   a. Inadequate cardiac contractions
   b. Abnormal cardiac structures
   c. Improper blood flow through the heart
   d. All of the above

6. What is a stress echocardiogram?
   a. A procedure that creates images of the heart
   b. A procedure that is performed after the patient has exercised
   c. A procedure that is performed when the patient's heart is placed under stress using a medication
   d. All of the above

7. What does a high cardiac calcium score mean?
   a. Confirmed diagnoses of coronary artery disease
   b. A possible plaque buildup on the coronary artery wall
   c. Little or no plaque buildup on the coronary artery wall
   d. None of the above

8. Why should the patient avoid taking Coumadin prior to a cardiac procedure?
   a. The patient's bleeding time increases.
   b. There is a high risk of bleeding.
   c. The healthcare provider will need to take precautions to control bleeding that might occur during the procedure.
   d. All of the above.

9. What is a first-pass scan?
   a. Captures images of blood going through the heart and lungs for the first time
   b. A type of cardiac blood pool scan
   c. A procedure that uses a radioactive tracer
   d. All of the above

10. Why would a healthcare provider order a transesophageal echocardiogram?
    a. To assess cardiac contraction under stress
    b. To generate an image of the heart that is not obstructed by bone
    c. To assess cardiac blood flow to the extremities
    d. To assess cardiac blood flow to the lungs

Answers

1. a. To measure the ejection fraction of the heart
2. a. To assess cardiac contractions
3. a. An ambulatory electrocardiogram
4. d. All of the above
5. d. All of the above
6. d. All of the above
7. b. A possible plaque buildup on the coronary artery wall
8. d. All of the above
9. d. All of the above
10. b. To generate an image of the heart that is not obstructed by bone

CHAPTER 20

Lung Tests and Procedures

The lungs exchange carbon dioxide and oxygen in the hemoglobin in red blood cells. In order to do so effectively, the lungs must be able to expand and retract and blood must flow freely to the lungs. When the patient experiences signs and symptoms of lung disorder and disease, the healthcare provider tests the lungs and orders procedures to evaluate the respiratory system.

The healthcare provider can examine the respiratory tract using a bronchoscopy and removes samples of suspicious tissue for microscopic examination. The capacity and function of the lungs are measured using several pulmonary function tests. Blood flow to the lungs is monitored by a lung scan and by performing a pulmonary angiogram to identify restriction or blockage of blood flow to the lungs. This is also performed using CT imaging.

The patient may experience difficulty breathing when excess fluid builds in the plural space, inhibiting the expansion of the lung. A thoracentesis is sometimes performed, which removes the excess fluid.

Diseases such as lung cancer can destroy part of or the entire lung, requiring the healthcare provider to surgically remove a portion (wedge resection), lobes of the lung (lobectomy), or the entire lung (pneumonectomy).

In this chapter, you will learn about these tests and procedures.

Learning Objectives

1. Lung Scan
2. Pulmonary Function Tests

Key Words

Body plethysmograph test
Carbon monoxide diffusing capacity
Exercise stress tests
Expiratory reserve volume (ERV)
Forced expiratory volume (FEV)
Forced vital capacity (FVC)
Functional residual capacity (FRC)
Gas diffusion tests
Inhalation challenge test
Lobectomy
Maximum voluntary ventilation (MVV)
Peak expiratory flow (PEF)
Perfusion scan
Pneumonectomy
Pulmonary emboli
Slow vital capacity (SVC)
Spirometry test
Total lung capacity (TLC)
Ventilation scan
V/Q scan

Lung Scan 1

A lung scan is performed to detect pulmonary emboli that imbed blood flow in the lungs. Following are the three types of lung scans:

- Perfusion: In a perfusion scan, a radioactive tracer is injected into a blood vessel of the patient. An image is taken of the lungs as the tracer circulates in the lungs. A pulmonary embolus is suspected in areas of the lung where the tracer is not seen.
- Ventilation: In a ventilation scan, the patient inhales gas that contains a radioactive tracer. An image is taken of the lungs. A pulmonary embolus is suspected in areas of the lung that are not receiving the tracer.

- V/Q: A V/Q scan consists of both the perfusion scan and the ventilation scan. The ventilation scan is performed first. This is the most commonly performed lung scan.

WHAT IS BEING EXAMINED?

- The presence of the pulmonary emboli

HOW IS THE TEST PERFORMED?

- The patient removes his clothing.
- The patient lies on a table.
- Perfusion scan
  - The radioactive trace is injected into the patient's arm.
- Ventilation scan
  - A mask is placed over the patient's mouth and nose.
  - The patient is asked to take a deep breath to inhale a mixture of oxygen and tracer gas and hold it for 10 seconds as the image is taken.
- The scanning camera is placed over the patient's chest.
- The patient must remain still when each image is taken.
- The scanning camera may be repositioned during the procedure.
- The patient may be repositioned during the procedure.

RATIONALE FOR THE TEST

- Assess blood flow to the lungs.

NURSING IMPLICATIONS

- Determine if the patient
  - Has signed a consent form
  - Takes anticoagulants such as Coumadin, aspirin, heparin, or Plavix
  - Is pregnant
  - Can lie still on the back

- Has pulmonary disease
- Has cardiac disease
- Can hold the breath for 10 seconds, if the ventilation scan is performed
- Has taken Pepto-Bismol or barium prior to the procedure

UNDERSTANDING THE RESULTS

- The lung scan takes about 30 minutes to perform. The results are usually known immediately.
- Normal test results indicate
  - Normal blood flow throughout the lungs
- Abnormal test results indicate
  - Obstructed blood flow in a portion of the lungs
  - Pulmonary emboli identified

TEACH THE PATIENT

- Explain
  - Why the procedure is being performed.
  - That the patient will need to sign a consent form.
  - What the patient will experience during the procedure.
  - That the patient will not feel any pain during the procedure except for a pinch from the needle used to administer the tracer, if the perfusion scan is performed.
  - That the patient must tell the healthcare provider if the patient is taking anticoagulants.
  - That the patient may be asked to change positions during the procedure.
  - That the patient might have swelling at the injection site that is relieved by placing warm compresses on the site.
  - After using the toilet following the procedure, the patient should promptly flush the toilet and wash hands thoroughly with soap and water since the tracer exits the body through urine and stool.
  - If the patient is breast-feeding, any breast milk produced up to 2 days following the procedure should be discarded.

Pulmonary Function Tests 2

There are a number of pulmonary function tests used to assess how well the patient's lungs perform. They are as follows:

- Gas diffusion: Measures the amount of gasses that cross the alveoli per minute. These include arterial blood gases and the carbon monoxide diffusing capacity.
- Spirometry: Measures the volume and capacity of the lungs.
- Exercise stress: Measures the effect the exercise has on the lungs.
- Body plethysmograph: Measures the volume and capacity of the lungs.
- Inhalation challenge: Assesses the patient's airway responses to allergens.

WHAT IS BEING EXAMINED?

- Function of the lungs

HOW IS THE TEST PERFORMED?

- A nose clip may be placed on the patient's nose to prevent the patient from breathing through the nose, depending on the test that is being performed.
- A mouthpiece connected to the measuring device is placed in the patient's mouth.
- Gas diffusion tests
  - Arterial blood gases
    - A needle is inserted into an artery in the patient's arm.
    - A sample of blood is removed from an artery.
    - Pressure is placed on the site for approximately 5 minutes to stop any bleeding.
    - A bandage is then placed over the site.
  - Carbon monoxide diffusing capacity
    - A mask is connected to a container of a mixture of air and a small amount of carbon monoxide.
    - The patient breathes the gas mixture.

- A mouthpiece connected to the measuring device is placed in the patient's mouth.
- The patient exhales. The amount of carbon monoxide in the patient's breath is measured by the device. This is referred to as the diffusing capacity of the patient's lungs.

- Spirometry test
  - A nose clip is placed on the patient's nose to prevent breathing through the nose.
  - A mouthpiece connected to the spirometer is placed in the patient's mouth.
  - The patient inhales, and then exhales with force to measure the forced vital capacity (FVC). The exhaled air flow is also measured halfway through exhalation to measure the forced expiratory flow 25% to 75%. The peak exhale flow is also measured, which is referred to as the peak expiratory flow (PEF).
  - The patient inhales, then exhales with force. The amount of air exhaled is measured each second for 3 seconds to determine the forced expiratory volume (FEV).
  - The patient is asked to take a deep breath and then exhale to measure the maximum voluntary ventilation (MVV).
  - The patient is asked to take a deep breath and then slowly exhale to measure the slow vital capacity (SVC) and the total lung capacity (TLC).
  - The patient is asked to take a normal breath and exhale to measure the functional residual capacity (FRC).
  - The patient is asked to inhale normally and then exhale with force (RV). The amount exhaled is subtracted from the functional residual capacity (FRC) to calculate the expiratory reserve volume (ERV).
  - The patient's expiratory reserve volume (ERV) is calculated by subtracting the reserve volume (RV) from the functional residual capacity (FRC).
  - The radioactive trace is injected into the patient's arm.
- Exercise stress test
  - The patient undergoes a spirometry test.
  - The patient exercises.
  - The patient undergoes another spirometry test.
  - Results of both spirometry tests are compared to assess the patient's lung function before and after exercising.

- Inhalation challenge test
    - The patient undergoes a spirometry test.
    - The patient places a face mask over nose and mouth.
    - The face mask is attached to a nebulizer.
    - An allergen is gradually added to the mixture in the nebulizer.
    - The patient inhales a fine mist that contains the allergen from the nebulizer.
    - The patient is monitored for bronchospasm that may be triggered by the allergen.
    - The patient undergoes another spirometry test.
    - The results of both spirometry tests are compared to determine the effect that the antigen had on the patient.
- Body plethysmography test
    - The patient sits inside an airtight plethysmograph booth.
    - The booth is filled with a mixture of oxygen and helium or 100% oxygen.
    - Instruments attached to the booth measure pressure changes within the booth as the patient breathes.

RATIONALE FOR THE TEST

- Assess the function of the patient's lungs.
- Monitor the progress of lung therapy.

NURSING IMPLICATIONS

- Determine if the patient
    - Has eaten a heavy meal 8 hours before the test
    - Is able to have nose pinched
    - Is able to breathe through a mouthpiece
    - Is able to breathe normally
    - Is able to follow directions during the test
    - Has removed dentures if they prevent a tight seal around the patient's mouth
    - Has smoked 6 hours before the test

- Has ingested caffeine 6 hours before the test
- Has exercised 6 hours before the test
- Is wearing comfortable clothes if the patient is taking the exercise stress test
- Has taken medication that affects the respiratory tract
- Has taken a sedative before the test
- Has cardiovascular disease
- Has allergies
- Is pregnant

UNDERSTANDING THE RESULTS

- The lung scan takes about 30 minutes to perform. The results are usually known immediately.
- Normal test results indicate
  - Normal for age, height, sex, weight, and race
- Abnormal test results indicate
  - Decreased pulmonary function based on the patient's age, height, sex, weight, and race

TEACH THE PATIENT

- Explain
  - Why the procedure is being performed.
  - What the patient will experience during the procedure.
  - That the patient will not feel any pain during the test except if undergoing the arterial blood gas test, during which the patient will feel a pinch when the needle is inserted into the patient's arm.
  - That the patient should avoid taking a sedative prior to the test.
  - That the patient may be asked to stop taking respiratory medication 24 hours before the test. Taking medication may be resumed after the test.
  - That the patient may be given instructions during the test.
  - That the patient should not eat a big meal 8 hours before the test.
  - That the patient should wear comfortable clothes the day of the test, especially undergoing an exercise stress test.

- After using the toilet following the procedure, the patient should promptly flush the toilet and wash hands thoroughly with soap and water since the tracer exits the body through urine and stool.
- If the patient is breast-feeding, any breast milk produced up to 2 days following the procedure should be discarded.
- That the patient should tell the healthcare provider if feeling lightheaded when breathing rapidly during the test. The patient will be given time to adjust his/her breathing, which normally relieves the lightheadedness.

Summary

There are several tests used to assess pulmonary function. These tests measure the carbon dioxide and oxygen exchange and the lungs' capability to expand and retract. In addition, these tests also examine blood flow to the lungs. Blood flow to the lungs is monitored by a lung scan and by performing a pulmonary angiogram to identify restriction or blockage of blood flow to the lungs.

There are a number of procedures that can be performed to restore pulmonary function or remove diseased tissue. For example, a bronchoscope is used to view the respiratory tract and remove obstructions or take tissue samples. A thoracotomy is a surgical procedure performed to remove a portion or the entire diseased lung. A thoracentesis is performed to drain excess pleural fluid from the pleural space enabling the patient's lungs to fully expand.

Quiz

1. What is a V/Q scan?
   a. A common lung scan in which a perfusion scan is performed and then a ventilation scan is performed
   b. A common lung scan in which a ventilation scan is performed and then a perfusion scan is performed
   c. A scan that measures the volume and quantity of a lung
   d. None of the above

2. What test is performed if the healthcare provider suspects a pulmonary embolus?
   a. Bronchoscopy
   b. Lung scan
   c. Pulmonary function test
   d. Thoracotomy

3. What is the reason for administering the V/Q scan?
   a. To assess for a pulmonary embolus
   b. To assess the capacity and function of the lung
   c. To assess vital capacity
   d. To assess for cardiac inflections

4. What do normal results for a pulmonary function test depend on?
   a. Patient's age, height, sex, weight, and race
   b. Laboratory standard
   c. If the patient wears dentures
   d. If the patient is a smoker

5. What does the gas diffusion test measure?
   a. Oxygen in the blood
   b. The amount of gases that cross the alveoli per minute
   c. CO_2 in the blood
   d. The volume and capacity of the lungs

6. The inhalation challenge test assesses the patient's airway responses to allergens.
   a. True
   b. False

7. A perfusion scan involves
   a. A radioactive tracer being injected into a blood vessel
   b. Contrast material being injected into a blood vessel
   c. The patient inhaling a gas
   d. None of the above

8. The spirometry test measures
   a. The amount of gases that cross the alveoli per minute
   b. The patient's airway responses to allergens
   c. Volume and capacity of the lungs
   d. None of the above

9. Why would a patient be asked to refrain from taking a sedative prior to a pulmonary function test?
   a. A sedative may invalidate the test results because it might slow down respiration.
   b. The patient will fall asleep during the test.
   c. The patient may vomit during the test.
   d. None of the above.

10. How does use of an incentive spirometer assist the patient's respiratory system?
    a. It helps the patient expand the lungs.
    b. It helps the patient reduce lung expansion.
    c. It prevents antigens from entering the respiratory tract.
    d. None of the above.

Answers

1. b. A common lung scan, in which a ventilation scan is performed and then a perfusion scan is performed.
2. b. Lung scan.
3. a. To assess for a pulmonary embolus.
4. a. Patient's age, height, sex, weight, and race.
5. b. The amount of gases that cross the alveoli per minute.
6. a. True.
7. a. A radioactive tracer being injected into a blood vessel.
8. c. Volume and capacity of the lungs.
9. a. A sedative may invalidate the test results because it might slow down respiration.
10. a. It helps the patient expand the lungs.

CHAPTER 21

Tests and Procedures for Females

Female patients routinely undergo breast and cervical examinations for signs of cysts, growths, abnormal tissue, structural abnormalities, and infection. In this chapter you will learn about tests and procedures that are performed to test for disorders and repair of disorders.

When a mammogram reveals a suspicious growth, the healthcare provider usually orders a breast ultrasound to closely examine the growth and then possibly a breast biopsy. If the tissue sample is cancerous, the healthcare provider may perform a mastectomy. A mastectomy may be performed even if there is no sign of breast cancer. These are explored in this chapter.

The patient may decide to have her breasts altered for therapeutic or cosmetic reasons. You will learn about procedures that augment, reduce, and lift the breast in this chapter.

There are a number of tests used to examine the vulva, vagina, cervix, uterus, and fallopian tubes. Many of these tests enable the healthcare provider to take a tissue sample or perform a biopsy on abnormal tissue. If the tissue sample is identified to be cancerous, the cancerous organ is removed. You will learn about these tests and procedures in this chapter.

Learning Objectives

1. Breast Cancer Gene Test (BRCA)
2. Breast Ultrasound
3. Pap Smear
4. Vaginosis Tests
5. Sperm Penetration Tests

Key Words

BRCA1
BRCA2

Breast Cancer Gene Test 1

Scientists have discovered two genes called the breast cancer genes (BRCA1, BRCA2) that if mutated are associated with breast and ovarian cancer. A woman who carries this mutated gene and has a family history of breast or ovarian cancer may have a higher than normal chance of developing these cancers. However, she also has a chance of not developing these cancers. The presence of the breast cancer gene does not mean that the woman will develop breast ovarian cancer. The breast cancer gene test determines if the patient's BRCA1 and BRCA2 genes are mutated. If so, some patients may decide to have a mastectomy and/or oophorectomy to prevent these cancers from developing. Patients who test positive may also be advised to take tamoxifen to inhibit this gene.

Hint

Male patients who have a family history of breast cancer and prostate cancer may also have this gene. These patients can also develop breast cancer and prostate cancer.

WHAT IS BEING EXAMINED?

- Determine if the patient has the breast cancer gene.

HOW IS THE PROCEDURE PERFORMED?

- The healthcare provider removes a blood sample from a vein in the patient's arm.
- The blood sample is sent to the laboratory for examination.

RATIONALE FOR THE TEST

- Assess the patient's chances of developing breast or ovarian cancer.
- Provide information to enable the patient to decide whether or not to take preventive measures.

NURSING IMPLICATIONS

- Determine if the patient
  - Has signed a consent form
  - Has been taken anticoagulant medication such as Plavix, Coumadin, heparin, or aspirin
  - Has undergone genetic counseling to understand the significance of this test

UNDERSTANDING THE RESULTS

- The procedure takes less than 1 hour. Results are ready within 2 weeks of the procedure.
- Normal (negative) test results indicate
  - The BRCA1 and BRCA2 genes are not mutated.
- Uncertain (variant of uncertain significance VUS)
  - It is undetermined if the BRCA1 and BRCA2 genes are mutated.
- Abnormal (positive) test results indicate
  - One or both BRCA1 and BRCA2 genes are mutated.
  - Positive for endometrial hyperplasia.
  - Cancer cells were present in the sample.

Caution

A normal results does not mean that the patient will not develop breast cancer or ovarian cancer.

TEACH THE PATIENT

- Explain
  - Why the test is being performed.
  - What the patient will experience during the test.
  - That the patient will have to sign a consent form.
  - That the patient should undergo genetic counseling before the test is administered to fully understand the significance of the test.
  - A negative result does not mean that the patient will not develop breast cancer or ovarian cancer.
  - A positive result does not mean that the patient will develop breast cancer or ovarian cancer. It means there is a higher probability if the patient also has a strong family trait of breast cancer or ovarian cancer.
  - That the patient should undergo genetic counseling if the test result is positive or uncertain to fully understand the significance of these results.

Breast Ultrasound 2

A breast ultrasound is used to examine suspicious findings of a mammogram because ultrasound technology can differentiate between a cyst and solid tissue. Furthermore, the breast ultrasound can examine areas of the breast that are difficult to view on a mammogram.

WHAT IS BEING EXAMINED?

- Suspicious findings of a mammogram

HOW IS THE PROCEDURE PERFORMED?

- The patient removes her clothing from above the waist.
- The patient either sits or lies on a table.

- A conductive gel is placed on the ultrasound transducer.
- The ultrasound transducer is pressed on and moved around the breast.
- Images of breast tissue are displayed on a computer screen.
- Any conductive gel remaining on the breast is removed.

RATIONALE FOR THE TEST

- To determine if a lump in the breast is a cyst or a solid mass
- To assess suspicious results from a mammogram
- To assess the underlying cause of swelling and pain in the breast
- To guide the healthcare provider's decision to take a breast biopsy
- To assess breast implants

NURSING IMPLICATIONS

- Determine if the patient
  - Can sit or lie on her back

UNDERSTANDING THE RESULTS

- The procedure takes less than 1 hour. Preliminary results are immediate. Complete results are available within 2 days.
- Normal test results indicate
  - Normal breast tissue
- Abnormal test results indicate
  - The lump is identified as a cyst.
  - The lump is identified as a solid mass.

TEACH THE PATIENT

- Explain
  - Why the procedure is being performed.
  - What the patient will experience during the procedure.
  - The patient may feel pain if she has a fibrocystic breast.

Pap Smear 3

A Pap smear is a procedure that removes sample cells from the cervix to assess if there are any abnormal cells. The sample is sent to the laboratory for microscopic identification. Further examination is necessary if the sample is positive, indicating abnormal cells on the cervix.

Caution

A negative Pap smear result does not mean that the patient is free from cervical cancer. It means that no abnormal cells were contained in the tissue sample. Abnormal cells might exist in areas of the cervix that was not tested.

WHAT IS BEING EXAMINED?

- The cervix

HOW IS THE PROCEDURE PERFORMED?

- The patient removes her clothing from below the waist.
- The patient lies on a table with her feet in stirrups.
- A speculum is placed into the vagina to spread the vaginal walls.
- The healthcare provider inserts a cytobrush to gather samples from several areas of the cervix and from the endocervical canal.
- The tissue sample is sent to the laboratory for identification.

RATIONALE FOR THE TEST

- To assess cervical tissue
- To identify abnormal tissue that might be cancerous

NURSING IMPLICATIONS

- Determine if the patient
  - Has been taking anticoagulant medication such as Plavix, Coumadin, heparin, or aspirin
  - Can lie on her back
  - Is pregnant
  - Is using birth control

- Has undergone cervical surgery
- Is not having her menstrual period
- Is 8 to 12 days after her menstrual period
- Has not used a tampon 24 hours prior to the procedure
- Has not douched 24 hours prior to the procedure
- Has not administered vaginal medications 24 hours prior to the procedure
- Did not have a pelvic, cervical, or vaginal infection within 6 weeks prior to the procedure

CAUTION

Assess if the patient has been the victim of rape. If so, the patient may not feel comfortable having a pelvic examination or a Pap smear performed.

UNDERSTANDING THE RESULTS

- The procedure takes less than 30 minutes. Results are available within 2 weeks.
- Normal test results indicate
  - Normal cervical tissue
- Abnormal test results indicate
  - Abnormal cervical tissue
- Inconclusive test results indicate
  - The same did not contain a sufficient amount of cells to make a determination if cells were normal.

TEACH THE PATIENT

- Explain
  - Why the procedure is being performed.
  - What the patient will experience during the procedure.
  - Not to use a tampon 24 hours prior to the procedure.
  - Not to douche 24 hours prior to the procedure.
  - Not to use vaginal medications 24 hours prior to the procedure.
  - That the patient cannot have the procedure performed during her menstrual period.

- That the patient cannot have the procedure performed if she had a pelvic, cervical, or vaginal infection within 6 weeks prior to the procedure.
- There may be a small amount of vaginal bleeding following the procedure.

Vaginosis Tests 4

Vaginosis is inflammation of the vulva and vagina caused by an infection or a reaction to an irritant, resulting in a painful vaginal discharge and itching. The most common causes of vaginosis are

- *Candida albicans:* This is a yeast infection that causes lumpy white discharge and itching.
- *Trichomonas vaginalis:* This causes a foamy, yellow-green, odorous vaginal discharge.
- Bacterial vaginosis: This causes a milky thick vaginal discharge that gives off a fishy odor.

Vaginosis tests are performed by the healthcare provider taking a sample of the vaginal discharge and sending the sample to a laboratory for examination. The four vaginosis tests are

- Whiff test: Potassium hydroxide solution is dropped on the sample. If a fishy odor emanates, then the patient has bacterial vaginosis.
- KOH slide: The sample is mixed with potassium hydroxide solution. Only the yeast remains on the slide, indicating that the patient has a yeast infection.
- Wet mount: The sample is mixed with saline on one slide. The laboratory technician then identifies, through microscopic examination, the organism causing the infection.
- Vaginal pH: The pH level of the sample is tested. A pH level greater than 4.5 indicates bacterial vaginosis.

WHAT IS BEING EXAMINED?

- The cause of vaginosis

HOW IS THE PROCEDURE PERFORMED?

- The patient removes her clothing from below the waist.
- The patient lies on a table with her feet in stirrups.

- A speculum is placed into the vagina to spread the vaginal walls.
- The healthcare provider inserts a swab to gather a sample of the discharge.
- The sample is sent to the laboratory for identification.

RATIONALE FOR THE TEST

- To assess the cause of vaginal itching, inflammation, and discharge
- To identify the treatment of vaginosis

NURSING IMPLICATIONS

- Determine if the patient
  - Has taken anticoagulant medication such as Plavix, Coumadin, heparin, or aspirin
  - Can lie on her back
  - Is pregnant
  - Is using birth control
  - Has undergone cervical surgery
  - Is 8 to 12 days after her menstrual period
  - Has had sexual intercourse 24 hours before the procedure
  - Has used a tampon 24 hours prior to the procedure
  - Has douched 24 hours prior to the procedure
  - Has administered vaginal medications 24 hours prior to the procedure
  - Is having her menstrual period

CAUTION

Assess if the patient has been the victim of rape. If so, the patient may not feel comfortable having a sample of the discharge taken.

UNDERSTANDING THE RESULTS

- The procedure takes less than 30 minutes. Results are available within 2 days.
- Normal test results indicate
  - Normal vaginal discharge: No vaginosis

- Abnormal test results indicate
    - Presence of a microorganism that is causing vaginosis

TEACH THE PATIENT

- Explain
    - Why the procedure is being performed.
    - What the patient will experience during the procedure.
    - That the patient will have to sign a consent form.
    - Not to use a tampon 24 hours prior to the procedure.
    - Not to douche 24 hours prior to the procedure.
    - Not to use vaginal medications 24 hours prior to the procedure.
    - Cannot have the procedure performed during her menstrual period.
    - There will be a small amount of vaginal bleeding following the procedure.
    - Should not have sexual intercourse 24 hours before the procedure.

Sperm Penetration Tests 5

Sperm penetration tests are performed, when a woman is having difficulty becoming pregnant, to determine if the sperm can move through the cervical mucus and into the fallopian tubes. The two types of sperm penetration tests are

- Sperm penetration assay: This test mixes sperm with hamster eggs to see if the sperm can penetrate the egg. The result is measured as a sperm capacitation index.
- Sperm mucus penetration: This test determines if sperm can move through the cervical mucus.

WHAT IS BEING EXAMINED?

- Sperm penetration capability

HOW IS THE PROCEDURE PERFORMED?

- The patient signs a consent form.
- Sperm penetration assay.
    - Ejaculation should not occur for 2 days before the semen sample is taken.

  - Ejaculation should occur within 5 days before the semen sample is taken.
  - Before the semen sample is taken, the patient should urinate.
  - The patient's hands and penis should be washed.
  - Do not use lubricants or condoms when collecting the sample.
  - Do not collect a semen sample after withdrawal from intercourse.
  - Place the semen sample in a sterile cup.
  - Keep the sample at body temperature.
  - Keep the sample from direct sunlight.
  - Deliver the sample to the laboratory immediately.
  - The laboratory will mix the semen with hamster eggs to determine the sperm capacitation index.
- Sperm mucus penetration test
  - The woman must be ovulating in order to perform this test.
  - Determine if she is ovulating by collecting a urine sample in the mid to late morning.
  - The woman should avoid drinking fluids before giving a urine sample.
  - The urine sample is analyzed for the presence of luteinizing hormone (LH). This hormone indicates that she is ovulating.
  - The woman is given a pelvic examination during which a sample of cervical mucus is collected.
  - The man provides a semen sample (see sperm penetration assay above).
  - The cervical mucus and the semen samples are combined at the laboratory to determine if sperm can move through the cervical mucus.

RATIONALE FOR THE TEST

- To determine the underlying cause of infertility

NURSING IMPLICATIONS

- Determine if the male patient
  - Has ejaculated for 2 days before the semen sample is taken
  - Has ejaculated within 5 days before the semen sample is taken
  - Has urinated before the semen sample was taken
  - Washed his hands and penis before the semen sample was taken

  - Used lubricants or condoms when collecting the semen sample
  - Collected a semen sample after withdrawal from intercourse
  - Placed the semen sample in a sterile cup
  - Kept the sample at body temperature
  - Kept the sample from direct sunlight
  - Delivered the sample to the laboratory immediately
  - Has signed a consent form
- Determine of the female patient
  - Was ovulating during the test

UNDERSTANDING THE RESULTS

- The procedure takes less than 1 hour. Results are known within 2 days.
- Normal test results indicate
  - Sperm penetrated the hamster egg.
  - Sperm moved through the cervical mucus.
- Abnormal test results indicate
  - Sperm could not penetrate the hamster egg.
  - Sperm could not move through the cervical mucus.

TEACH THE PATIENT

- Explain
  - Why the procedure is being performed.
  - What the patient will experience during the procedure.
  - That the patient will not feel any pain during the procedure.
  - Must sign a consent form.
  - How to collect the semen sample.

Summary

In this chapter you learned about commonly performed tests and procedures that assess a woman's breast and cervix for structural defects and disease. The tests involved visual examination of the patient's vulva, vagina, cervix, uterus, and fallopian tubes and looking through the skin using X-ray and ultrasound.

Any suspicious area of the breast or cervix is examined closely by taking a tissue sample or a biopsy of an abnormal growth to identify it. If the sample is cancerous, the healthcare provider will perform one of several procedures to remove either the affected area or affected organs.

Quiz

1. What does it mean when a woman has BRCA1 and BRCA2?
   a. The woman may have a higher than normal chance of developing breast cancer or ovarian cancer.
   b. The woman does not have cancer.
   c. The woman has cancer.
   d. The woman will develop breast cancer or ovarian cancer.

2. What does a negative Pap smear mean?
   a. The patient is free from cervical cancer.
   b. No abnormal cells were contained in the tissue sample.
   c. The patient has abnormal cells.
   d. The patient has cancer.

3. What test might a healthcare provider order if a woman is having difficulty becoming pregnant?
   a. Sperm penetration tests
   b. Pap smear
   c. The Whiff test
   d. The KOH test

4. A breast ultrasound is commonly ordered
   a. In place of a mammogram
   b. If there are suspicious results on a mammogram
   c. Always in combination with a mammogram
   d. Only when performing breast augmentation

5. What might a healthcare provider prescribe if a woman has the BRCA1 and BRCA2 genes?
   a. Calcium chloride
   b. Potassium chloride
   c. Tamoxifen
   d. Sodium chloride

6. What does it mean if a woman's husband has the BRCA1 and BRCA2 genes?
   a. He can pass these genes to the woman during sexual intercourse.
   b. The woman is at a high risk for cancer.
   c. The couple should refrain from unprotected sexual intercourse.
   d. The man might be at high risk for developing breast cancer and/or prostate cancer.

7. A common yeast infection that causes vaginosis is
   a. *Candida albicans*
   b. *Trichomonas vaginalis*
   c. Bacterial vaginosis
   d. None of the above

8. What cause of vaginosis causes a fishy odor?
   a. *Candida albicans*
   b. *Trichomonas vaginalis*
   c. Bacterial vaginosis
   d. None of the above

9. A breast ultrasound can differentiate between a cyst and solid tissue.
   a. True
   b. False

10. How would you respond if the patient who is scheduled for a Pap smear today mentions that her menstrual period ended 7 days ago?
    a. Prepare the patient for the test.
    b. Reschedule the test for the next day.
    c. Reschedule the test for the next week.
    d. Reschedule the test for 2 weeks.

Answers

1. a. The woman may have a higher than normal chance of developing breast cancer or ovarian cancer.
2. b. No abnormal cells were contained in the tissue sample.
3. a. Sperm penetration tests.
4. b. If there are suspicious results on a mammogram.
5. c. Tamoxifen.
6. d. The man might be at high risk for developing breast cancer and/or prostate cancer.
7. a. *Candida albicans.*
8. c. Bacterial vaginosis.
9. a. True.
10. b. Reschedule the test for the next day.

CHAPTER 22

Maternity Tests

There are several tests that are performed during pregnancy and shortly after childbirth to assess the health of the fetus and newborn. In a high-risk pregnancy, the healthcare provider might perform a chorionic villus sampling or amniocentesis early on in the pregnancy to determine if the fetus has a genetic disorder or other health issues.

Amniocentesis, for example, may be suggested in high-risk pregnancies that have a high risk of birth defects. This condition is confirmed by performing a cordocentesis where a sample of blood is taken from the umbilical cord while in the womb.

Later in the pregnancy, the healthcare provider performs a biophysical profile of the fetus to determine the overall health of the fetus. This is when an assessment is made of the fetal heart rate, breathing and body movements, muscle tone, and the volume of amniotic fluid.

It is also around this same period when the mother may undergo a contraction stress test. The contraction stress test determines if the fetus is healthy enough to survive the reduced oxygen levels that are common with natural childbirth.

In some pregnancies, the woman might experience an incompetent cervix that could result in the cervix opening prior to the 37th week of gestation, causing a premature birth. In this situation, the healthcare provider is likely to perform a cervical cerclage, which temporarily closes the cervix until the mother enters labor.

If the birth is premature, the healthcare provider may perform a cranial ultrasound to determine if there were complications caused by the premature birth.

The newborn is typically administered the sweat test that helps determine if the newborn has a high level of chloride in his sweat, which may be an indication of cystic fibrosis.

You will learn about all these tests in this chapter.

Learning Objectives

1. Biophysical Profile (BPP)
2. Contraction Stress Test
3. Cranial Ultrasound
4. Amniocentesis
5. Cordocentesis
6. Sweat Test
7. Chorionic Villus Sampling (CVS)
8. Karyotyping
9. Cervical Cerclage (Weak Cervix)
10. Galactosemia Test

Key Words

Amniotic fluid volume
Body movement
Breathing movement
Chorioamnionitis
Chorionic villi
Fetal anemia
Fetal lung development
Fetal monitor
Fetal ultrasound
Intraventricular hemorrhage (IVH)
Macroduct technique
Muscle tone
Nonstress test
Oxytocin
Periventricular leukomalacia (PVL)
Pilocarpine
Transabdominal
Transcervical

Biophysical Profile

The biophysical profile (BPP) test assesses the health of the fetus and is commonly performed in the last trimester of the pregnancy, although the healthcare provider may perform this test earlier and frequently in high-risk pregnancies. The BPP test consists of a nonstress test and a fetal ultrasound. Each element of the test is graded according to Table 22.1.

WHAT IS BEING EXAMINED?

- Fetal assessment

HOW IS THE PROCEDURE PERFORMED?

- The patient refrains from smoking 2 hours before the test.
- The patient must have a full bladder except if she is near term. If the patient is unable to fill her bladder, a urinary catheter is inserted into the urethra and saline will be infused into the bladder.
- The patient lies on a table with her abdomen exposed.

Table 22.1 Biophysical Profile Elements

	Nonstress Test	Breathing Movement	Body Movement	Muscle Tone	Amniotic Fluid Volume
Normal (2 points)	Two or more heart rate increases with a rate of 15 minutes or greater while the fetus moves	One or more breathing movements of 60 seconds	Three or more arm, leg, or body movements	Flexed arms and legs; head rests on chest; one or more extensions and flexions	1 cm of amniotic fluid in the uterine cavity
Abnormal (0 points)	One or more heart rate increases or the rate is not 15 minutes or greater while the fetus moves	Breathing movements of less than 60 seconds	Less than three arm, leg, or body movements	Arms, legs, and spine are extended; open hand; fetus not returning to normal position; extension and flexion is slow	Less than 1 cm of amniotic fluid in the uterine cavity

- Nonstress test
  - Conductive gel is placed on her abdomen.
  - Two belts of the fetal monitor are placed on her abdomen to measure fetal heart rate and contractions.
  - The fetal monitor constantly records the fetus during the test. The test results appear on a strip of paper.
  - The patient is asked to press a button connected to the fetal monitor each time the fetus moves and if there is a contraction.
  - The two belts of the fetal monitor are removed and the gel is wiped from the patient's abdomen.
- Fetal ultrasound
  - Conductive gel is applied to her abdomen.
  - A transducer is pressed against and moved about the abdomen.
  - Images of the fetus appear on a computer screen.
  - Conductive gel is wiped off the abdomen.

RATIONALE FOR THE TEST

- Assess
  - Fetus movement
  - Fetal heart rate
  - Volume of amniotic fluid
  - Muscle tone
  - Breathing rate

NURSING IMPLICATIONS

- Determine if the patient
  - Can lie on her back
  - Is able to lie still
  - Is able to follow instructions
  - Has smoked 2 hours before the test
  - Has a full bladder
  - Is hypo- or hyperglycemic
  - Is using alcohol or narcotics

UNDERSTANDING THE RESULTS

- The procedure takes less than 30 minutes. Results are known immediately.
- Normal test results indicate
  - A score of 8 or greater
- Undetermined test results indicate
  - A score of 5 to 8. The patient should be retested.
- Abnormal test results indicate
  - A score of less than 5. The healthcare provider will order different tests to further assess the fetus and mother.

TEACH THE PATIENT

- Explain
  - Why the test is being performed.
  - What the patient will experience during the test.
  - That the patient will not feel any pain during the test.
  - That the patient must not smoke 2 hours before the test.
  - That the patient must have a full bladder unless she is near term.

Contraction Stress Test 2

The healthcare provider usually performs a BPP test of the fetus to assess breathing, movement, muscle tone, and the volume of amniotic fluid. If the BPP test indicates suspicious results, the healthcare provider might perform a contraction stress test.

The contraction stress test determines if the fetus will remain healthy during natural childbirth. Uterine contractions reduce oxygen to the fetus, which normally does not harm the fetus. However, some fetuses can become negatively affected by the lower oxygen level, so the healthcare provider might decide a caesarean birth.

During the contraction stress test, a fetal heart monitor is attached to the mother while the mother is administered oxytocin. Oxytocin is a hormone that induces uterine contractions. The fetal heart rate is expected to decelerate during a contraction and accelerate following the contraction. If the heart rate does not accelerate, then the fetus may not remain healthy during natural childbirth.

Hint

This test is not usually performed if the mother has in the past had a cesarean section, placenta previa, placenta abruptio, incompetent cervix, premature rupture of the amniotic membrane, been administered magnesium sulfate, or is pregnant with multiple fetuses.

WHAT IS BEING EXAMINED?

- The fetus's ability to remain healthy during natural childbirth

HOW IS THE PROCEDURE PERFORMED?

- The patient refrains from eating or drinking 8 hours before the test.
- The patient refrains from smoking 2 hours before the test.
- The patient empties her bladder.
- The patient signs a consent form.
- The patient lies on the table with back raised, lying slightly to the left side.
- The patient exposes her abdomen.
- Conductive gel is placed on the patient's abdomen.
- Two fetal monitor straps are placed on the abdomen to record contractions and the fetal heart rate.
- The patient's vital signs are monitored.
- A baseline is measured for 10 minutes of the fetal heart rate and contractions.
- A low dose of oxytocin is administered. The dose is increased until there are three contractions, each lasting more than 45 seconds over a period of 10 minutes.
- The fetal monitor continually records the fetal heart rate and contractions during the test.
- The healthcare provider continues to observe the fetal monitor, which shows the fetal heart and contractions have returned to the baseline values.

RATIONALE FOR THE TEST

- Assess
  - Ability of the fetus to remain healthy during natural childbirth
  - Health of the placenta

NURSING IMPLICATIONS

- Determine if the patient
  - Has signed a consent form
  - Has any allergies
  - Has been taking anticoagulant medication such as Plavix, Coumadin, heparin, or aspirin
  - Can lie on her back
  - Is able to lie still
  - Is able to follow instructions
  - Has eaten or drunk 8 hours before the test
  - Has smoked 2 hours before the test
  - Has emptied her bladder

UNDERSTANDING THE RESULTS

- The procedure takes less than 2 hours. Preliminary results are known immediately. Detailed results are known within a week.
- Normal (negative) test results indicate
  - The fetal heart rate returned to normal following the contractions.
- Abnormal (positive) test results indicate
  - The fetal heart rate showed late decelerations following the contractions.
  - Contractions were hyperstimulated lasting longer than 90 seconds.

Caution

The contraction stress test may have a false-positive result.

TEACH THE PATIENT

- Explain
  - Why the test is being performed.
  - What the patient will experience during the test.
  - That the patient will not feel any pain during the procedure, except for a pinch from the needle used to administer the oxytocin.
  - That the patient cannot eat 12 hours before the open biopsy.

- That the patient cannot smoke 2 hours before the test.
- That the patient should empty her bladder before the test.

Cranial Ultrasound 3

A cranial ultrasound is performed on premature newborns to assess complications that might have arisen during the premature birth. During a cranial ultrasound, images of the newborn's brain are captured, displayed on a computer screen, and stored on a computer. The healthcare provider might order several cranial ultrasounds weeks apart. Some complications such as intraventricular hemorrhage (IVH) can be detected during the first week of birth while other complications such as periventricular leukomalacia (PVL) might occur 8 weeks after birth. PVL is damaged tissue around the ventricles.

Hint

A cranial ultrasound is not performed after the child is 18 months of age because the cranium is fully formed and the fontanelle is closed.

WHAT IS BEING EXAMINED?

- Assess the premature newborn's brain for complications resulting from the premature birth.

HOW IS THE PROCEDURE PERFORMED?

- A conduction gel is placed on the newborn's head.
- A transducer is moved across the newborn's head, capturing images that appear on a computer monitor.

RATIONALE FOR THE TEST

- Assess
  - For hydrocephalus
  - For the risk of developing cerebral palsy
  - The cause of an enlarged head
  - For IVH
  - For PVL

NURSING IMPLICATIONS

- The mother can hold the newborn during the test.
- The newborn can be fed during the test.

UNDERSTANDING THE RESULTS

- The procedure takes less than 30 minutes. Preliminary results are known immediately. Detailed results are known within a week.
- Normal test results indicate
  - No complications found
- Abnormal test results indicate
  - Complications found

TEACH THE PATIENT

- Explain
  - Why the test is being performed.
  - What the newborn will experience during the test.
  - That the newborn will not feel any pain during the test.
  - That the mother can hold and feed the newborn during the test.

Amniocentesis 4

Amniotic fluid contains cells shed by the fetus. About the 16th week of gestation, the healthcare provider may perform amniocentesis, which is the removal of some amniotic fluid. Fetal cells contain the amniotic fluid which is analyzed to determine if the fetus has a birth defect. Amniocentesis is also performed during the third trimester, if there is a risk of premature birth, to determine fetal lung development and to assess if the mother has chorioamnionitis, which is an infection of the amniotic fluid.

Amniocentesis is ordered if an integrated test result is positive, indicating that the fetus has a high chance of having a birth defect. These tests include α-fetoprotein (AFP), estriol, inhibin A, and chorionic gonadotropin (hCG). It is also ordered if parents are carriers of a genetic trait that is likely to be passed on to the fetus. These include cystic fibrosis, Duchenne muscular dystrophy, sickle cell anemia, thalassemia, hemophilia, and Tay-Sachs disease.

Amniocentesis is also performed to determine if the fetus is Rh-positive when the mother has the Rh factor. This tests the amniotic fluid for increased bilirubin levels after the 20th week of gestation, which indicates that the fetal blood cells are being attacked by the mother's antibodies.

HINT

Amniocentesis does not identify all birth defects. There are many birth defects that are not revealed by amniocentesis.

WHAT IS BEING EXAMINED?

- The fetus

HOW IS THE PROCEDURE PERFORMED?

- The patient will sign a consent form.
- The patient must have an empty bladder.
- The patient lies on a table with her abdomen exposed.
- The insertion site is cleaned with an antiseptic.
- The insertion site is injected with a local anesthetic.
- Conductive gel is placed on the mother's abdomen.
- A fetal monitor is placed on the mother's abdomen to monitor the fetus during the procedure.
- The mother's vital signs are monitored during the procedure.
- The healthcare provider performs a fetal ultrasound to guide insertion of the needle.
- A needle is passed through the abdomen into the uterus. It is removed and reinserted if the fetus moves close to the needle.
- Two tablespoons of amniotic fluid are drawn up from the needle into a syringe.
- The needle is removed.
- A bandage covers the insertion site.

RATIONALE FOR THE TEST

- Assess
    - For birth defects
    - Fetal lung development

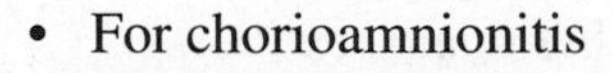

- For chorioamnionitis
- For Rh antibodies

NURSING IMPLICATIONS

- Determine if the patient
  - Signed a consent form
  - Has emptied her bladder
  - Can lie on her back
  - Is able to lie still

UNDERSTANDING THE RESULTS

- The procedure takes less than 30 minutes. Results are known within 2 weeks.
- Normal test results indicate
  - No signs of birth defect.
  - Fetal lungs are adequately developed.
  - No signs of chorioamnionitis.
- Abnormal test results indicate
  - Signs of birth defect.
  - Fetal lungs are not adequately developed.
  - The mother has chorioamnionitis.

Caution

A normal result from amniocentesis does not rule out a birth defect.

TEACH THE PATIENT

- Explain
  - Why the test is being performed.
  - What the patient will experience during the test.
  - That the patient will not feel any pain during the test, except a pinch when the local anesthetic is administered.
  - That the patient should breath slowly during the procedure to relax her abdominal muscle.

- That the patient may feel a cramp in the abdomen.
- That the patient may feel pulling on her abdomen when amniotic fluid is removed.
- That the patient must sign a consent form.
- That the patient must empty her bladder before the procedure.
- That the patient should avoid exercise for 24 hours following the test.
- That the patient should not have sexual intercourse for 24 hours following the test.
- That the patient should not douche or use tampons for 24 hours following the test.

Caution

The patient should call her healthcare provider if she notices fluid or bloody discharge from the insertion site or if there is swelling and redness at the insertion site. She should also call her healthcare provider if she experiences fever, pain, or cramping in her abdomen.

Cordocentesis 5

If amniocentesis or other tests reveal that the fetus might have anemia, the healthcare provider may order a cordocentesis to confirm the finding, typically in the second trimester. A cordocentesis is the sampling of fetal blood from the umbilical cord to determine if the fetus has a blood disorder or is Rh-positive. The blood sample is also used to assess the oxygen level in fetal blood.

WHAT IS BEING EXAMINED?

- Fetal blood

HOW IS THE PROCEDURE PERFORMED?

- The patient will sign a consent form.
- The patient must have an empty bladder.
- The patient lies on a table with her abdomen exposed.
- The insertion site is cleaned with an antiseptic.
- The insert site is injected with a local anesthetic.

- Conductive gel is placed on the mother's abdomen.
- A fetal monitor is placed on the mother's abdomen to monitor the fetus during the procedure.
- The mother's vital signs are monitored during the procedure.
- The healthcare provider performs a fetal ultrasound to guide insertion of the needle.
- A needle is passed through the abdomen into the umbilical cord. The needle is removed.
- A small amount of blood is drawn up from the needle into a syringe.
- The needle is removed.
- A bandage covers the insertion site.

HINT

The healthcare provider may administer medication to temporarily stop the fetus from moving and to prevent preterm labor and infection.

RATIONALE FOR THE TEST

- Assess
  - For fetal anemia
  - If the fetus is Rh-positive
  - Oxygen level in fetal blood

NURSING IMPLICATIONS

- Determine if the patient
  - Signed a consent form
  - Has emptied her bladder
  - Can lie on her back
  - Is able to lie still

UNDERSTANDING THE RESULTS

- The procedure takes less than 30 minutes. Results are known within 2 weeks.
- Normal test results indicate
  - Fetal blood is normal.

- Abnormal test results indicate
  - Fetal anemia.
  - The fetus is Rh-positive.
  - Abnormal oxygen level in the fetal blood.

TEACH THE PATIENT

- Explain
  - Why the test is being performed.
  - What the patient will experience during the test.
  - That the patient will not feel any pain during the test, except a pinch when the local anesthetic is administered.
  - That the patient should breathe slowly during the procedure to relax her abdominal muscle.
  - That the patient may feel a cramp in the abdomen.
  - That the patient must sign a consent form.
  - That the patient must empty her bladder before the procedure.
  - That the patient should avoid exercise for 24 hours following the test.
  - That the patient should not have sexual intercourse for 24 hours following the test.
  - That the patient should not douche or use tampons for 24 hours following the test.

CAUTION

The patient should call her healthcare provider if she notices fluid or bloody discharge from the insertion site or if there is swelling and redness at the insertion site. She should also call her healthcare provider if she experience fever, pain, or cramping in her abdomen.

Sweat Test 6

The sweat test is administered to a newborn between the ages of 2 days and 5 months old to assess if the newborn might have cystic fibrosis. Children who have cystic fibrosis have increased sodium chloride in their sweat. The sweat test measures the amount of chloride in sweat in newborns.

WHAT IS BEING EXAMINED?

- Chloride in sweat

HOW IS THE PROCEDURE PERFORMED?

- The right arm or thigh is washed and dried.
- A small gauze pad is soaked with pilocarpine.
- Another small gauze pad is soaked with salt water.
- Both are placed on the skin.
- Electrodes are placed over the gauze pads.
- A mild current flows through the electrodes forcing the medication into the skin. The newborn may feel a slight tingling.
- The gauze pads are removed after 10 minutes.
- The skin under the gauze pad that contains pilocarpine is red.
- The skin is cleaned with water and dried.
- The reddened skin is covered with a dry gauze pad.
- The dry gauze pad is covered with plastic to prevent evaporation.
- The dry gauze pad remains in place for 30 minutes to soak up sweat.
- The dry gauze pad is removed and placed in a collection bottle, which is sealed and sent to the laboratory where the dry gauze is weighed to determine the amount of chloride and/or sodium that is on the gauze.
- The skin beneath the dry gauze pad is cleaned and dried.
- The skin remains reddened for less than 5 hours following the test.

HINT

The healthcare provider may use the macroduct technique of collecting sweat, which uses a coil rather than gauze pads.

RATIONALE FOR THE TEST

- To assess for cystic fibrosis

NURSING IMPLICATIONS

- Determine if the patient
  - Is less than 4 weeks of age. Newborns less than 4 weeks of age usually do not produce enough sweat for the test.

- Is dehydrated.
- Is ill.
- Is taking steroids.
- Has a rash.

UNDERSTANDING THE RESULTS

- The procedure takes less than 1 hour. Results are known within 2 days.
- Normal test results indicate
    - Normal amount of chloride
- Borderline test results indicate
    - A high normal level of chloride
- Abnormal test results indicate
    - A high level of chloride

CAUTION

An abnormal result indicates that the newborn may have cystic fibrosis. Further testing is necessary to diagnose cystic fibrosis.

TEACH THE PATIENT

- Explain
    - Why the test is being performed.
    - What the newborn will experience during the test.
    - That the newborn will not feel any pain during the test, except for a slight tingle when current is applied to the electrode.
    - That any reddened skin will return to normal color within 5 hours of the test.

Chorionic Villus Sampling 7

The placenta contains chorionic villi, which are tiny growths that contain the same genetic material as the fetus. Chorionic villus sampling is a procedure in which a sampling of chorionic villi is biopsied between the 10th and 12th week of gestation

and is examined to determine if the fetus has a genetic disorder. There are two methods used to biopsy the chorionic villus. These are

- Transabdominal: A needle is inserted through the abdomen.
- Transcervical: A catheter is inserted through the cervix.

HINT

Chorionic villus sampling is performed earlier in gestation than amniocentesis.

WHAT IS BEING EXAMINED?

- Genetic material of the chorionic villus

HOW IS THE PROCEDURE PERFORMED?

- The patient signs a consent form.
- The patient lies on a table.
- The patient either removes her clothing below the waist or exposes her abdomen, depending on which biopsy method is used by the healthcare provider.
- Conductive gel is placed on her abdomen.
- A fetal monitor is attached to the abdomen to monitor the fetus.
- Transabdominal
  - The insertion site is cleaned with an anesthetic.
  - A local anesthetic is administered to the insertion site.
  - An ultrasound transducer is pressed on and moved around the abdomen.
  - Images of the fetus and uterus are displayed on a computer screen.
  - A needle is inserted through the abdomen.
  - A sample of the chorionic villi is removed.
  - The needle is removed.
  - The insertion site is bandaged.
- Transcervical
  - The patient places her feet in stirrups.
  - A speculum is inserted to spread apart the vaginal wall.
  - The cervix is cleaned with an antiseptic soap.

- An ultrasound transducer is pressed on and moved around the abdomen.
- Images of the fetus and uterus are displayed on a computer screen.
- A catheter is inserted through the cervix.
- A sample of the chorionic villi is removed.
- The needle is removed.

RATIONALE FOR THE TEST

- To assess for fetal genetic disorder

NURSING IMPLICATIONS

- Determine if the patient
  - Has a full bladder
  - Has signed a consent form
  - Has allergies

UNDERSTANDING THE RESULTS

- The procedure takes less than 1 hour. Results are known within 2 weeks.
- Normal test results indicate
  - Normal cells in the sample
- Abnormal test results indicate
  - Abnormal genetic material found

TEACH THE PATIENT

- Explain
  - Why the test is being performed.
  - What the patient will experience during the test.
  - That the patient will not feel any pain during the test, except for a slight pinch when the local anesthetic is administered.
  - That patient might feel cramping following the procedure.
  - That the patient may have vaginal spotting following the procedure.
  - That the patient may have a small amount of amniotic fluid leakage following the procedure.

Caution

The patient should call her healthcare provider if cramping, spotting or leakage of amniotic fluid occurs 48 hours after the procedure. A call should be made if she experiences fever, swelling at the insertion site or is dizzy.

Karyotyping 8

Karyotyping is a test that determines the number and quality of chromosomes in a cell and is used to detect possible genetic disorders. Tissue samples for karyotyping are typically taken during chorionic villus sampling or amniocentesis.

WHAT IS BEING EXAMINED?

- Chromosomes of fetal cells

HOW IS THE PROCEDURE PERFORMED?

- The patient should undergo genetic counseling to understand the test and its results.
- The sample is taken using chorionic villus sampling or amniocentesis.
- Chromosomes are removed from cells and examined under a microscope.

RATIONALE FOR THE TEST

- To assess for fetal genetic disorder

NURSING IMPLICATIONS

- Determine if the patient has undergone genetic counseling

UNDERSTANDING THE RESULTS

- The procedure takes less than 1 hour. Results are known within 2 weeks.
- Normal test results indicate
    - Normal number and quality of chromosomes.

- Abnormal test results indicate
    - Abnormal number and quality of chromosomes

TEACH THE PATIENT

- Explain
    - Why the test is being performed.
    - Why is it important to have genetic counseling to fully understand the test and its results.

Cervical Cerclage (Weak Cervix) 9

If the patient has an incompetent cervix, the cervix might open prior to the 37th week of gestation and could result in premature birth. The healthcare provider may perform a cervical cerclage, which is a procedure to close the cervix, to ensure that the cervix remain closed until after the 37th week of gestation.

CAUTION

When a cervical cerclage has been performed, the cervix must be manually opened before the patient goes into labor otherwise the healthcare provider may perform a cesarean section.

WHAT IS BEING EXAMINED?

- Closing of a incompetent cervix

HOW IS THE PROCEDURE PERFORMED?

- The patient removes her clothing.
- The patient lies on a table with her feet in stirrups.
- A breathing tube is inserted into the patient's throat.
- The patient is connected to an ECG.
- The patient is connected to a pulse oximeter to measure the oxygen content of the her blood.
- A speculum is placed into the vagina to spread the vaginal walls.

- If the amnion (amniotic sac) is protruding
  - A catheter is inserted through the cervix.
  - A bulb at the end of the catheter is inflated pushing the amnion into the pelvis.
- Incisions are made in the cervix.
- Tape is tied through the incisions, closing the cervix or stitches made to close the cervix.

RATIONALE FOR THE TEST

- To prevent premature opening of the cervix

NURSING IMPLICATIONS

- Determine if the patient
  - Has signed a consent form
  - Has any allergies
  - Has been taking anticoagulant medication such as Plavix, Coumadin, heparin, or aspirin
  - Can lie on her back
  - Is able to lie still
  - Has been treated for infection of the pelvis, cervix, or vagina
  - Has not eaten or drunk for 12 hours before the procedure
  - Does not have uterine contractions
  - Does not have vaginal bleeding
  - Does not have ruptured membranes

UNDERSTANDING THE RESULTS

- The procedure takes less than 1 hour. Results are immediate.
- Normal test results indicate
  - The cervix is closed.
- Abnormal test results indicate
  - The healthcare provider is unable to close the cervix.

TEACH THE PATIENT

- Explain
  - Why the procedure is being performed.
  - What the patient will experience during the procedure.
  - That the patient will have to sign a consent form.
  - That the patient should not eat or drink 12 hours before the procedure.
  - Should not exercise following the procedure.
  - Should not perform any heavy lifting following the procedure.

Caution

The patient should call her healthcare provider if she has a fever, has an odorous discharge from the vagina, experiences heavy vaginal bleeding, or has abdominal pain.

Galactosemia Test 10

Newborns who have galactosemia may experience seizures, brain damage, and mental retardation if their body is unable to convert galactose, which is found in breast milk and formula, into glucose. This is because they lack three enzymes needed for this process. The galactosemia test determines if these enzymes are present in the newborn's blood.

Hint

These enzymes can also be detected in a urine sample.

WHAT IS BEING MEASURED?

- Enzymes that convert galactose into glucose

HOW IS THE TEST PERFORMED?

- A blood specimen is collected from the heel.

RATIONALE FOR THE TEST

- To screen for galactosemia

NURSING IMPLICATIONS

- Assess if the patient
  - Has had a blood transfusion
  - Has relief from galactosemia. Galactosemia is a genetic disorder.

UNDERSTANDING THE RESULTS

- Test results are available within 2 days. The laboratory determines normal values based on calibration of testing equipment with a control test. Test results are reported as high, normal, or low based on the laboratory's control test.
- Normal (negative): The enzymes are present.
- Abnormal (positive): The enzymes are not present.

TEACH THE PATIENT

- Explain
  - Why blood sample is taken.
  - How the sample is taken.
  - That the newborn who has galactosemia must not ingest milk or milk by-products.

Summary

In this chapter, you learned of tests that assess the well-being of the fetus and of the newborn and procedures that prevent premature birth.

In a high-risk pregnancy, a chorionic villus sampling or amniocentesis is performed to determine if the fetus has a genetic disorder or other health issues such as anemia. If the fetus is suspected of being anemic, a cordocentesis is likely performed in which a sample of blood is taken from the umbilical cord while in the womb.

A BPP test of the fetus is performed later in the pregnancy to assess the fetal heart rate, breathing and body movements, muscle tone, and the volume of amniotic fluid. During this time the mother undergoes a contraction stress test to determine if the fetus is healthy enough to survive the reduced oxygen levels that are common with natural childbirth.

In cases of incompetent cervix, which could result in the cervix opening prior to the 37th week of gestation causing a premature birth, the healthcare provider may perform a cervical cerclage, which temporarily closes the cervix until the mother enters labor.

A premature birth can result in complications for the fetus. A cranial ultrasound might be performed to identify some complications.

Quiz

1. What is meant by a score of 7 on the biophysical profile (BPP) test?
    a. Undetermined results. The test should be repeated.
    b. The fetus is normal.
    c. There are possible problems with the fetus.
    d. None of the above.

2. Why is oxytocin administered during a contraction stress test?
    a. To lower blood pressure of the fetus
    b. To stop contractions
    c. To induce contractions
    d. To keep the fetus from moving

3. Why is a cranial ultrasound not performed after 18 months of age?
    a. The fontanelle is closed.
    b. The fontanelle remains open.
    c. The test is too painful for the baby to undergo.
    d. Complications from a premature birth would have already manifested.

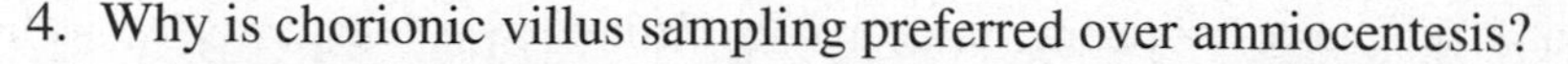

4. Why is chorionic villus sampling preferred over amniocentesis?
   a. Chorionic villus sampling can be performed earlier in the pregnancy than amniocentesis.
   b. Chorionic villus sampling can be performed later in the pregnancy than amniocentesis.
   c. Chorionic villus sampling is safer than amniocentesis.
   d. None of the above.

5. What might an increase in bilirubin levels from an amniocentesis taken after the 20th week indicate?
   a. The fetus is healthy.
   b. Fetal blood cells are attacking the mother's antibodies.
   c. Fetal blood cells are being attacked by the mother's antibodies.
   d. The fetus's lungs have matured.

6. When is a contraction stress test not performed?
   a. The mother had a past cesarean section.
   b. The mother has placenta previa.
   c. The mother has placcnta abruptio.
   d. All of the above.

7. Late decelerations in a contraction stress test means
   a. The fetus might not be healthy enough to withstand natural childbirth.
   b. The fetus is healthy enough to withstand natural childbirth.
   c. The mother is not healthy enough to withstand natural childbirth.
   d. None of the above.

8. During the BPP test performed early in the trimester, the mother should
   a. Have an empty bladder before the test
   b. Have a full bladder before the test
   c. Have had a bowel movement prior to the test
   d. None of the above

9. What is the purpose of using pilocarpine during a sweat test?
   a. Pilocarpine helps to draw sweat from the newborn.
   b. Pilocarpine prevents the newborn from sweating.
   c. Pilocarpine is not used in the sweat test.
   d. Pilocarpine protects the skin from electrodes that are placed on the skin during the test.

10. What happens when a patient who had a cervical cerclage goes into labor?
   a. The patient is administered pilocarpine to stop the contractions.
   b. The sutures open as part of natural childbirth.
   c. The cervix is manually opened or the patient undergoes a cesarean section.
   d. None of the above.

Answers

1. a. Undetermined results. The test should be repeated.
2. c. To induce contractions.
3. a. The fontanelle is closed.
4. a. Chorionic villus sampling can be performed earlier in the pregnancy than amniocentesis.
5. c. Fetal blood cells are being attacked by the mother's antibodies.
6. d. All of the above.
7. a. The fetus might not be healthy enough to withstand natural childbirth.
8. b. Have a full bladder before the test.
9. a. Pilocarpine helps to draw sweat from the newborn.
10. c. The cervix is manually opened or the patient undergoes a cesarean section.

CHAPTER 23

Chest, Abdominal, Urinary Tract Tests and Procedures

When there are suspected disorders of the upper gastrointestinal tract, the thyroid gland, the liver, gallbladder, kidneys, spleen, the urinary tract and other organs in the upper part of the body, the healthcare provider is likely to order a number of tests to uncover the underlying problem.

Some tests enable the healthcare provider to look down the esophagus to examine the stomach, duodenum, and the bile and pancreatic ducts and to take a biopsy or, in some cases, to remove an obstruction.

Other tests enable the healthcare provider to scan the liver, spleen, gallbladder, and kidneys by using contrast material to highlight the structure of the organ.

Images of the organ are captured with a camera and studied to uncover diseases and disorders.

There are also procedures that the healthcare provider can perform to temporarily or permanently repair a problem. It can be to remove a cancerous thyroid gland, remove a tumor from the bladder, or fix urinary incontinence.

You will learn about these and other tests and procedures in this chapter.

Learning Objectives

1. Upper Gastrointestinal (UGI) Series
2. Esophagus Test Series
3. Gallbladder Scan
4. Kidney Scan
5. Liver and Spleen Scan
6. Urinalysis
7. Urine Culture and Sensitivity Test
8. Renin Assay Test
9. Thyroid Scan
10. Thyroid and Parathyroid Ultrasound
11. Thyroid Hormone Tests
12. Thyroid-Stimulating Hormone (TSH) Test
13. Thyroid Surgery
14. Salivary Gland Scan
15. D-xylose Absorption Test
16. Stool Culture
17. Enterotest (Giardiasis String Test)
18. Stool Analysis
19. Fecal Occult Blood Test (FOBT)
20. Overnight Dexamethasone Suppression Test

Key Words

Bethanechol sensitivity test
Bladder scan
Bravo wireless
Cholecystokinin
Esophageal acidity test
Esophageal manometry
Free thyroxine (FT_4)
Function study
Malabsorption syndrome
Near-total thyroidectomy
Perfusion study
Radioactive iodine uptake (RAIU) test
Total thyroidectomy
Total thyroxine (T_4)
Triiodothyronine (T_3)
Whole-body thyroid scan

Upper Gastrointestinal Series 1

The upper gastrointestinal (UGI) series consists of a group of tests that assess the esophagus, stomach, and the duodenum. Prior to the series, the patient ingests barium contrast material and water. X-ray images of the esophagus, stomach, and duodenum are taken using a fluoroscope as the barium moves through the upper gastrointestinal tract. Images are displayed on a computer screen and stored for further review. If the healthcare provider sees anything suspicious, he might perform an endoscopy where an endoscope is inserted down the esophagus and into the stomach and duodenum to directly view the upper gastrointestinal tract.

Hint

The upper gastrointestinal series is also performed during a full gastrointestinal series, which also involves examination of the lower gastrointestinal tract.

WHAT IS BEING EXAMINED?

- The esophagus, stomach, and duodenum

HOW IS THE PROCEDURE PERFORMED?

- Twelve hours before the test, the patient is administered a laxative.
- The patient signs a consent form.

- The patient removes clothing.
- The patient lies on the back.
- An X-ray of the upper gastrointestinal tract is taken.
- The patient is sitting when drinking the barium contrast material.
- A series of X-rays are taken.
- The patient may be asked to drink additional amounts of barium contrast material during the test. After each swallow, additional X-rays are taken.

RATIONALE FOR THE TEST

- To assess the cause of stomach pain and indigestion
- To assess the cause of malabsorption syndrome

NURSING IMPLICATIONS

- Determine if the patient
  - Has signed a consent form
  - Has any allergies
  - Has been taking anticoagulant medications such as Plavix, Coumadin, heparin, or aspirin
  - Can lie on the back or stomach
  - Is able to lie still
  - Is able to follow instructions
  - Has eaten a low-fiber diet for 3 days before the test
  - Has not eaten or drunk 12 hours before the test
  - Has taken a laxative 12 hours before the test
  - Has removed dentures before the test
  - Is able to swallow the barium contrast material

UNDERSTANDING THE RESULTS

- The procedure takes less than 1 hour. Preliminary results are known immediately. Comprehensive results are ready within 2 weeks of the procedure.

- Normal test results indicate
  - Normal structure of the upper gastrointestinal tract
- Abnormal test results indicate
  - Abnormal structure of the upper gastrointestinal tract

TEACH THE PATIENT

- Explain
  - Why the procedure is being performed.
  - What the patient will experience during the procedure.
  - That the patient will not feel any pain during the procedure.
  - Should eat a low-fiber diet for 3 days before the test.
  - Must not eat or drink 12 hours before the test.
  - Should take a laxative 12 hours before the test.
  - Will be swallowing the barium contrast material several times during the test. Barium contrast material is sweet flavored.
  - The patient might feel bloated following the test.
  - The patient might feel nausea following the test.
  - Feces will appear white for 3 days following the test.
  - The healthcare provider may administer a stool softener following the test to prevent constipation from the barium contrast material.

CAUTION

The patient should call the healthcare provider if he/she has not had a bowel movement within 3 days following the test.

Esophagus Test Series 2

The esophagus test series consists of two tests that assess the esophagus and esophageal sphincters. These tests are

- Esophageal manometry: Measures esophageal muscle contractions
- Esophageal acidity test: Measures the pH of the esophagus

WHAT IS BEING EXAMINED?

- The esophagus

HOW IS THE PROCEDURE PERFORMED?

- The patient signs a consent form.
- The patient removes clothing.
- The patient lies on a table with head slightly raised.
- Esophageal manometry
  - The patient swallows a transducer that is attached to a tube.
  - The tube contains pressure-sensing holes.
  - The patient is asked to swallow several times during the test. Pressure in the esophagus is measured with each swallow.
  - A graph depicting the pressure is recorded and assessed by the healthcare provider.
- Esophageal acidity test (bravo wireless)
  - The patient swallows a capsule that contains a transmitter. The transmitter measures the acidity of the patient's stomach and transmits it to a receiver carried by the patient.
  - The patient keeps a diary of his/her activities.
  - The patient is asked to press a button on the pager when he/she experiences a symptom.
  - The transmitter passes through bowel movement within a week.
- Esophageal acidity test (nonwireless)
  - An anesthetic is sprayed in the back of patient's throat and in his/her nose.
  - A tube containing a probe that measures acidity is passed through the patient's nose and down the esophagus.
  - The probe remains in place for 24 hours as it records the acidity level in the esophagus while patient goes about normal daily activities.
  - The patient keeps a diary of his/her activities and symptoms.
  - The patient must avoid activities that may cause the monitor to become wet.
  - The probe is removed after 24 hours and the results are analyzed by the healthcare provider.

RATIONALE FOR THE TEST

- To assess the cause of gastroesophageal reflux disease (GERD)
- To assess the cause of chest pain

NURSING IMPLICATIONS

- Determine that the patient
  - Has signed a consent form
  - Has not eaten or drunk for 12 hours before the test
  - Has not smoked or ingested alcohol 24 hours before the test
  - Does not have esophageal varices
  - Does not have heart failure
  - Has been taking anticoagulant medications such as Plavix, Coumadin, heparin, or aspirin
  - Can lie on the back
  - Can swallow
  - Has not taken Prevacid, Prilosec, Protonix, Nexium, Aciphex, Zantac, Pepcid, Axid, Tagamet, or antacids prior to and during the test
  - Has not taken corticosteroids
  - Has not taken blood pressure medication
  - Has not taken medication to treat bladder and intestinal muscle spasms
  - Has not taken medication for Parkinson disease
  - Has not taken theophylline

UNDERSTANDING THE RESULTS

- The procedure takes less than 1 hour. The esophageal acidity test takes 24 hours. Results are known within 2 weeks of the test.
- Normal test results indicate
  - Normal structure and function of the esophagus or normal pH level
- Abnormal test results indicate
  - Abnormal structure or function of the esophagus or acidic pH level

TEACH THE PATIENT

- Explain
  - Why the procedure is being performed.
  - What the patient will experience during the procedure.
  - That the patient will not feel any pain during the procedure.
  - That the patient should not eat or drink for 12 hours before the test.
  - Should not have smoked or ingested alcohol 24 hours before the test.
  - Should not take Prevacid, Prilosec, Protonix, Nexium, Aciphex, Zantac, Pepcid, Axid, Tagamet, or antacids prior to and during the test.

Gallbladder Scan 3

A gallbladder scan assesses the function of the gallbladder and is used to identify blockages in the bile ducts. A radioactive tracer is injected into the patient's vein. The tracer is removed from the blood by the liver, which places and adds the tracer to bile that flows into the gallbladder and into the duodenum. A camera takes an image of the tracer as the tracer flows through the liver to the duodenum.

WHAT IS BEING EXAMINED?

- The gallbladder

HOW IS THE PROCEDURE PERFORMED?

- The patient signs a consent form.
- The patient removes clothing.
- The patient lies on a table.
- The injection site is cleaned.
- The patient is injected with the radioactive tracer.
- The camera is moved into position.
- Images are taken immediately, then every 10 minutes for 90 minutes.
- The patient is administered cholecystokinin to stimulate the gallbladder.
- The patient may be administered morphine sulfate to diagnose gallbladder inflammation.

- The patient should drink lots of fluid to flush the tracer from his body.
- The patient should flush the toilet quickly after urinating and defecating, since the tracer is excreted in urine and feces for 2 days following the test.

RATIONALE FOR THE TEST

- To assess the gallbladder
- To assess the cause of upper right abdominal pain

NURSING IMPLICATIONS

- Determine if the patient
  - Has signed a consent form
  - Has not eaten or drunk for 12 hours before the test
  - Has allergies
  - Is pregnant
  - Has recently had a barium test performed within 4 days prior to the test
  - Has been taking anticoagulant medications such as Plavix, Coumadin, heparin, or aspirin
  - Has taken Pepto-Bismol

UNDERSTANDING THE RESULTS

- The procedure takes less than 2 hours and results are known within 2 days.
- Normal test results indicate
  - Gallbladder is functioning normally.
  - Normal shape and size of the gallbladder.
- Abnormal test results indicate
  - Possible liver disease if the tracer does not flow into the gallbladder.
  - Gallbladder is inflamed or blocked.
  - The bile duct is narrowed or blocked.

TEACH THE PATIENT

- Explain
  - Why the procedure is being performed.
  - What the patient will experience during the procedure.

- That the patient will not feel any pain during the procedure except for a pinch when the tracer is injected into the vein.
- That the patient must sign a consent form.
- That the patient must not eat or drink for 12 hours before the test.
- That the patient should not take anticoagulant medications such as Plavix, Coumadin, heparin, or aspirin for 2 weeks prior to the test.
- That the patient should not take Pepto-Bismol for 48 hours before the test.
- That the patient should drink lots of fluids to flush the tracer following the test.
- That the patient should flush the toilet immediately after urinating or defecating.
- That the patient should wash hands thoroughly after urinating or defecating.
- That the patient should apply a warm compress to the injection site if there is swelling.
- If the patient is breast-feeding, breast milk must be discarded for 2 days after the scan.

Kidney Scan 4

A kidney scan assesses the function of the kidneys. A radioactive tracer is injected in the patient's vein. The tracer moves through the blood vessels in the kidneys. A camera takes images of the tracer as the tracer flows through the kidney, illustrating where blood flows unobstructed and where blood flow is blocked. There are two types of kidney scans. These are

- Function study: This measures the time that the tracer takes to pass through the kidneys and enter the bladder as part of urine.
- Perfusion study: This assesses blood flow through the kidneys.

Hint

A kidney scan is an alternative to the intravenous pyelogram (IVP) test.

WHAT IS BEING EXAMINED?

- The kidneys

HOW IS THE PROCEDURE PERFORMED?

- The patient signs a consent form.
- The patient removes clothing.
- Patients lie on a table, although they may be asked to sit or stand during the test.
- The injection site is cleaned.
- The patient is injected with the radioactive tracer.
- The camera is moved into position.
- Function study
  - Images are taken every 3 minutes for 30 minutes.
  - The patient may be administered a diuretic to increase kidney function.
- Perfusion study
  - Movement of the tracer through the kidneys is recorded on a renogram.
- The patient should drink lots of fluid to flush the tracer from body.
- The patient should flush the toilet quickly after urinating or defecating, since the tracer is excreted in urine and feces for 2 days following the test.

RATIONALE FOR THE TEST

- To assess blood flow through the kidneys
- To assess the function of the kidneys

NURSING IMPLICATIONS

- Determine if the patient
  - Has signed a consent form.
  - Has drunk three glasses of water before the test.
  - Can lie still.
  - Has allergies.
  - Is pregnant.
  - Has recently had a barium test performed within 4 days prior to the test.
  - Has been taken anticoagulant medications such as Plavix, Coumadin, heparin, or aspirin.

- Has taken Pepto-Bismol.
- Has taken antihypertensives.
- If the patient is breast-feeding, breast milk must be discarded for 2 days after the scan.

UNDERSTANDING THE RESULTS

- The procedure takes less than 2 hours. Preliminary results are known within 2 days.
- Normal test results indicate
  - Kidney is functioning normally
  - Normal kidney size and shape
- Abnormal test results indicate
  - Abnormal kidney function.
  - Abnormal kidney size and shape.
  - There is narrowing or blockage of blood vessels.

TEACH THE PATIENT

- Explain
  - Why the procedure is being performed.
  - What the patient will experience during the procedure.
  - That the patient will not feel any pain during the procedure except for a pinch when the tracer is injected into his vein.
  - That the patient must sign a consent form.
  - That the patient must drink three glasses of water before the test.
  - That the patient should not take anticoagulant medications such as Plavix, Coumadin, heparin, or aspirin for 2 weeks prior to the test.
  - That the patient should not take Pepto-Bismol for 48 hours before the test.
  - That the patient should drink lots of fluids to flush the tracer following the test.
  - That the patient flushes the toilet immediately after urinating or defecating.
  - That the patient should wash hands thoroughly after urinating or defecating.

- That the patient should apply a warm compress to the injection site if there is swelling.
- If the patient is breast-feeding, breast milk must be discarded for 2 days after the scan.

Liver and Spleen Scan 5

A liver and spleen scan assesses the function of the liver and spleen. A radioactive tracer is injected into the patient's vein. The tracer moves through the blood vessels in the liver and spleen. A camera takes an image of the tracer as it flows through the liver and spleen, illustrating where blood flows unobstructed and where blood flow is blocked.

WHAT IS BEING EXAMINED?

- The liver and spleen

HOW IS THE PROCEDURE PERFORMED?

- The patient signs a consent form.
- The patient will empty the bladder.
- The patient removes clothing.
- The patient lies on a table.
- The injection site is cleaned.
- The patient is injected with the radioactive tracer.
- The camera is moved into position.
- Several images are taken.
- The patient should drink lots of fluid to flush the tracer from body.
- The patient should flush the toilet quickly after urinating or defecating, since the tracer is excreted in urine and feces for 2 days following the test.

RATIONALE FOR THE TEST

- To assess blood flow through the liver and spleen
- To assess if cancer metastasized to the liver

- To assess the spleen after an injury
- To assess treatment for cancer

NURSING IMPLICATIONS

- Determine if the patient
  - Has signed a consent form.
  - Can lie still.
  - Has allergies.
  - Is pregnant.
  - Has recently had a barium test performed within 4 days prior to the test
  - Has been taking anticoagulant medication such as Plavix, Coumadin, heparin, or aspirin.
  - Has taken Pepto-Bismol.
  - Has emptied the bladder.
  - If the patient is breast-feeding, breast milk must be discarded for 2 days after the scan.

UNDERSTANDING THE RESULTS

- The procedure takes less than 2 hours. Results are known within 2 days.
- Normal test results indicate
  - Liver and spleen are functioning normally.
  - Normal size and shape of the liver and spleen.
- Abnormal test results indicate
  - Abnormal liver function
  - Abnormal spleen function
  - Abnormal size and shape of the spleen
  - Abnormal size and shape of the liver

TEACH THE PATIENT

- Explain
  - Why the procedure is being performed.
  - What the patient will experience during the procedure.

- That the patient will not feel any pain during the procedure except for a pinch when the tracer is injected into the vein.
- That the patient must sign a consent form.
- That the patient must have an empty bladder.
- That the patient should not take anticoagulant medications such as Plavix, Coumadin, heparin, or aspirin for 2 weeks prior to the test.
- That the patient should not take Pepto-Bismol for 48 hours before the test.
- That the patient should drink lots of fluids to flush the tracer following the test.
- That the patient should flush the toilet immediately after urinating or defecating.
- That the patient should wash hands thoroughly after urinating or defecating.
- That the patient should apply a warm compress to the injection site if there is swelling.
- If the patient is breast-feeding, breast milk must be discarded for 2 days after the scan.

Urinalysis 6

Waste material carried by blood is filtered by kidneys and excreted as urine. A urinalysis is performed to determine the characteristics of the urine and to determine the existence and amount of substances in the urine. Urine characteristics:

- Clarity: How clear is the urine?
- Color: What is the color of the urine?
- Specific gravity: The balance between water and substances in the urine.
- Odor: The aroma of urine.
- pH: How acidic or alkaline is the urine?

There are several methods used to capture the urine sample. These are

- Clean-catch, midstream, one-time urine collection: Urine is collected after the patient begins to urinate.
- Double-voided urine collection: Urine is collected the second time that the patient voids.
- 24-hour urine collection: Urine is collected over a 24-hour period.

WHAT IS BEING EXAMINED?

- Urine

HOW IS THE PROCEDURE PERFORMED?

- The patient washes hands.
- The patient opens the collection cup without touching the inside of the cup.
- Clean the urethral area with an antiseptic.
- Do not touch the cup to the urethra or any skin when collecting the sample.
- Do not contaminate the urine sample with feces, pubic hair, or other substances.
- If the sample becomes contaminated, then begin again with a new collection cup.
- Double-voided urine collection.
  - The patient urinates.
  - The patient drinks a lot of water.
  - The patient waits until he/she has the feeling to urinate.
- Urinate for 5 seconds making sure no skin aside from the urethra touches the urine.
- Move the collection cup into the urine stream.
- Do not touch the collection cup to the urethra or any skin.
- Remove the collection cup and continue urinating.
- Place the lid on the collection cup.
- Refrigerate the collection cup until the sample is sent to the laboratory.
- 24-hour urine collection
  - The patient empties bladder the first thing in the morning.
  - The patient collects urine the next and subsequent times that he/she urinates for the 24-hour period.
  - The patient must note the time when urinates. This begins the 24-hour period.
  - Urine is kept in a gallon container that must be refrigerated between collections.
  - The patient must note the last time when urinated at the end of the 24-hour period.
  - If patients urinate within the 24-hour period and do not collect the urine, then they must begin the 24-hour urine collection again.

RATIONALE FOR THE TEST

- To assess for kidney and other disorders

NURSING IMPLICATIONS

- Determine if the patient
  - Has collected the urine properly
  - Has exercised prior to the test
  - Is menstruating
  - Has eaten blackberries, beets, or foods that might color urine
  - Has been exposed to sun
  - Is on extended bed rest
  - Has recently undergone a test that used contrast material
  - Has ingested nicotine or caffeine
  - Has had a recent illness or is under stress
  - Keeps warm during testing if testing catecholamine levels
  - Is pregnant
  - Has had severe vomiting
  - Has taken Dilantin, Pyridium, vitamin B, diuretics, rifampin, Trimpex, ascorbic acid, Probalan, Benemid, allopurinol, insulin, lithium, laxatives, steroids, antacids, NSAIDs, antibiotics, potassium supplements, growth hormones or parathyroid hormone, aspirin, antidepressants (including tricyclic), nitroglycerin, tetracycline, theophylline, cold and sinus medication
  - Has recently taken sodium-based medication

UNDERSTANDING THE RESULTS

- The collection takes a few seconds except for the 24-hour collection. Results are known within 2 days.
- The laboratory determines normal values based on calibration of testing equipment with a control test. Test results are reported as high, normal, or low based on the laboratory's control test.
- Normal
  - Color: pale to dark amber
  - Clarity: clear

    - Odor: Nutty
    - Specific gravity: 1.005 to 1.030
    - pH: 4.6 to 8.0
    - Protein: None
    - Glucose: None
    - Ketones: None
    - Leukocytes: None
    - Erythrocytes: None
    - Casts: None
    - Microorganism: None
    - Crystals: Slight
    - Squamous cells: None
    - Uric acid: 250 to 800 mg (24-hour urine collection)
    - Calcium: 100 to 250 mg (24-hour urine collection)
    - Catecholamines: < 100 mcg (24-hour urine collection)
    - Epinephrine: < 20 mcg (24-hour urine collection)
    - Norepinephrine: < 100 mcg (24-hour urine collection)
    - Dopamine: 65 to 400 mcg (24-hour urine collection)
    - Normetanephrine: 10 to 80 mcg (24-hour urine collection)
    - Metanephrine: < 1.3 mg (24-hour urine collection)
    - Vanillylmandelic acid: < 6.8 mg (24-hour urine collection)
    - Cortisol: < 100 mcg (24-hour urine collection)
    - Phosphate: 0.9 to 1.3 g (24-hour urine collection)
    - Potassium: 25 to 100 mEq/L (24-hour urine collection)
    - Sodium: 40 to 220 mEq (24-hour urine collection)
    - Microalbumin: < 30 mg (24-hour urine collection)
- Abnormal
    - Color: Clear (chronic kidney disease, diabetes), dark (dehydration), or red (blood)
    - Clarity: Cloudy (microorganism)
    - Odor: Fruity (diabetes), foul (urinary tract infection), maple syrup (maple syrup urine disease)

- Specific gravity: < 1.005 (overhydration, kidney disease, diuretics) or > 1.030 (dehydration, hyperglycemia, vomiting, diarrhea)
- pH: < 4.6 (diabetes, dehydration, excess alcohol, aspirin overdose) or > 8.0 (vomiting, urinary tract infection, kidney disease)
- Protein: Present (infection, high blood pressure, kidney disorder, glomerulonephritis)
- Glucose: Present (diabetes, liver disease, kidney disorder, adrenal gland disorder)
- Ketones: Present (diabetes, starvation, anorexia, bulimia, alcoholism)
- Leukocytes: Present (urinary tract infection, glomerulonephritis, kidney stones, kidney disorder)
- Erythrocytes: Present (urinary tract infection, glomerulonephritis, kidney stones, kidney disorder)
- Casts: Present (inflammation, kidney disorder, lead poisoning, heart failure)
- Microorganism: Present (urinary tract infection)
- Crystals: Large amount (kidney stone, kidney disorder)
- Squamous cells: Present (contaminated urine sample)
- Uric acid: < 250 or > 800 mg (24-hour urine collection) (kidney stones, gout)
- Calcium: < 100 or > 250 mg (24-hour urine collection) (hyperparathyroidism, osteoporosis, excess vitamin D, excess dietary calcium, dehydration)
- Catecholamines: > 100 mcg (24-hour urine collection) (pheochromocytoma)
- Epinephrine: > 20 mcg (24-hour urine collection) (infection, stress, trauma)
- Norepinephrine: > 100 mcg (24-hour urine collection) (infection, stress, trauma)
- Dopamine: < 65 or > 400 mcg (24-hour urine collection) (infection, stress, trauma)
- Normetanephrine: < 10 or > 80 mcg (24-hour urine collection) (infection, stress, trauma)
- Metanephrine: > 1.3 mg (24-hour urine collection) (pheochromocytoma)
- Vanillylmandelic acid: > 6.8 mg (24-hour urine collection) (pheochromocytoma)

- Cortisol: > 100 mcg (24-hour urine collection) (Cushing syndrome)
- Phosphate: < 0.9 or > 1.3 g (24-hour urine collection) (kidney disorder, hyperparathyroidism, osteomalacia)
- Potassium: < 25 or > 100 mEq/L (24-hour urine collection) (kidney disorder)
- Sodium: < 40 or > 220 mEq (24-hour urine collection) (indicate high blood pressure, heart disease, or kidney disorder)
- Microalbumin: > 30 mg (24-hour urine collection) (kidney disorder or diabetic nephropathy)

TEACH THE PATIENT

- Explain
  - Why the test is being performed.
  - What the patient will experience during the test.
  - That the patient will not feel any pain during the test.
  - How to perform the test.
  - That the patient should not exercise prior to the test.
  - That the patient should not take the test if she is menstruating.
  - That the patient should avoid eating blackberries, beets, or foods that might color urine.
  - That the patient should avoid usual exposure to sun.
  - That the patient should avoid extended bed rest.
  - That the patient should avoid ingesting nicotine or caffeine.
  - That the patient should keep warm during the test if catecholamine levels are being tested.
  - That the patient should avoid taking Dilantin, Pyridium, vitamin B, diuretics, rifampin, Trimpex, ascorbic acid, Probalan, Benemid, allopurinol, insulin, lithium, laxatives, steroids, antacids, NSAIDs, antibiotics, potassium supplements, growth hormones or parathyroid hormone, aspirin, antidepressants (including tricyclic), nitroglycerin, tetracycline, tricyclic, theophylline, cold and sinus medications, 2 weeks before the test.
  - That the patient should avoid taking sodium-based medication 2 weeks before the test.

Urine Culture and Sensitivity Test 7

A urine culture is ordered when the patient is suspected of having a urinary tract infection. A urine collection is placed in an environment conducive to the growth of microorganisms for 3 days. The urine is examined to identify the presence and the type of microorganism. Once the microorganism is identified, a sensitivity test is performed to determine the medication that kills the microorganism.

WHAT IS BEING EXAMINED?

- Existence of microorganisms in urine

HOW IS THE PROCEDURE PERFORMED?

- The patient washes hands.
- The patient opens the collection cup without touching the inside of the cup.
- Clean the urethral area with an antiseptic.
- Do not touch the cup to the urethra or any skin when collecting the sample.
- Do not contaminate the urine sample with feces, pubic hair, or other substances.
- If the sample becomes contaminated, then begin again with a new collection cup.
- Urinate for 5 seconds making sure no skin aside from the urethra touches the urine.
- Move the collection cup into the urine stream.
- Do not touch the collection cup to the urethra or any skin.
- Remove the collection cup and continue urinating.
- Place the lid on the collection cup.
- Refrigerate the collection cup until the sample is sent to the lab.

HINT

A sterile sample may be taken from a Foley catheter.

RATIONALE FOR THE TEST

- To assess the existence and type of microorganism in a patient with a urinary tract infection.
- To assess the medication to use to treat the urinary tract infection.

NURSING IMPLICATIONS

- Determine if the patient
  - Has collected the urine properly
  - Has taken antibiotics, vitamin C, or diuretics

UNDERSTANDING THE RESULTS

- The collection takes a few seconds. Results are known within 3 days.
- Normal (negative) test results indicate
  - No microorganism is present in the urine collection.
- Abnormal (positive) test results indicate
  - Microorganisms are present in the urine collection.

TEACH THE PATIENT

- Explain
  - Why the test is being performed.
  - What the patient will experience during the test.
  - That the patient will not feel any pain during the test.
  - How to perform the test.
  - Should not take antibiotics, vitamin C, or diuretics before the test.

Renin Assay Test 8

The renin assay test is performed along with the aldosterone test to determine the underlying cause of hypertension. Renin is an enzyme produced by the kidneys. Aldosterone is a hormone produced by the adrenal glands. Together these work to balance the sodium and potassium levels within the patient. A high renin level might indicate a kidney disorder. A low renin level might indicate Conn syndrome. A low renin level and a high aldosterone level might indicate an adrenal gland tumor.

WHAT IS BEING EXAMINED?

- Renin level in blood

HOW IS THE PROCEDURE PERFORMED?

- See How to Collect Blood Specimen from a Vein in Chapter 1.

RATIONALE FOR THE TEST

- To assess the underlying cause of hypertension

NURSING IMPLICATIONS

- Determine that the patient
  - Has not taken β-blockers, ACE inhibitors, diuretics, aspirin, corticosteroids. and estrogen 4 weeks prior to the test
  - Has not eaten natural black licorice for 2 weeks prior to the test
  - Has not ingested caffeine 24 hours prior to the test
  - Has eaten a low-sodium diet 3 days prior to the test
  - Has not eaten or drunk 8 hours prior to the test
  - Is not pregnant

UNDERSTANDING THE RESULTS

- Test results are available quickly. The laboratory determines normal values based on calibration of testing equipment with a control test. Test results are reported as high, normal, or low based on the laboratory's control test.
- Normal test results indicate
  - Normal levels of renin and aldosterone
- Abnormal test results indicate
  - High renin levels might indicate a kidney disorder.
  - Low renin levels might indicate Conn syndrome.
  - Low renin levels and high aldosterone levels might indicate an adrenal gland tumor.

TEACH THE PATIENT

- Explain
  - Why the test is being performed.
  - What the patient will experience during the test.

- That the patient will not feel any pain during the test except a pinch when the blood sample is taken.
- That the patient should not take β-blockers, ACE inhibitors, diuretics, aspirin, corticosteroids, and estrogen 4 weeks prior to the test.
- That the patient should not eat natural black licorice for 2 weeks prior to the test.
- That the patient should not ingest caffeine 24 hours prior to the test.
- That the patient should eat a low-sodium diet 3 days prior to the test.
- That the patient should not eat or drink 8 hours prior to the test.

Thyroid Scan 9

A thyroid scan assesses the function of the thyroid gland. There are two types of thyroid scans. These are

- Radioactive iodine uptake (RAIU) test: This assesses the absorption of a radioactive tracer by the thyroid gland.
- Whole-body thyroid scan: This test assesses whether or not thyroid cancer has metastasized.

WHAT IS BEING EXAMINED?

- The thyroid

HOW IS THE PROCEDURE PERFORMED?

- A blood sample is taken to measure the amount of thyroid hormones in the blood.
- The patient signs a consent form.
- The patient removes clothing above the waist.
- The patient removes dentures.
- The patient removes jewelry from the upper part of the body.
- The patient either swallows the radioactive tracer 24 hours before the test or is administered technetium 2 hours prior to the test.
- The patient lies on a table.

- The table is tipped backward.
- The gamma scintillation camera takes images of the thyroid gland.
- Images are also taken again 24 hours after the initial images were taken.
- The patient should drink lots of fluid to flush the tracer from body.
- The patient should discard breast milk for 2 days following the test if she is breast-feeding.
- The patient should flush the toilet quickly after urinating or defecating, since the tracer is excreted through urine and feces for 2 days following the test.

RATIONALE FOR THE TEST

- To assess the thyroid function
- To assess the treatment for thyroid disease

NURSING IMPLICATIONS

- Determine that the patient
  - Has signed a consent form
  - Has not eaten or drunk 2 hours before the test
  - Has not taken antithyroid medication, thyroid hormones, Cordarone, Pacerone, iodine, or kelp a week before the test
  - Has eaten a low-iodine diet
  - Has allergies
  - Is not pregnant
  - Is breast-feeding
  - Has recently had a radioactive iodine test performed within 4 weeks prior to the test
  - Has been taking anticoagulant medications such as Plavix, Coumadin, heparin, or aspirin

UNDERSTANDING THE RESULTS

- The procedure takes less than 1 hour. Results are known within a week.
- Normal test results indicate
  - Thyroid is functioning normally.
  - Normal shape and size of the thyroid.

- Abnormal test results indicate
  - Abnormal shape or size of the thyroid.
  - Nodules are found on the thyroid.
  - Iodine is in other tissues indicating that thyroid cancer may have metastasized.

TEACH THE PATIENT

- Explain
  - Why the test is being performed.
  - What the patient will experience during the test.
  - That the patient will not feel any pain during the procedure except for a pinch when the tracer is injected into the vein.
  - That the patient may feel flushed when technetium is injected, if it is administered.
  - The patient swallows the capsule with water that contains the tracer if the capsule is used.
  - Must sign a consent form.
  - Must not eat or drink for 2 hours before the test.
  - Should not take anticoagulant medications such as Plavix, Coumadin, heparin, or aspirin for 2 weeks prior to the test.
  - Drink lots of fluids to flush the tracer following the test.
  - Flush the toilet immediately after urinating or defecating.
  - Wash hands thoroughly after urinating or defecating.
  - If the patient is breast-feeding, breast milk must be discarded for 2 days after the scan.
  - Should not take antithyroid medication, thyroid hormones, Cordarone, Pacerone, iodine, or kelp a week before the test.
  - Should eat a low-iodine diet.

Thyroid and Parathyroid Ultrasound 10

The thyroid and parathyroid ultrasound is used to assess the size and shape of the thyroid gland and the parathyroid glands, which are located behind the thyroid gland.

Hint

This test is not used to assess the function of these glands.

WHAT IS BEING EXAMINED?

- The thyroid and parathyroid glands

HOW IS THE PROCEDURE PERFORMED?

- The patient signs a consent form.
- The patient removes clothing above the waist.
- The patient lies on a table with neck stretched.
- Conductive gel is placed on the patient's neck.
- A transducer is moved over the thyroid gland.
- Images are displayed on a computer screen and saved in a computer.
- The gel is wiped from the patient's neck.

RATIONALE FOR THE TEST

- To assess the size and shape of the thyroid and parathyroid glands.

NURSING IMPLICATIONS

- Determine if the patient
  - Has signed a consent form
  - Removed jewelry from the neck
  - Removed his clothing above the waist
  - Is able to lie still

UNDERSTANDING THE RESULTS

- The procedure takes less than 1 hour. Results are known within a week.
- Normal test results indicate
  - Normal shape and size of the thyroid
  - Normal shape and size of the parathyroid glands

- Abnormal test results indicate
  - Abnormal shape or size of the thyroid
  - Abnormal shape or size of the parathyroid glands

TEACH THE PATIENT

- Explain
  - Why the test is being performed.
  - What the patient will experience during the test.
  - That the patient will not feel any pain during the test.
  - That the patient must sign a consent form.

Thyroid Hormone Tests 11

The thyroid gland produces two hormones. These are thyroxine (T_4) and triiodothyronine (T_3). The thyroid hormone tests measure the level of thyroid hormones in the patient's blood. There are three thyroid hormone tests. These are

- Free thyroxine (FT_4): This test determines the amount of thyroxine that is not bound to globulin.
- Total thyroxine (T_4): This test determines the total amount of thyroxine that is attached to globulin and that is not bound to globulin.
- Triiodothyronine (T_3): This test determines the total amount of triiodothyronine that is attached to globulin and not bound to globulin.

WHAT IS BEING EXAMINED?

- The thyroid hormone level in blood

HOW IS THE PROCEDURE PERFORMED?

- See How to Collect Blood Specimen from a Vein in Chapter 1.

RATIONALE FOR THE TEST

- To assess for hyperthyroidism and hypothyroidism

- To assess for the cause of abnormal thyroid-stimulating hormone (TSH) test results
- To assess for treatment of hyperthyroidism and hypothyroidism

NURSING IMPLICATIONS

- Determine that the patient
  - Has not taken birth control pills, corticosteroids, estrogen, Dilantin, Tegretol, amiodarone, propranolol, or lithium 4 weeks prior to the test
  - Has not been taking anticoagulant medications such as Plavix, Coumadin, heparin, or aspirin
  - Is not pregnant
  - Has not undergone a test that used contrast material 4 weeks prior to the test

UNDERSTANDING THE RESULTS

- Test results are available quickly. The laboratory determines normal values based on calibration of testing equipment with a control test. Test results are reported as high, normal, or low based on the laboratory's control test.
- Normal test results indicate
  - Normal level of thyroxine (T_4) and triiodothyronine (T_3)
- Abnormal test results indicate
  - High levels might indicate hyperthyroidism, Graves disease, goiter, thyroiditis, excess intake of thyroid medication.
  - Low levels might indicate hypothyroidism, pituitary gland disorder, thyroiditis.

TEACH THE PATIENT

- Explain
  - Why the test is being performed.
  - What the patient will experience during the test.
  - That the patient will not feel any pain during the test except a pinch when the blood sample is taken.
  - That the patient should not take birth control pills, corticosteroids, estrogen, Dilantin, Tegretol, amiodarone, propranolol, lithium, or anticoagulant medications such as Plavix, Coumadin, heparin, or aspirin 4 weeks prior to the test.

Thyroid-Stimulating Hormone Test 12

The hypothalamus produces thyrotropin-releasing hormone (TRH), which causes the pituitary gland to produce the thyroid-stimulating hormone (TSH), causing the thyroid to produce thyroid hormones. The TSH test determines the underlying cause of thyroid disorder.

WHAT IS BEING EXAMINED?

- The TSH level in blood

HOW IS THE PROCEDURE PERFORMED?

- See How to Collect Blood Specimen from a Vein in Chapter 1.

RATIONALE FOR THE TEST

- Assess the underlying cause of hyperthyroidism and hypothyroidism.

NURSING IMPLICATIONS

- Determine that the patient
  - Has not taken lithium, Tapazole, propylthiouracil, or corticosteroids 4 weeks prior to the test
  - Has not taken anticoagulant medication such as Plavix, Coumadin, heparin, or aspirin
  - Is not pregnant
  - Has not undergone a test that used contrast material 4 weeks prior to the test

UNDERSTANDING THE RESULTS

- Test results are available quickly. The laboratory determines normal values based on calibration of testing equipment with a control test. Test results are reported as high, normal, or low based on the laboratory's control test.
- Normal test results indicate
  - Normal level TSH

- Abnormal test results indicate
  - High levels might indicate hypothyroidism, Hashimoto thyroiditis, pituitary gland tumor, insufficient dose of thyroid hormone.
  - Low levels might indicate hyperthyroidism, Graves disease, goiter, pituitary gland disorder, excess dose of thyroid hormone.

TEACH THE PATIENT

- Explain
  - Why the test is being performed.
  - What the patient will experience during the test.
  - That the patient will not feel any pain during the test except a pinch when the blood sample is taken.
  - That the patient should not take lithium, Tapazole, propylthiouracil, or corticosteroids 4 weeks prior to the test.

Thyroid Surgery 13

Thyroid surgery is performed to remove a portion or all of the thyroid gland to treat thyroid cancer or to remove a benign nodule that is interfering with the patient's swallowing or breathing. There are three types of thyroid surgical procedures. These are

- Near-total thyroidectomy: One lobe, the isthmus, and part of the other lobe of the thyroid gland are removed to treat Graves disease.
- Total thyroidectomy: The whole thyroid gland is removed in addition to neighboring lymph nodes to treat thyroid cancer.
- Thyroid lobectomy: A lobe of the thyroid gland is removed. The healthcare provider may or may not remove the isthmus that connects the lobes of the thyroid gland. This is commonly performed to assess the lobe for cancer.

WHAT IS BEING EXAMINED?

- Surgical removal of part or all of the thyroid gland

HOW IS THE PROCEDURE PERFORMED?

- The patient signs a consent form.

- The patient removes clothing.
- The patient will lie on a table with head tipped backward.
- An intravenous line is inserted into the patient's arm.
- The patient is administered a general anesthetic.
- A breathing tube is placed down the patient's throat.
- The patient is connected to an ECG during the procedure.
- The patient is connected to a pulse oximeter.
- The insertion site is cleaned with an anesthetic.
- An incision is made in the skin above the thyroid gland.
- Part or all of the thyroid is removed.
- A drain might be inserted into the neck and connected to a collection device. The drain is emptied frequently and removed once drainage has stopped.
- The incision is closed with stitches and the site is covered with colloidin, which is a clear protective waterproof glue.

RATIONALE FOR THE TEST

- To assess for cancerous tissues
- To remove a cancerous thyroid gland
- To treat Graves disease

NURSING IMPLICATIONS

- Determine if the patient
  - Has signed a consent form
  - Removed dentures
  - Has any allergies
  - Has been taking anticoagulant medication such as Plavix, Coumadin, heparin, or aspirin
  - Is pregnant
  - Has recently had a radioactive iodine test performed within 4 weeks prior to the surgery
  - Can lie on the back
  - Is able to lie still
  - Has eaten or drunk 12 hours before the surgery

UNDERSTANDING THE RESULTS

- The procedure takes less than 1 hour. Results are ready within 2 weeks of the procedure.
- Normal test results indicate
  - Thyroid is removed
- Abnormal test results indicate
  - Thyroid tissue was cancerous

CAUTION

A normal result does not mean that the patient is cancer free. It means that no cancerous cells were found in the tissue sample.

TEACH THE PATIENT

- Explain
  - Why the procedure is being performed.
  - What the patient will experience during the procedure.
  - That the patient will not feel any pain during the procedure except for a pinch from the needle used to the insert the intravenous saline lock.
  - That the patient cannot eat 12 hours before the surgery.
  - The incision site may be sore for 3 days following the surgery.
  - A bruise may develop around the site.
  - That the patient should avoid any stretching or pulling of neck muscles for 24 hours after the procedure.
  - The patient should gargle with warm salt water to ease a sore throat that follows the surgery.
  - The patient's neck will feel tender, stiff, and swollen following the procedure. Tenderness resolves in a week. The swelling resolve in 8 weeks.
  - The patient might have a small scar on the neck.
  - The patient must sign a consent form.
  - The patient must remove dentures.
  - The patient must stop taking anticoagulant medication such as Plavix, Coumadin, heparin, or aspirin.

- The patient might be required to undergo lifetime thyroid hormone replacement therapy and calcium replacement therapy, depending on the surgical procedure.
- Support the back of patient's neck when raising the head.

CAUTION

The patient should call the healthcare provider if he/she has fever, experiences redness and drainage from the incision site after the drain has been removed, or has pain a week or more following the procedure. The patient should also call the healthcare provider if experiencing muscle cramps, hoarseness, tingling around the lips, hands, and feet, which are signs of hypocalcemia.

Salivary Gland Scan 14

A salivary gland scan assesses the function of the salivary glands to determine the underlying cause of xerostomia (dry mouth) or swelling.

WHAT IS BEING EXAMINED?

- The salivary glands

HOW IS THE PROCEDURE PERFORMED?

- The patient signs a consent form.
- The patient removes clothing above the waist.
- The patient removes dentures.
- The patient removes jewelry from the upper part of the body.
- A radioactive tracer is injected into the patient's vein.
- The patient lies on a table.
- The table is tipped backward.
- The gamma scintillation camera takes images of the salivary glands.
- The patient sucks a lemon, which normally causes the glands to release more saliva.
- The patient should drink lots of fluid to flush the tracer from the body.

- The patient should discard breast milk for 2 days following the test if she is breast-feeding.
- The patient should flush the toilet quickly after urinating or defecating, since the tracer is excreted in urine and feces for 2 days following the test.

RATIONALE FOR THE TEST

- To assess the salivary glands function
- To assess the underlying cause of swollen salivary glands
- To assess the underlying cause of dry mouth

NURSING IMPLICATIONS

- Determine if the patient
  - Has signed a consent form
  - Has allergies
  - Is pregnant
  - Is breast-feeding
  - Has recently had a radioactive iodine test performed within 4 weeks prior to the test
  - Has been taking anticoagulant medications such as Plavix, Coumadin, heparin, or aspirin

UNDERSTANDING THE RESULTS

- The procedure takes less than 1 hour. Results are known within a week.
- Normal test results indicate
  - Salivary glands are functioning normal.
- Abnormal test results indicate
  - Abnormal shape or size of the salivary glands.
  - Nodules are found on the salivary glands.
  - Blockage is found on the salivary glands.

TEACH THE PATIENT

- Explain
  - Why the test is being performed.

- What the patient will experience during the test.
- That the patient will not feel any pain during the procedure except for a pinch when the tracer is injected into the vein.
- The patient may feel flushed when tracer is injected.
- The patient must sign a consent form.
- The patient should not take anticoagulant medications such as Plavix, Coumadin, heparin, or aspirin for 2 weeks prior to the test.
- The patient should drink lots of fluids to flush the tracer following the test.
- The patient should flush the toilet immediately after urinating or defecating.
- The patient should wash hands thoroughly after urinating and defecating.
- If the patient is breast-feeding, breast milk must be discarded for 2 days after the scan.

D-xylose Absorption Test 15

Patients with malabsorption syndrome are unable to absorb certain nutrients into their blood from the intestinal tract. This can result in malnutrition and chronic diarrhea. The D-xylose absorption test assesses whether or not these signs are a result of malabsorption syndrome by asking the patient to drink a solution of D-xylose and then measuring the amount of D-xylose in the patient's blood and urine.

WHAT IS BEING MEASURED?

- Absorption of D-xylose

HOW IS THE TEST PERFORMED?

- A sample of urine and blood are taken.
- The patient drinks a solution containing D-xylose.
- A blood sample is taken 2 hours after the patient drinks the solution.
- A 24-hour urine sample is collected once the patient drinks the solution.
- Urine and blood samples are then sent to the laboratory for analysis.

RATIONALE FOR THE TEST

- Assess the intestine's ability to absorb D-xylose.

NURSING IMPLICATIONS

- Determine if the patient
  - Has eaten fruits, jellies, pastries, or other foods high in pentose for 24 hours prior to the test
  - Has taken aspirin, indomethacin, cardiac medication, or antibiotics
  - Has eaten or drunk, except for water, for 12 hours prior to the test
  - Has emptied the bladder prior to the test
  - Has an infection
  - Has rested prior to and during the test
  - Has a digestive disorder that reduces the time it takes the stomach to empty

UNDERSTANDING THE RESULTS

- The results are available quickly. The laboratory determines normal values based on calibration of testing equipment with a control test. Test results are reported as high, normal, or low based on the laboratory's control test.
- Normal test results indicate
  - The patient's intestines absorb D-xylose without a problem.
- Low value test results indicate
  - Malabsorption syndrome
  - Celiac disease
  - Whipple disease
  - Crohn disease
- High value results indicate
  - Hodgkin disease
  - Scleroderma

TEACH THE PATIENT

- Explain
  - Why the test is administered.
  - How the test is performed.
  - How to collect a 24-hour urine sample.
  - That the patient will not feel pain except a pinch when blood samples are taken.
  - Should not eat fruits, jellies, pastries, or other foods high in pentose for 24 hours prior to the test.
  - Should not take aspirin, indomethacin, cardiac medication, or antibiotics.
  - Should not eat or drink, except for water, 12 hours prior the test.
  - Should empty his bladder prior to the test.
  - Should rest prior to and during the test.

Stool Culture 16

A patient may exhibit diarrhea and other signs of an infection. The healthcare provider orders a stool culture to determine if the underlying cause is a microorganism. A sample of the patient's stool is sent to the laboratory where it is placed in an environment that encourages microorganisms to grow. After 3 days, laboratory technicians determine if a microorganism is present and if so, which microorganism.

HINT

The healthcare provider typically orders a sensitivity test of the sample along with the stool culture. The sensitivity test determines the medication that kills the microorganism.

WHAT IS BEING MEASURED?

- Microorganism in a stool sample

HOW IS THE TEST PERFORMED?

- Urinate prior to defecating.

- Do not mix urine and stool.
- The patient is asked to defecate in a basin in the toilet or bedpan. Do not defecate in the toilet, since this might contaminate the sample.
- Place on clean gloves.
- A sample of stool is placed in a clean container. It is critical that the stool sample is not contaminated with hair, toilet paper, and other elements that might produce a false-positive test result.
- Cap and label the container.
- Wash hands immediately to prevent the spread of the microorganism.
- The container is immediately sent to the laboratory.

RATIONALE FOR THE TEST

- Assess if there is a microorganism in the stool sample.
  - Identify the microorganism in the stool sample.

NURSING IMPLICATIONS

- Determine if the patient
  - Has her menstrual period
  - Has taken antibiotics or medication to treat an infection
  - Has had a test that used contrast material within 10 days of the test
  - Has contaminated the stool sample

UNDERSTANDING THE RESULTS

- The results are available within 3 days.
- Normal test results indicate
  - No disease causing microorganism
- Abnormal test results indicate
  - Disease causing microorganism identified
- Indeterminate
  - Insufficient sample size
  - Contaminated sample

TEACH THE PATIENT

- Explain
  - Why the test is administered.
  - How the test is performed.
  - How to collect the stool sample.
  - Not to take the sample during her menstrual period.
  - Not to take antibiotics or medication to treat an infection.
  - The patient will not feel pain.

Enterotest (Giardiasis String Test) 17

A patient who has severe diarrhea might have giardiasis. Giardiasis is caused by an intestinal parasite called *Giardia intestinalis,* which is found in water, food, or soil that are contaminated with feces. The enterotest determines if the patient has giardiasis by sampling fluid in the duodenum. The patient swallows a gelatin capsule that is attached to a string. The string is taped to the outside of the patient's mouth while the capsule dissolves in the stomach and the duodenum. The string is then removed and examined under a microscope.

WHAT IS BEING MEASURED?

- Fluid in the duodenum

HOW IS THE TEST PERFORMED?

- The patient swallows a gelatin capsule that is attached to a string.
- The string is taped to the outside of the patient's cheek.
- The patient waits 6 hours.
- The string is removed and sent to the laboratory for microscopic analysis to determine the presence of *G. intestinalis.*

RATIONALE FOR THE TEST

- Assess for the presence of *G. intestinalis.*

NURSING IMPLICATIONS

- Determine if the patient
  - Can swallow
  - Is able to have the string attached to the outside of his mouth

UNDERSTANDING THE RESULTS

- The result is available within 3 days.
- Normal test results indicate
  - No *G intestinalis* is present.
- Abnormal test results indicate
  - *Giardia intestinalis* is present.

TEACH THE PATIENT

- Explain
  - Why the test is administered.
  - How the test is performed.
  - How to collect the stool sample.
  - The patient should not take antibiotics or medication to treat an infection.
  - That the patient will not feel pain although the string might feel uncomfortable.

Stool Analysis 18

Stool analysis is the examination of the patient's feces to identify digestive tract disorders. The patient's stool sample is examined for color, volume, consistency, odor, and the presence of blood, fat, mucus, fiber, bile, and glucose.

WHAT IS BEING MEASURED?

- Stool sample

HOW IS THE TEST PERFORMED?

- Urinate prior to defecating.
- Do not mix urine and stool.

- The patient is asked to defecate in a basin in the toilet or bedpan. Do not defecate in the toilet, since this might contaminate the sample.
- Place on clean gloves.
- A sample of stool is placed in a clean container. It is critical that the stool sample isn't contaminated with hair, toilet paper, and other elements that might produce a false-positive test result.
- Cap and label the container.
- Wash hands immediately to prevent the spread of the microorganism.
- The container is immediately sent to the laboratory.

RATIONALE FOR THE TEST

- Assess
  - Digestive tract disorder
  - Liver disorder
  - Pancreatic disorder
  - Colon cancer
  - Absorption disorder

NURSING IMPLICATIONS

- Determine if the patient
  - Has her menstrual period
  - Has taken aspirin, Aleve, ibuprofen, antacids, NSAIDs, laxatives, antidiarrheal medication, antibiotics, or medication to treat an infection for 2 weeks prior to the test
  - Has had a test that used contrast material within 10 days of the test.
  - Has contaminated the stool sample
  - Has eaten cauliflower, bananas, red meat, cantaloupe, parsnips, turnips, or beets prior to the test
  - Has taken vitamin C prior to the test
  - Has drunk alcohol prior to the test
  - Has had an enema prior to the test

UNDERSTANDING THE RESULTS

- The results are available within 3 days.
- Normal test results indicate
  - Stool is well-formed, soft, and brown
  - No mucus or blood in the stool
  - Minimum amount of glucose
- Abnormal test results indicate
  - Stool is loose, hard, or a color other than brown
  - Mucus or blood in the stool
  - Significant amount of glucose, fat, trypsin, or elastase in stool

TEACH THE PATIENT

- Explain
  - Why the test is administered.
  - How the test is performed.
  - How to collect the stool sample.
  - Not to take the sample during her menstrual period.
  - Not to take aspirin, Aleve, ibuprofen, antacids, NSAIDs, laxatives, antidiarrheal medication, antibiotics, or medication to treat an infection for 2 weeks prior to the test.
  - Not to eat cauliflower, bananas, red meat, cantaloupe, parsnips, turnips, or beets prior to the test.
  - Not to take vitamin C prior to the test.
  - Not to drink alcohol prior to the test.
  - Not to have an enema prior to the test.
  - The patient will not feel pain.

Fecal Occult Blood Test 19

Blood in the stool is not always visible. The fecal occult blood test (FOBT) examines the stool for blood that is not visible to the naked eye. This is referred to as occult blood. Although the presence of occult blood is linked to colon cancer, there are many other causes of occult blood in the stool.

WHAT IS BEING MEASURED?

- Occult blood in the stool

HOW IS THE TEST PERFORMED?

- Urinate prior to defecating.
- Do not mix urine and stool.
- The patient is asked to defecate in a basin in the toilet or bedpan. Do not defecate in the toilet, since this might contaminate the sample.
- Place on clean gloves.
- A sample of stool is placed on a card that is chemically treated. It is critical that the stool sample is not contaminated with hair, toilet paper, and other elements that might produce a false-positive test result.
- A reaction agent is placed on the stool sample. If the sample turns blue, then there is blood in the stool sample.
- Wash hands immediately.

HINT

The healthcare provider may collect a stool sample during a digital rectum examination rather than ask the patient to defecate.

RATIONALE FOR THE TEST

- Assess for blood in the stool.

NURSING IMPLICATIONS

- Determine if the patient
  - Has her menstrual period
  - Has contaminated the stool sample
  - Has been taking anticoagulant medications such as Plavix, Coumadin, aspirin, or heparin.

UNDERSTANDING THE RESULTS

- The results are available immediately.
- Normal test results indicate
  - No blood in the stool sample

- Abnormal test results indicate
    - Anal fissures
    - Peptic ulcer
    - Polyps
    - Hemorrhoids
    - Gastroesophageal reflux disease
    - Ulcerative colitis
    - Use of anticoagulation medication
    - Crohn disease

TEACH THE PATIENT

- Explain
    - Why the test is administered.
    - How the test is performed.
    - How to collect the stool sample.
    - Not to take the sample during her menstrual period.
    - Stop taking anticoagulant medications such as Plavix, Coumadin, aspirin, or heparin.
    - The patient will not feel pain.

Overnight Dexamethasone Suppression Test 20

The overnight dexamethasone suppression test is used to assess if the patient has Cushing syndrome. The pituitary gland secretes adrenocorticotropic hormone (ACTH) based on the amount of cortisol in the patient's blood. Adrenocorticotropic hormone signals the adrenal glands to secrete cortisol. In Cushing syndrome, cortisol is secreted regardless of the secretion of the adrenocorticotropic hormone level. The overnight dexamethasone suppression test requires the patient to take dexamethasone, which is a corticosteroid. This increases the cortisol level in the patient's blood and, therefore, signals the pituitary gland not to secrete adrenocorticotropic hormone. As a result, the adrenal glands should not secrete cortisol. In the morning, the patient's cortisol level should be relatively low. If not, then the patient might have Cushing syndrome.

WHAT IS BEING MEASURED?

- Cortisol level in blood

HOW IS THE TEST PERFORMED?

- At 11.00 PM, the patient is administered dexamethasone by mouth with milk or antacid.
- A blood sample is taken at 8.00 AM.
- The level of cortisol in the blood sample is measured.

RATIONALE FOR THE TEST

- Assess
  - For Cushing syndrome.

NURSING IMPLICATIONS

- Determine if the patient
  - Has eaten or drunk 12 hours prior to drawing the blood sample
  - Has taken methadone, aspirin, MAOIs, lithium, diuretics, 48 hours before the blood sample is drawn

UNDERSTANDING THE RESULTS

- The results are available quickly. The laboratory determines normal values based on calibration of testing equipment with a control test. Test results are reported as high, normal, or low based on the laboratory's control test.
- Normal value results indicate
  - A relatively low level of cortisol in the blood sample
- High value results indicate
  - Cushing syndrome
  - Cancer
  - Diabetes
  - Hyperthyroidism

TEACH THE PATIENT

- Explain
    - Why the test is administered.
    - How the test is performed.
    - Do not eat or drink 12 hours prior to drawing the blood sample.
    - Do not take methadone, aspirin, MAOIs, lithium, diuretics, 48 hours before the blood sample is drawn.

Summary

When a patient reports abnormal feeling in the chest, abdomen, or urinary tract, the healthcare provider can hone in on the problem by performing one or more tests and procedures.

In this chapter, you learned about tests for the upper gastrointestinal tract, thyroid gland, liver, gallbladder, kidneys, spleen, urinary tract, and other organs in the upper part of the body.

You learned how the healthcare provider looks down the esophagus to examine the stomach, duodenum, and the bile and pancreatic ducts and how to take a biopsy or, in some cases, remove an obstruction.

And you learned how scans of the liver, spleen, gallbladder, and kidney using contrast material highlight structural problems within the organ.

You also learned about procedures that can temporarily or permanently repair a problem, whether it is removing a cancerous thyroid gland, or a tumor from the bladder, or fixing urinary incontinence.

Quiz

1. What test is used to measure esophageal muscles?
    a. Esophageal manometry
    b. Esophageal acidity test
    c. Nonwireless
    d. Bravo wireless

2. What procedure is used to examine the pancreatic ducts and the bile ducts?
   a. Cholecystokinin stimulation
   b. Endoscopic retrograde cholangiopancreatogram
   c. Perfusion study
   d. Function study

3. What is the purpose of the urodynamic test?
   a. To assess bladder function
   b. To assess the position of the urethra
   c. To assess urethral pressure
   d. None of the above

4. Cloudy urine might indicate
   a. The patient is dehydrated.
   b. The patient has taken Pepto-Bismol.
   c. Bacterial infection.
   d. The patient is overhydrated.

5. During a 24-hour urine collection, urine from the first time the patient urinated should
   a. Be refrigerated
   b. Be stored in a gallon container
   c. Should be discarded
   d. All of the above

6. Why might a whole body thyroid scan be ordered?
   a. To assess for thyroid cancer metastasis
   b. To assess for hyperthyroidism
   c. To assess for hypothyroidism
   d. None of the above

7. A total thyroxine test is used to determine
   a. The total amount of thyroxine that is missing from the patient
   b. The amount of thyroxine that is not bound to globulin
   c. The amount of thyroxine that is attached to globulin and that is not bound to globulin
   d. All of the above

8. How is the cause of xerostomia determined?
   a. Using the thyroid gland scan
   b. Using the salivary gland scan
   c. Using the liver scan
   d. Using the bladder scan

9. What is the purpose of the renin assay test?
   a. To determine the underlying cause of hypertension
   b. To determine the underlying cause of hypotension
   c. To determine the underlying cause of renal disease
   d. To determine the underlying cause of liver disease

10. Prior to taking the urine culture and sensitivity test the patient should
   a. Take antibiotics
   b. Not take antibiotics
   c. Avoid eating for 48 hours prior to the test
   d. None of the above

Answers

1. a. Esophageal manometry.
2. b. Endoscopic retrograde cholangiopancreatogram.
3. a. To assess bladder function.
4. c. Bacterial infection.
5. c. Should be discarded.

6. a. To assess for thyroid cancer metastasis.
7. c. The amount of thyroxine that is attached to globulin and that is not bound to globulin.
8. b. Using the salivary gland scan.
9. a. To determine the underlying cause of hypertension.
10. b. Not take antibiotics.

CHAPTER 24

Bone and Muscle Tests

Aching bones and muscles might be from a cause other than overexercising. It could be a sign of an underlying disorder that needs immediate medical attention. Healthcare providers are able to assess the reason for the patient's discomfort by testing the patient's bones and muscles.

Healthcare providers have an assortment of tests and procedures that are used to investigate signs of a disorder. An image of the bone can be taken by using a bone scan where a radioactive tracer highlights the structure of the bone or a myelogram that uses contrast material to bring out the structure of the spine into view. An arthrogram is used to create an image of soft tissues and structures of a joint. A bone mineral density test is ordered to determine the thickness of bone, looking for the first sign of osteoporosis.

When X-ray, ultrasound, and other imaging technologies indicate something is abnormal, the healthcare provider may perform arthroscopy to look directly into the joint or a bone biopsy, or bone marrow aspiration to take samples of suspicious tissue.

Once a diagnosis is made, there are a number of procedures that the healthcare provider can perform to either fix the disorder or reduce the impact of the disorder on the patient's daily activities. If the patient has a herniated disc, the healthcare provider can perform a discectomy to remove all or part of the disc. The healthcare provider can fuse together bones or use arthroplasty to reconstruction a joint. Discomfort from spinal stenosis can be reduced by performing a decompressive laminectomy that removes the lamina that is causing pressure on the nerve.

You will learn about these tests and procedures in this chapter.

Learning Objectives

1. Bone Mineral Density (BMD)
2. Bone Scan

Key Words

Conductive velocity
Dual-energy X-ray absorptiometry (DEXA)
Dual-photo absorptiometry (DPA)
Peripheral dual-energy X-ray absorptiometry (P-DEXA)
Quantitative computed tomography (QCT)
Spinal stenosis
T-score
Z-score

Bone Mineral Density 1

The bone mineral density (BMD) test measures the density of bone to assess if the patient has osteopenia or osteoporosis. A low bone density might result in an increased risk for fracture. There are five ways to measure BMD. These are

- Ultrasound: This test uses sound waves to determine the density of bone. However, this method does not assess hip and the spine, which are bones that commonly fracture because of low BMD.
- Dual-photo absorptiometry (DPA): This method uses a low-dose radioactive tracer to measure bone density in all bones including the hip and spine.

- Quantitative computed tomography (QCT): This method measures the density of the vertebra; however, this is less accurate than the DPA, DEXA, and P-DEXA methods.
- Dual-energy X-ray absorptiometry (DEXA): This method uses two X-ray beams to measure bone density and can measure up to 2% bone loss, making it the most accurate way to measure BMD.
- Peripheral dual-energy X-ray absorptiometry (P-DEXA): This method measures bone density in arms and legs, but cannot be used to measure the bone density of the hip or spine.

WHAT IS BEING EXAMINED?

- BMD

HOW IS THE PROCEDURE PERFORMED?

- The patient signs a consent form.
- The patient lies on a table.
- The ultrasound, CT, or X-ray device scans the site.
- The insertion site is cleaned with an anesthetic.

RATIONALE FOR THE TEST

- To assess for osteoporosis
- To assess the progression of osteoporosis
- To assess the impact of long-term treatment with corticosteroids

NURSING IMPLICATIONS

- Determine if the patient
  - Has signed a consent form
  - Is pregnant
  - Is able to lie still
  - Has had broken bones
  - Has arthritis
  - Has had bone replacement
  - Has had a barium test within 2 weeks of the test

UNDERSTANDING THE RESULTS

- The procedure takes less than 1 hour. Results are ready within 2 weeks of the procedure.
- The patient is given a T-score.
  - The T-score compares the patient's BMD with the average healthy 30-year-old's T-score.
  - A negative T-score indicates that the patient's BMD is less than the T-score of the average healthy 30 years old.
  - A positive T-score indicates that the patient's BMD is greater than the T-score of the average healthy 30 years old.
- The patient is given a Z-score. The Z-score compares the patient to those of the same age, sex, and race.
  - A negative Z-score indicates that the patient's BMD is less than the Z-score of the average person of his/her age, sex, and race.
  - A positive Z-score indicates that the patient's BMD is greater than the Z-score of the average person of his/her age, sex, and race.
- Normal test results indicate
  - T-Score
    - Normal is within +/−1 T-score of the average healthy 30-year-old's T-score.
  - Z-Score
    - Normal is within +/−1 Z-score of the average person of his/her age, sex, and race.
- Abnormal test results indicate
  - T-Score
    - 2.5 or less than T-score of the average healthy 30-year-old's T-score indicates osteoporosis.
  - Z-Score
    - 2.5 or less than Z-score of the average person of his/her age, sex, and race.

TEACH THE PATIENT

- Explain
  - Why the procedure is being performed.

- What the patient will experience during the procedure.
- That the patient will not feel any pain.

Bone Scan 2

A bone scan assesses the infection, trauma, and metastasized cancer growth to the bone. A radioactive tracer is injected in the patient's vein. The tracer is removed from the blood into the bone. A gamma camera takes an image of the tracer as the tracer is absorbed. Lack of absorption indicates bone infarction and possibly cancer. Areas of high absorption might indicate an infection, tumor, or fracture.

WHAT IS BEING EXAMINED?

- The bone

HOW IS THE PROCEDURE PERFORMED?

- The patient signs a consent form.
- The patient removes his/her clothes.
- The patient lies on a table.
- The injection site is cleaned.
- The patient is injected with the radioactive tracer.
- The gamma camera is moved into position.
- Images are taken immediately, then periodically for up to 5 hours.
- The patient should drink lots of fluid to flush the tracer from his/her body.
- The patient should flush the toilet quickly after urinating and defecating since the tracer is excreted in urine and feces for 2 days following the test.

RATIONALE FOR THE TEST

- Determine if cancer has metastasized.
- Assess for bone infection.

NURSING IMPLICATIONS

- Determine if the patient
  - Has signed a consent form
  - Has emptied his/her bladder
  - Has allergies
  - Is pregnant
  - Is breast-feeding
  - Has recently had a barium test performed within 4 days prior to the test
  - Has not been taking anticoagulant medication such as Plavix, Coumadin, heparin, or aspirin for 2 weeks prior to the test
  - Has not taken Pepto-Bismol for 48 hours before the test

UNDERSTANDING THE RESULTS

- The procedure takes less than 1 hour. Results are known within 2 days.
- Normal test results indicate
  - The bone is normal.
- Abnormal test results indicate
  - Cancer has metastasized to the bone.
  - The bone is inflamed.
  - A bone infarction is identified.

TEACH THE PATIENT

- Explain
  - Why the procedure is being performed.
  - What the patient will experience during the procedure.
  - That the patient will not feel any pain during the procedure except for a pinch when the tracer is injected into his/her vein.
  - That the patient must sign a consent form.
  - That the patient should not take anticoagulant medication such as Plavix, Coumadin, heparin, or aspirin for 2 weeks prior to the test.
  - That the patient should not take Pepto-Bismol for 48 hours before the test.
  - That the patient must empty his/her bladder before the test.
  - That the patient should drink lots of fluids to flush the tracer following the test.

- That the patient should flush the toilet immediately after urinating and defecating.
- That the patient should wash his/her hands thoroughly after urinating and defecating.
- That the patient should apply a warm compress to the injection site if there is swelling.
- If the patient is breast-feeding, breast milk must be discarded for 2 days after the scan.

Summary

Healthcare providers have an assortment of tests and procedures that are used to investigate signs of a disorder. Initially they will order an image be taken of the suspected structure. These include a myelogram that uses contrast material to highlight structures in the spine, or a bone scan that uses a radioactive tracer to make the structure of the bone appear on a video screen. The healthcare provider may also order an arthrogram to visualize the soft tissues and structures of a joint.

The thickness of bone is a factor that determines the risk of spontaneous stress factors, especially with the elderly who are susceptible to osteoporosis. The healthcare provider assesses the thickness of bone by ordering a BMD test.

When imaging technology indicates something might be abnormal, the healthcare provider usually performs arthroscopy to look directly into the joint or a bone biopsy, or bone marrow aspiration to take samples of suspicious tissue.

The healthcare provider can perform procedures to either fix the disorder or reduce the impact of the disorder on the patient's daily activities. For example, a herniated disc is repaired by performing a discectomy to remove all or part of the disc. The healthcare provider can fuse together bones or use arthroplasty to reconstruct a joint. Discomfort from spinal stenosis can be reduced by performing a decompressive laminectomy that removes the lamina that is causing pressure on the nerve.

Quiz

1. What parts of the body are measured using P-DEXA?
   a. Arms and legs
   b. Spine and arms
   c. Hip and spine
   d. Hip and legs

2. What can produce a false-positive result in the bone mineral density (BMD) test?
   a. The patient has arthritis.
   b. The patient has a T-score of 1.
   c. The patient has a Z-score of 1.
   d. The patient has an X-score of less than 1.

3. What does a negative T-score mean?
   a. The patient's BMD is greater than the T-score of the average healthy 30 years old.
   b. The patient's BMD is equal to the T-score of the average healthy 30 years old.
   c. The patient's BMD is less than the T-score of the average healthy 30 years old.
   d. None of the above.

4. A normal T-score is +/−1 of the average healthy 30-year-old's T-score.
   a. True
   b. False

5. What does the lack of absorption indicate in a bone scan?
   a. The patient is elderly.
   b. Bone infarction and possibly cancer.
   c. The patient is healthy.
   d. Infection.

6. What is a Z-score?
   a. A score that compares the patient's conductive velocity with that of people his/her own age, sex, and race.
   b. A score that compares the patient's BMD with that of a healthy 30 years old.
   c. A score that compares the patient's BMD with that of people his/her own age, sex, and race.
   d. None of the above.

7. What is the most accurate method of measuring BMD?
   a. Dual-photo absorptiometry
   b. Quantitative computed tomography
   c. Dual-energy X-ray absorptiometry
   d. Ultrasound

8. Why should a patient flush the toilet quickly after urinating following a bone scan?
   a. The radioactive tracer is excreted in urine.
   b. To prevent infection.
   c. To reduce the spread of bone demineralization.
   d. All of the above.

9. The patient who has given birth 4 weeks ago arrives for a bone scan. How would you respond?
   a. Ask the patient if she is breast-feeding.
   b. Ask the patient if the infant was normal.
   c. Assess the patient's blood calcium level.
   d. Ask the patient to drink a liter of water.

10. Why would a healthcare provider not use ultrasound to test the bone density of the patient's hip?
    a. Ultrasound cannot be used to measure bone density of the hip.
    b. The patient is unable to lie still for the test.
    c. Ultrasound is used to determine the density of bone in the spine.
    d. Ultrasound is used to measure only high bone density.

Answers

1. a. Arms and legs.
2. a. The patient has arthritis.
3. c. The patient's BMD is less than the T-score of the average healthy 30 years old.

4. b. False.
5. b. Bone infarction and possibly cancer.
6. c. A score that compares the patient's BMD with that of people of his/her own age, sex, and race.
7. c. Dual-energy X-ray absorptiometry.
8. a. The radioactive tracer is excreted in urine.
9. a. Ask the patient if she is breast-feeding.
10. a. Ultrasound cannot be used to measure bone density of the hip.

CHAPTER 25

Tests for Males

There are a number of medical tests and procedures that are specifically designed to diagnose and treat disorders that affect men. There are a group of tests and procedures focused on fertility.

When a man is unable to impregnate a woman, the healthcare provider orders tests to assess if there is an underlying problem with the man's reproductive organs. The initial test is a semen analysis that assesses the man's semen and sperm. Depending on the results, a testicular scan or testicular ultrasound is ordered to determine if there is a structural disorder. One such structural disorder is varicocele, which is a large vein that blocks blood flow to the testicles. This is relieved by performing a varicocele repair.

The healthcare provider may follow up with a testicular examination or an erectile dysfunction test. If the erectile dysfunction test returns positive results, the healthcare provider may perform a penile implant procedure where a device is inserted to cause an erection.

Some men desire to become infertile by having their vas deferens cut or blocked by a vasectomy. This prevents sperm from mixing with semen, resulting in no sperm in the ejaculate. A vasectomy in some instances can be reversed by performing a vasovasostomy.

Men are susceptible to developing an enlarged prostate gland, which could be caused by prostate cancer. Prostatic cancer cells are in part fueled by testosterone, which is produced by the testicles. The healthcare provider might perform an orchiectomy, which is the surgical removal of one or both testicles. This reduces the level of testosterone in the patient's body.

Alternatively, the healthcare provider may perform a prostatectomy, which is the removal of the prostate gland. However, this procedure may leave the patient with erectile dysfunction and urinary incontinence.

In this chapter, you will learn about these tests and procedures.

Learning Objectives

1. Erectile Dysfunction Tests
2. Testicular Ultrasound
3. Testicular Scan
4. Testicular Examination
5. Semen Analysis

Key Words

Color duplex Doppler
Epididymis
Fructose level
Intracavernosal injection
Liquefaction time
Nocturnal penile tumescence (NPT)
Prostaglandin E1
Semen volume
Snap gauge
Sperm count
Sperm morphology
Sperm motility
Transillumination
Vas deferens

Erectile Dysfunction Tests 1

Erectile dysfunction is commonly caused by psychological, blood vessel, and nerve disorders. There are three tests that are commonly ordered to assess erectile dysfunction. These are

- Color duplex Doppler: This test assesses blood flow through the penis using an ultrasound.
- Nocturnal penile tumescence (NPT): This test assesses if the patient has erections during sleep.
- Intracavernosal injection: This test injects prostaglandin E1 into the base of the penis to cause an erection.

WHAT IS BEING EXAMINED?

- Penile erection

HOW IS THE PROCEDURE PERFORMED?

- The patient signs a consent form.
- The patient removes clothing below the waist.
- The patient lies on a table.
- Color duplex Doppler.
  - Conductive gel is placed on the penis.
  - The Doppler ultrasound transducer is moved across the penis.
  - An image on a video screen shows the direction and velocity of blood flowing through the penis.
- NPT
  - A snap gauge is placed around the penis when the patient goes to sleep.
  - An erection during sleep will snap the film, indicating that the patient had an erection while sleeping.
  - Alternatively, a transducer is placed on the penis when the patient goes to sleep and it records the number of erections, the length of time of the erection, and how rigid the erection was during sleep.
- Intracavernosal injection
  - Prostaglandin E1 is injected into the base of the penis causing an erection.
  - A measurement is taken of the duration and how rigid the erection is.

RATIONALE FOR THE TEST

- To assess the underlying cause of erectile dysfunction

NURSING IMPLICATIONS

- Determine if the patient
  - Has signed a consent form
  - Is stressed or depressed
  - Has not taken Cialis, Levitra, or Viagra prior to the test
  - Has any allergies
  - Has not taken sleeping pills 2 days before the test, if the patient is undergoing the NPT test
  - Has not ingested alcohol 2 days before the test, if the patient is undergoing the NPT test
  - Has not had an erection lasting more than 4 hours
  - Has not ingested nicotine within 2 hours of the test
  - Has not been taking anticoagulant medication such as Plavix, Coumadin, heparin, or aspirin
  - Can lie still

UNDERSTANDING THE RESULTS

- The procedure takes less than 1 hour except for the NPT test, which is performed overnight. Results are immediate.
- Normal test results indicate
  - Normal blood flow in the penis
  - Normal erection immediately and overnight
- Abnormal test results indicate
  - Abnormal blood flow in the penis
  - Abnormal or no erection

TEACH THE PATIENT

- Explain
  - Why the procedure is being performed.
  - What the patient will experience during the procedure.
  - That the patient will not feel any pain during the procedure except for a pinch from the needle if intracavernosal injection is performed,
  - That the patient will have to sign a consent form.

- That the patient must not take Cialis, Levitra, or Viagra prior to the test.
- That the patient must not have taken sleeping pills 2 days before the test, if the patient is undergoing the NPT test.
- That the patient must not ingest alcohol 2 days before the test, if the patient is undergoing the NPT test.
- That the patient must not ingest nicotine within 2 hours of the test.
- That the patient must stop taking anticoagulant medication such as Plavix, Coumadin, heparin, or aspirin until after the test is completed.

Testicular Ultrasound 2

A testicular ultrasound is a procedure used to produce an image of the testicles, scrotum, epididymis, and vas deferens to detect if there is any structural dysfunction.

WHAT IS BEING EXAMINED?

- Testicles, scrotum, epididymis, and vas deferens

HOW IS THE PROCEDURE PERFORMED?

- The patient removes clothing below the waist.
- The patient lies on a table.
- Conductive gel is placed on the scrotum.
- A transducer is moved across the scrotum.
- An image of the testicles, scrotum, epididymis, and vas deferens is seen on a video screen.

RATIONALE FOR THE TEST

- To assess a mass on the testicles
- To identify the underlying cause of pain in the scrotum
- To assess structures within the scrotum

NURSING IMPLICATIONS

- Determine if the patient can lie still.

UNDERSTANDING THE RESULTS

- The procedure takes 1 hour. Results are known within 2 days.
- Normal test results indicate
  - Normal structures in the scrotum
- Abnormal test results indicate
  - Inflammation.
  - Twisted spermatic cord.
  - Growth.
  - Pyocele, spermatocele, or hematocele in the scrotum.
  - Hernia is present.

TEACH THE PATIENT

- Explain
  - Why the procedure is being performed.
  - What the patient will experience during the procedure.
  - That the patient will not feel any pain during the procedure.

Testicular Scan 3

A testicular scan assesses the function of the testicles and is used to identify blockages. A radioactive tracer is injected in the patient's vein. The tracer flows into the testicles. A camera takes an image of the tracer as the tracer flows through the testicles.

WHAT IS BEING EXAMINED?

- The testicles

HOW IS THE PROCEDURE PERFORMED?

- The patient signs a consent form.
- The patient removes clothing below the waist.

- The patient lies on a table.
- The penis is taped to the abdomen.
- The testicles are supported by a towel.
- The injection site is cleaned.
- The patient is injected with the radioactive tracer.
- The camera is moved into position.
- Two images are taken, one every 15 minutes.

RATIONALE FOR THE TEST

- To assess the testicles

NURSING IMPLICATIONS

- Determine if the patient
  - Has signed a consent form
  - Has allergies
  - Has had a barium test performed 4 days prior to the test
  - Has taken aspirin

UNDERSTANDING THE RESULTS

- The procedure takes less than 1 hour. Results are known within 2 days.
- Normal test results indicate
  - No blockages in the testicles
  - Normal shape and size of the testicles
- Abnormal test results indicate
  - Abnormal shape and size of the testicles.
  - Testicles are inflamed or blocked.

TEACH THE PATIENT

- Explain
  - Why the procedure is being performed.
  - What the patient will experience during the procedure.

- That the patient will not feel any pain during the procedure except for a pinch when the tracer is injected into his vein.
- That the patient must sign a consent form.
- That the patient should not take anticoagulant medication such as Plavix, Coumadin, heparin, or aspirin for 2 weeks prior to the test.
- That the patient must drink lots of fluids to flush the tracer following the test.
- That the patient should flush the toilet immediately after urinating and defecating.
- That the patient should wash hands thoroughly after urinating and defecating.
- That the patient should apply a warm compress to the injection site if there is swelling.

Testicular Examination

A testicular examination is performed to assess the patient's testicles, scrotum, and penis for testicular atrophy, growths, and abnormalities.

WHAT IS BEING EXAMINED?

- The testicles, scrotum, and penis

HOW IS THE PROCEDURE PERFORMED?

- The patient empties his bladder prior to the examination.
- The patient removes clothing below the waist.
- The patient lies on a table and then stands during the examination.
- The scrotum and testicles are palpated.
- Transillumination is performed to highlight any growth found during palpation. This is where a light is placed behind the testicles, making the outline of the growth visible.
- The healthcare provider assesses the testicles for consistency, weight, size, and texture.
- Lymph nodes in the groin and thigh are palpated.

RATIONALE FOR THE TEST

- To assess the testicles, scrotum, and penis

NURSING IMPLICATIONS

- Determine if the patient
  - Can lie still
  - Can stand

UNDERSTANDING THE RESULTS

- The procedure takes less than 1 hour. Results are known immediately.
- Normal test results indicate
  - Normal shape and size of the testicles, scrotum, and penis
  - Normal shape and size of the lymph nodes
- Abnormal test results indicate
  - Abnormal shape and size of the testicles, scrotum, and penis
  - Abnormal shape and size of the lymph nodes

TEACH THE PATIENT

- Explain
  - Why the examination is being performed.
  - What the patient will experience during the examination.
  - That the patient will not feel any pain during the examination.
  - That the patient should empty his bladder before the examination.

Semen Analysis 5

Semen analysis is performed to assess the volume of semen and number of quality sperm that is produced in an ejaculation to determine the underlying cause of infertility. There are eight factors that are analyzed. These are

- Semen volume: This is the amount of semen in an ejaculation.
- Liquefaction time: This is the time it takes for the semen to liquefy.

- Sperm morphology: This is the number of normally shaped sperm.
- Sperm motility: This is the percentage of sperm that shows forward movement.
- Sperm count: This is the number of sperm in a milliliter of semen in one ejaculation.
- Fructose level: This is the amount of fructose in semen to provide energy for sperm.
- pH: This measures the pH level of the semen.
- White blood cell count: This measures the number of white blood cells in semen, which is normally zero.

WHAT IS BEING EXAMINED?

- Semen and sperm

HOW IS THE PROCEDURE PERFORMED?

- The patient signs a consent form.
- Ejaculation should not occur for 2 days before the semen sample is taken.
- Ejaculation should occur within 5 days before the semen sample is taken.
- Before the semen sample is taken, the patient should urinate.
- The patient's hands and penis should be washed.
- The patient should not use lubricants or condoms when collecting the sample.
- The patient should not collect a semen sample after withdrawing from intercourse.
- The patient should place the semen sample in a sterile cup.
- The patient should keep the sample at body temperature.
- The patient should keep the sample away from direct sunlight.
- The patient should deliver the sample to the laboratory immediately.

RATIONALE FOR THE TEST

- To determine the underlying cause of infertility
- To assess the vasectomy
- To assess the vasovasostomy

NURSING IMPLICATIONS

- Determine if the patient
  - Has not taken tegument, sulfasalazine, testosterone, estrogen, or nitrofurantoin
  - Has not taken Echinacea, St. John's wort, caffeine, cocaine, alcohol, marijuana, or tobacco
  - Has not ejaculated for 2 days before the semen sample is taken
  - Has ejaculated within 5 days before the semen sample is taken
  - Has urinated before the semen sample was taken
  - Had washed his hands and penis before the semen sample was taken
  - Did not use lubricants or condoms when collecting the semen sample
  - Did not collect a semen sample after withdrawing from intercourse
  - Placed the semen sample in a sterile cup
  - Kept the sample at body temperature
  - Kept the sample away from direct sunlight
  - Deliver the sample to the laboratory immediately
  - Has signed a consent form

UNDERSTANDING THE RESULTS

- The procedure takes less than 1 hour. Results are known within 2 days.
- Normal test results indicate
  - Normal semen volume.
  - Liquefaction time is less than 60 minutes.
  - 70% of sperm have normal shape.
  - 70% of sperm show forward mobility.
  - Sperm count is more than 20 million sperm per milliliter of semen or zero if a vasectomy was performed.
  - Semen pH is between 7.1 and 8.0.
  - 300 mg of fructose in 100 mL of semen.
  - No white blood cells in the semen.
- Abnormal test results indicate
  - Low semen volume.
  - Liquefaction time is 60 minutes or greater.

- Less than 70% of sperm have normal shape.
- Less than 70% of sperm show forward mobility.
- Sperm count is less than 20 million sperm per milliliter of semen or greater than zero if a vasectomy was performed.
- Semen pH is less than 7.1 or greater than 8.0.
- Less than 300 mg of fructose in 100 mL of semen.
- White blood cells present in the semen.

TEACH THE PATIENT

- Explain
  - That the patient will not feel any pain during the procedure.
  - Why the procedure is being performed.
  - What the patient will experience during the procedure.
  - That the patient must sign a consent form.
  - How to collect the semen sample.
  - That the patient must not take tegument, sulfasalazine, testosterone, estrogen, or nitrofurantoin.
  - That the patient must not take Echinacea, St. John's wort, caffeine, cocaine, alcohol, marijuana, or tobacco.

Summary

In this chapter you learned about tests and procedures specifically designed to diagnose and treat disorders that affect men.

You saw how semen analysis determines the quality of semen and sperm and how a testicular scan or testicular ultrasound gives the healthcare provider a view inside the structural defects of the patient's testicles, which do not show up on a testicular examination. The healthcare provider can remove structural disorders by performing procedures such as varicoceles repair.

Erectile dysfunction is a problem that affects some patients. You learned about the various erectile dysfunction tests that are used to diagnose this condition and about the penile implant procedure that restores the erection.

The healthcare provider can perform a vasectomy to cut or block the vas deferens, preventing sperm from mixing with semen resulting in no sperm in the ejaculate. A vasectomy in some instances can be reversed by performing a vasovasostomy.

An enlarged prostate gland could be caused by prostate cancer. An orchiectomy can be performed to remove one or both testicles resulting in a reduction of testosterone, which fuels prostate cancer cells. If this fails to reduce the prostate gland, then the healthcare provider may perform a prostatectomy—the removal of the prostate gland.

Quiz

1. What does the NPT test measure?
   a. Blood flowing through the penis
   b. If the patient has erections during sleep
   c. The success of a vasectomy
   d. None of the above

2. What is sperm morphology?
   a. The amount of semen in an ejaculation
   b. The percentage of perm showing forward movement
   c. The number of normally shaped sperm
   d. None of the above

3. How does the snap gauge work?
   a. It measures the size of an erection.
   b. The snap gauge consisting of a film is placed around the penis. The film snaps when the patient has erection.
   c. It measures the length of time of an erection.
   d. All of the above.

4. Why is transillumination performed?
   a. To highlight growths found during a testicular examination
   b. To visualize the prostate gland through the urethra
   c. To scan the prostate gland
   d. To scan the bladder

5. What is liquefaction time?
    a. The time necessary for semen to liquefy
    b. The time necessary for semen to dehydrate
    c. The time necessary for sperm to liquefy
    d. The time necessary for sperm to dehydrate
6. What might cause low sperm motility?
    a. Sample not kept at body temperature
    b. Delay in delivering the sample to the laboratory
    c. Sample not kept away from direct sunlight
    d. All of the above
7. What might cause a low sperm count?
    a. The sample was taken within 2 days of the last ejaculation.
    b. Having sexual intercourse 10 days before the sample is taken.
    c. Unable to have an erection.
    d. Unable to have an erection while sleeping.
8. Why is the fructose level in semen measured?
    a. Fructose prevents sperm motility.
    b. Fructose provides energy for sperm.
    c. Fructose reduces sperm count.
    d. Fructose increases semen volume.
9. What is the purpose of the Color duplex Doppler test?
    a. To assess the velocity and direction of blood flowing through the penis
    b. To assess size of an erection
    c. To assess if the patient had an erection during sleep
    d. To assess if the patient has urine in his bladder
10. What is the purpose of an intracavernosal injection?
    a. To increase sperm count
    b. To prevent an erection
    c. To cause an erection
    d. None of the above

Answers

1. b. If the patient has erections during sleep.
2. c. The number of normally shaped sperm.
3. b. The snap gauge consisting of a film is placed around the penis. The film snaps when the patient has erection.
4. a. To highlight growths found during a testicular examination.
5. a. The time necessary for semen to liquefy.
6. d. All of the above.
7. a. The sample was taken within 2 days of the last ejaculation.
8. b. Fructose provides energy for sperm.
9. a. To assess the velocity and direction of blood flowing through the penis.
10. c. To cause an erection.

CHAPTER 26

Skin Tests

Skin is the largest and the most visible organ in the body and is susceptible to wrinkles, blemishes, growths including both nonmelanoma and melanoma, and infection. Healthcare providers perform an assortment of tests and procedures to diagnose and treat skin conditions.

Lesions, warts, and blemishes that make skin unsightly and unhealthy can be removed by performing one of a number of procedures. For example, undesired tissue can be frozen with cryosurgery and removed, or layers of skin can be peeled away using chemicals or micrographic surgery. Layers can also be removed using curettage, electrosurgery, or the dermabrasion procedure where a rotating burr scrapes scars and aging skin to encourage new skin growth.

Melanoma and nonmelanoma skin cancer affects many patients. The healthcare provider can cure or minimize discomfort of this condition by performing a skin excision where the tumor is surgically removed.

Infected skin can be treated once the microorganism that causes the infection is identified. The healthcare provider is able to identify the microorganism by performing a wound culture, where a sample of the infected tissue is placed in an environment that is favorable for the growth of microorganism (culture) which is then identified. Once the laboratory determines the medication which will fight the microorganism, the medication is administered to find out whether it actually kills the microorganism.

Skin is also a perfect site for testing of allergic reactions. There are several tests that healthcare providers administer to the skin to identify allergens that cause the patient to develop an allergic reaction. The skin is also the site of the Mantoux skin test to determine if the patient has ever been exposed to *Mycobacterium tuberculosis.*

You will learn about these tests and procedures in this chapter.

Learning Objectives

1. Chemical Peel
2. Allergy Skin Testing
3. Dermabrasion
4. Mantoux Skin Test
5. Wound Culture

Key Words

Bacillus Calmette-Guérin (BCG) vaccination
Curettage
Exfoliates
Glycolic acid
Intradermal test
Liquid nitrogen
Phenol
Retin-A
Skin patch test
Skin prick test
Tretinoin cream
Trichloroacetic acid (TCA)

Chemical Peel 1

A chemical peel is a procedure that exfoliates injured or dead skin, enabling new skin to replace it. There are three types of chemical peels. These are

- Superficial: This procedure removes the surface layer of the skin for a smoother, brighter appearance and uses glycolic acid or dry ice for the peel.

- Medium: This procedure is commonly used to smooth fine wrinkles, remove blemishes, and treat pigment problems. It uses trichloroacetic acid (TCA) for the peel.
- Deep: This procedure is commonly used to correct coarse wrinkles and blotches on the face only and uses phenol for the peel. New skin might be lighter in tone because the skin loses some ability to produce pigment.

Caution

A chemical peel might change the patient's skin tone and result in scarring and cold sores. It might cause flaking and dryness.

WHAT IS BEING EXAMINED?

- Skin

HOW IS THE PROCEDURE PERFORMED?

- The patient moisturizes his/her skin twice a day for 3 weeks prior to the procedure.
- The patient is administered acyclovir 2 weeks prior to the procedure to prevent a viral infection.
- The patient signs a consent form.
- The patient removes his/her clothes from the affected area.
- The patient lies on a table.
- The healthcare provider may test the peel on skin at a less visible location.
- The site is cleaned with an antiseptic.
- Superficial peel
  - The chemical is applied on to the site using an applicator.
  - The chemical remains on the site for several minutes.
  - Alcohol or water is applied to the site to neutralize the chemical.
  - The chemical is removed by wiping the skin.
  - Discomfort is relieved by using a fan.
- Medium peel
  - The patient is administered a sedative.
  - The chemical is applied on to the site using an applicator.

- The chemical remains on the site for several minutes longer than in a superficial peel.
- Alcohol or water is applied to the site to neutralize the chemical.
- The chemical is removed by wiping the skin.
- Discomfort is relieved by using a cool compress.

- Deep peel
  - The patient is administered a general anesthetic.
  - The patient is connected to an ECG to monitor his/her heart.
  - The patient is connected to a pulse oximeter to monitor the oxygen level in the blood.
  - A breathing tube is inserted into the patient's mouth.
  - The chemical is applied on to a small area of the site using an applicator.
  - The chemical remains on the site for several minutes.
  - Alcohol or water is applied to the site to neutralize the chemical.
  - The chemical is removed by wiping the skin.
  - The healthcare provider waits 15 minutes before repeating treatment to another small area of the site in order to prevent the building of chemical on the skin.
  - The healthcare provider may cover the area with an ointment or tape. The ointment is removed with water 24 hours following the procedure. The tape is removed 2 days following the procedure.
- The patient might be told to be administered Retin-A on the site after the peel to speed up healing.

RATIONALE FOR THE TEST

- Cosmetic improvement of the skin

NURSING IMPLICATIONS

- Determine if the patient
  - Has signed a consent form
  - Has not been taking anticoagulant medication such as Plavix, Coumadin, heparin, or aspirin
  - Can lie still

- Has moisturized his/her skin twice a day for 3 weeks prior to the procedure
- Has administered acyclovir 2 weeks prior to the procedure to prevent a viral infection

UNDERSTANDING THE RESULTS

- The procedure takes less than 2 hours. Result is realized in 6 months following the procedure.
- Normal test results indicate
  - The wrinkle or the blemish is removed.
- Abnormal test results indicate
  - Scarring occurred

TEACH THE PATIENT

- Explain
  - Why the procedure is being performed.
  - What the patient will experience during the procedure.
  - That the patient will not feel any pain during the procedure.
  - That the patient must moisturize his/her skin twice a day for 3 weeks prior to the procedure.
  - That the patient must be administered acyclovir 2 weeks prior to the procedure to prevent a viral infection.
  - That the patient might have to administer Retin-A on the site to encourage healing.
  - That the patient must apply tretinoin cream every night on the site beginning 3 weeks following the procedure.
  - That the patient must clean the site frequently following the procedure.
  - That the patient must avoid exposure to the sun until the site heals.
  - That the patient must use sun block every day since new skins can be damaged by the sun.
  - For superficial peel, the skin is pink but eventually fades to your natural skin tone. Normal activities can be resumed immediately.

- For medium peel, it takes 7 days to heal during which the site is swollen and reddish. A crust forms over the site, which will fall off within 10 days following the procedure. Normal activities can resume in a week.
- For deep peel, it takes 2 weeks for regrowth to occur. The skin is reddening for 2 months. Healing takes less than 6 months. The patient will be prescribed pain medication for discomfort, antibiotics for infection, and corticosteroids for swelling. The patient can reduce crusting at the site by showering several times a day. Normal activities can resume in 2 weeks.

CAUTION

The patient should call his/her healthcare provider if he/she has fever and drainage from the site following the procedure. The patient should also call his/her healthcare provider if the site remains red for longer than 2 months following the procedure.

Allergy Skin Testing 2

An allergen is a substance that causes an immune reaction. Allergy skin testing is performed to identify allergens. There are three types of skin testing for allergens. These are

- Skin patch test: This test is used to identify allergens that cause contact dermatitis and requires the placement of a pad that contains an allergen solution on the skin for 72 hours. An allergic reaction occurs if the patient is allergic to the allergen.
- Skin prick test: This test requires that a drop of an allergen solution be placed on the patient's skin. The skin is scratched allowing the allergen solution to penetrate. A wheal occurs if the patient is allergic to the allergen.
- Intradermal test: This test requires that a small amount of an allergen solution be injected into the dermal layer of the skin. A wheal occurs if the patient is allergic to the allergen.

WHAT IS BEING EXAMINED?

- Skin

HOW IS THE PROCEDURE PERFORMED?

- The patient signs a consent form.
- The patient removes his/her clothes from the affected area.
- The patient lies on a table.
- The site is cleaned with an antiseptic.
- Skin prick test and intradermal test
    - The allergen is introduced into the skin.
    - The healthcare provider examines the site for wheals in 15 minutes.
- Skin patch test
    - The allergen-containing pad is applied to the site.
    - The pad is removed in 72 hours.
    - The site is assessed for redness and wheals.

RATIONALE FOR THE TEST

- To identify a source of a patient's allergic reaction

NURSING IMPLICATIONS

- Determine if the patient
    - Has signed a consent form
    - Has not been taking anticoagulant medication such as Plavix, Coumadin, heparin, or aspirin
    - Can lie still
    - Does not develop an itch at the site
    - Has not been taking antihistamines
    - Has not been taking antidepressants

UNDERSTANDING THE RESULTS

- The procedure takes less than 1 hour. Result is immediate except for the skin patch test, which takes 72 hours.
- Normal (negative) test results indicate
    - No change to the site

- Abnormal (positive) test results indicate
  - Wheals appear at the site.

TEACH THE PATIENT

- Explain
  - Why the procedure is being performed.
  - What the patient will experience during the procedure.
  - That the patient will not feel any pain during the procedure except for a pinch if the intradermal test or the skin prick test is performed.
  - The patient must keep the site clean and dry until the site is assessed by the healthcare provider.
  - The patient should not scratch the site.
  - The patient should not take antihistamines 2 weeks prior to the test.
  - The patient should not take antidepressants 2 weeks prior to the test.

CAUTION

The patient should call the healthcare provider if experiencing itchiness while the skin patch test is being performed.

Dermabrasion 3

Dermabrasion is a procedure that removes the upper layers of damaged skin caused by scars and aging to encourage new skin growth. The procedure is performed by using a rotating burr to remove damaged tissue.

WHAT IS BEING EXAMINED?

- Skin

HOW IS THE PROCEDURE PERFORMED?

- The patient signs a consent form.
- The patient removes his/her clothes from the affected area.
- The patient will lie on a table.
- The site is cleaned with an antiseptic.

- A local anesthetic is administered to the patient.
- The skin is then hardened by using liquid nitrogen or ice packs.
- A rotating burr is used to scrap the skin.
- An antibiotic-treated dressing is applied to the site.

RATIONALE FOR THE TEST

- To remove scars
- To remove growths from the upper layer of the skin
- To remove wrinkles

NURSING IMPLICATIONS

- Determine if the patient
  - Has signed a consent form
  - Has not been taking anticoagulant medication such as Plavix, Coumadin, heparin, or aspirin
  - Can lie still
  - Has not used isotretinoin
  - Has not had a face lift
  - Has not had active herpes
  - Has not had an immune disorder

UNDERSTANDING THE RESULTS

- The procedure takes less than 1 hour. Result is immediate.
- Normal test results indicate
  - The skin is removed.
- Abnormal test results indicate
  - Not all of the skin is removed.

TEACH THE PATIENT

- Explain
  - Why the procedure is being performed.
  - What the patient will experience during the procedure.

- That the patient will not feel any pain during the procedure except for a pinch when the local anesthetic is administered.
- The patient must keep the site clean and dry until it heals.
- The patient should not scratch or remove the scab that forms over the site.
- The site will heal within 6 weeks following the procedure.
- There might be a scar or skin tone change to the site.
- The procedure may not produce the desired result.
- Apply sunscreen every day.

Caution

The patient should call his/her healthcare provider if he/she has fever and drainage from the site following the procedure.

Mantoux Skin Test 4

The Mantoux skin test determines if the patient has ever been exposed to *Mycobacterium tuberculosis.* The healthcare provider injects the purified protein derivative (PPD), with the *M tuberculosis* antigen, into the patient's forearm. If the patient develops a wheal within 48 hours indicating a positive immune response, then he/she either is or had been infected with *M tuberculosis.* A chest X-ray is taken to diagnose a current infection of *M tuberculosis* should there be a positive immune response to the test.

WHAT IS BEING EXAMINED?

- Antibodies to the *M tuberculosis* antigen

HOW IS THE PROCEDURE PERFORMED?

- The patient signs a consent form.
- The patient exposes his/her forearm.
- The site is cleaned with an antiseptic.
- A needle is inserted slightly under the upper layer of the skin.

- Purified protein derivative is injected.
- The site remains uncovered.

RATIONALE FOR THE TEST

- To identify if the patient is or was ever exposed to *M tuberculosis*

NURSING IMPLICATIONS

- Determine if the patient
  - Has signed a consent form.
  - Has not been taking anticoagulant medication such as Plavix, Coumadin, heparin, or aspirin.
  - Has not had the bacillus Calmette-Guérin (BCG) vaccination. Patients who have had this vaccination will likely test positive.
  - Has not taken corticosteroids.
  - Has not had an immune-compromised disease.
  - Has not had a measles, mumps, rubella, chickenpox, or polio vaccine 6 weeks prior to the test.
  - Is older than 3 months of age.
  - Has not had a prior positive PPD and should not have another PPD.

UNDERSTANDING THE RESULTS

- The procedure takes less than 10 minutes. Result is known in 48 hours.
- Normal (negative) test results indicate
  - No change to the site
- Abnormal (positive) test results indicate
  - Wheals appear at the site.

CAUTION

A negative test result does not mean that the patient is not infected with M tuberculosis. *It can take up to 10 weeks following the infection for the immune system to develop antibodies.*

TEACH THE PATIENT

- Explain
  - Why the procedure is being performed.
  - What the patient will experience during the procedure.
  - That the patient will not feel any pain during the procedure except for a pinch when the needle is inserted beneath the skin.
  - That the patient must keep the site clean and dry.
  - That the patient should not scratch the site.
  - The patient must return in 48 hours to have the healthcare provider assess the test results.

Wound Culture 5

A wound culture is ordered when the patient is suspected of having a skin infection. A tissue sample of the infected area is taken and placed in an environment conducive to the growth of microorganisms for 3 days. The tissue sample is then examined to identify the presence and the type of microorganism. Once the microorganism is identified, a sensitivity test is usually performed to determine the medication that kills the microorganism.

WHAT IS BEING EXAMINED?

- Existence of microorganisms in infected skin

HOW IS THE PROCEDURE PERFORMED?

- The healthcare provider swabs the wound with a sterile swab to collect a tissue sample. More than one sample might be taken if the healthcare provider suspects that the microorganism may be aerobic (grows in the presence of oxygen) or anaerobic (grows without the presence of oxygen) bacteria.
- The sterile swab is placed in either an aerobic or anaerobic culture tube and sent to the laboratory.
- The sample(s) is then placed in a culture dish in an environment conductive to growing microorganism.
- After 3 days, tests are performed to identify the microorganism.

- Medication is applied to a portion of the culture to determine which medication kills the microorganism. Preliminary results are known quickly, but final results can take 72 hours or longer.

RATIONALE FOR THE TEST

- To assess the existence and type of microorganism in a patient with a skin infection
- To assess the medication used to treat the skin infection

NURSING IMPLICATIONS

- Determine if the patient has not taken antibiotics or applied an antiseptic to the wound prior to the test.

UNDERSTANDING THE RESULTS

- The collection takes a few seconds. Results are known within 3 days.
- Normal (negative) test results indicate
  - No microorganism present in the tissue sample
- Abnormal (positive) test results indicate
  - Microorganisms are present in the tissue sample.

TEACH THE PATIENT

- Explain
  - Why the test is being performed.
  - What the patient will experience during the test.
  - That the patient will not feel any pain during the test.
  - That the patient should not use antibiotics or apply antiseptic to the wound before the test.

Summary

Healthcare providers perform an assortment of tests and procedures to diagnose and treat skin conditions.

Undesired skin tissue can be frozen with cryosurgery and removed. Layers of skin can be peeled away using chemicals, Mohs micrographic surgery, curettage, electrosurgery, or by using the dermabrasion procedure where a rotating burr scrapes scars from aging skin to encourage new skin growth.

The healthcare provider can cure or minimize discomfort of melanoma and nonmelanoma skin cancer by performing a skin excision from where the tumor is surgically removed.

The microorganism that causes a skin infection is identified by performing a wound culture. In a wound culture, a sample of the infected tissue is placed in an environment conducive for growing microorganisms. The microorganism is then identified and medication is administered to the microorganism to determine which medication kills the microorganism.

Skin is used to test for allergic reactions. There are several tests that healthcare providers administer to the skin to identify allergens that cause the patient to develop an allergic reaction. The skin is also the site of the Mantoux skin test to determine if the patient has ever been exposed to *M tuberculosis.*

Quiz

1. Why is it important to know if the patient ever received BCG vaccination prior to administering the Mantoux skin test?
   a. Patients who had this vaccination are prevented from taking the Mantoux skin test result.
   b. Patients who received this vaccination will likely have a positive Mantoux skin test.
   c. Patient who had this vaccination will experience seizures if given the Mantoux skin test.
   d. Patient who had this vaccination will experience anaphylactic shock if given the Mantoux skin test.

2. What is TCA used for?
   a. Superficial chemical peel
   b. Medium chemical peel
   c. Deep chemical peel
   d. None of the above

3. Chemical peel changes skin tone.
   a. True
   b. False

4. How does the dermabrasion improve the appearance of skin?
   a. It removes the outer layer of skin, enabling new skin to grow.
   b. It removes the inner layer of skin, enabling new skin to grow.
   c. It uses a chemical to remove the out layer of skin, enabling new skin to grow.
   d. All of the above.

5. Why is not the Mantoux skin test used to diagnose tuberculosis?
   a. A positive result indicates that the patient has developed antibodies to *M tuberculosis* antigen possibly from a previous exposure to *M tuberculosis*.
   b. A negative result indicates that the patient has not developed antibodies to *M tuberculosis* antigen; however, the immune system can take up to 10 weeks to develop the antibodies following the infection.
   c. Diagnosis is made using an X-ray.
   d. All of the above.

6. What allergy test requires that an allergen-containing pad be applied to the patient's skin?
   a. Skin patch test
   b. Skin prick test
   c. Intradermal test
   d. The Q patch test

7. Why is the patient usually administered acyclovir 2 weeks prior to a chemical peel?
   a. To prevent a fungal infection
   b. To prevent a bacterial infection
   c. To prevent a viral infection
   d. To reduce bleeding during the procedure

8. Why would a patient be administered Retin-A following a chemical peel?
   a. To reduce bleeding following the procedure
   b. To prevent an infection
   c. To encourage healing
   d. None of the above

9. What would you recommend if the site of a chemical peel remained red for 3 months following the procedure?
   a. Rush the patient to the hospital.
   b. Contact the healthcare provider immediately.
   c. Place a topical antibiotic on the site.
   d. Do nothing.

10. What is the allergy test called where a drop of allergen solution is placed on the skin and the skin is scratched?
    a. Skin patch test
    b. Skin prick test
    c. Intradermal test
    d. The Q patch test

Answers

1. b. Patients who received this vaccination will likely have a positive Mantoux skin test result.
2. b. Medium chemical peel.
3. a. True.
4. a. It removes the outer layer of skin, enabling new skin to grow.
5. d. All of the above.
6. a. Skin patch test.
7. c. To prevent a viral infection.
8. c. To encourage healing.
9. b. Contact the healthcare provider immediately.
10. b. Skin prick test.

CHAPTER 27

Sinus, Ears, Nose, Throat (ENT) Tests and Procedures

Snoring, headaches, frequent infections, and turning the volume on the TV high could be signs of an underlying problem that might be prevented or resolved by performing one of several procedures that you will learn about in this chapter.

Snoring is annoying to those who have to listen to it and can also be a symptom of something more ominous such as obstructed sleep apnea. The healthcare provider can fix this problem by performing an uvulopalatopharyngoplasty or radiofrequency palatoplasty that focus on the underlying cause of snoring.

There can be a number of reasons why a patient has a headache. One common cause is sinusitis. The healthcare provider can perform a number of tests to assess the sinus, including a sinus endoscopy, sinus X-ray, and a sinus aspiration where a sample of sinus is sent to the laboratory for a culture and sensitivity test. If problem is a blockage of the sinus rather than an infection, the healthcare provider might perform sinus surgery to remove the blockage.

The headache may be caused by something more involved than sinusitis. The healthcare provider might order an electroencephalogram, which records electrical activity in the brain and can help the healthcare provider diagnose the underlying cause of the headache and other neural symptoms.

Frequent infections can be challenging to resolve. The healthcare provider might order a throat culture, sputum culture, or sputum cytology study to identify the cause of the infection and the mediation that will kill the infecting microorganism. In addition, the healthcare provider may perform a procedure to prevent reinfection. The most common are a tonsillectomy tympanostomy tube, and a tympanocentesis.

Decreased hearing can be caused by a number of factors, including a buildup of cerumen in the ear canal, disorders of the eardrum, or a neurological problem. The healthcare provider can perform tympanometry and audiometric tests to determine the cause of hearing loss.

These and other tests and procedures are discussed in this chapter.

Learning Objectives

1. Electroencephalogram (EEG)
2. Throat Culture
3. Sputum Culture
4. Sputum Cytology
5. Sinus X-ray
6. Tympanometry
7. Audiometric Testing

Key Words

Acoustic immittance
Auditory brain stem response (ABR)
Otitis media
Otoacoustic emissions (OAE)
Pure-tone audiometry
Saliva
Speech reception/word recognition
Sputum
Tuning fork test
Tympanometer
Vestibular test
Whispered speech test

Electroencephalogram 1

Electroencephalogram (EEG) measures electrical activities of the brain using electrodes attached to the patient's head. Electrical activities are recorded and displayed on a video screen and stored in a computer.

HINT

The healthcare provider may order an ambulatory EEG, which monitors electrical activities of the brain while the patient goes about normal daily activities.

WHAT IS BEING EXAMINED?

- Electrical activities in the brain

HOW IS THE PROCEDURE PERFORMED?

- The patient signs a consent form.
- The patient lies on a table.
- Twenty-five electrodes are attached to the patient's head using a paste.
- Wires connect the electrodes to the electroencephalogram.
- Electrical activity is measured.
- A strobe light might be displayed as the electrical activity is measured, if the stroboscopic simulation test is performed.
- Sleep might be induced with a sedative, if the sleep study is performed.
- The patient might be asked to hyperventilate.

RATIONALE FOR THE TEST

- To determine the underlying cause of narcolepsy
- To determine the underlying cause of seizures
- To assess if the patient has dementia
- To assess if the patient is brain-dead

NURSING IMPLICATIONS

- Determine if the patient
  - Has signed a consent form

- Has taken tranquilizers, barbiturates, sedatives, seizure medication, or muscle relaxant prior to the test
- Has ingested caffeine 8 hours before the test
- Has not shampooed his/her hair the morning of the test
- Has not applied creams, sprays, lotion, or oil on the scalp prior the test
- Has not slept prior to the test, if the test is studying sleeping disorders
- Has not moved during the test
- Has normal body temperature

UNDERSTANDING THE RESULTS

- The procedure takes less than 2 hours unless a sleep study is performed, which might take 8 hours. Results are available within 2 weeks.
- Normal test results indicate
  - Normal electrical activities
- Abnormal test results indicate
  - Abnormal electrical activities

TEACH THE PATIENT

- Explain
  - Why the procedure is being performed.
  - What the patient will experience during the procedure.
  - That the patient will not feel any pain during the procedure.
  - That the patient should not take tranquilizers, barbiturates, sedatives, seizure medication, or muscle relaxants prior to the test.
  - That the patient should not ingested caffeine 8 hours before the test.
  - That the patient should not shampoo his/her hair the morning of the test.
  - That the patient should not apply creams, sprays, lotion, or oil on the scalp prior the test.
  - That the patient should not sleep prior to the test, if the test is studying sleeping disorders.

Throat Culture 2

A throat culture is ordered when the patient has a throat infection. A tissue sample of the infected area is taken and placed in an environment conducive to the growth of microorganisms for 3 days. The tissue sample is then examined to identify the presence and the type of microorganism. Once the microorganism is identified, a sensitivity test is usually performed to determine the medication that kills the microorganism.

WHAT IS BEING EXAMINED?

- Existence of microorganisms in the infected throat

HOW IS THE PROCEDURE PERFORMED?

- The healthcare provider swabs the throat with a sterile swab to collect a tissue sample. More than one sample might be taken.
- The sterile swab is placed in a culture tube and sent to the laboratory.
- The sample(s) is then placed in a culture dish in an environment conducive to growing microorganism.
- After 3 days, tests are performed to identify the microorganism.
- Medication is applied to a portion of the culture to determine which medication kills the microorganism.

RATIONALE FOR THE TEST

- To assess the existence and type of microorganism in a patient with a throat infection.
- To assess the medication to use to treat the throat infection.

NURSING IMPLICATIONS

- Determine if the patient has taken antibiotics or applied an antiseptic to the throat prior to the test.

UNDERSTANDING THE RESULTS

- It takes a few seconds to collect the specimen. Results are known within 3 days.

- Normal (negative) test results indicate
  - No microorganism is present in the sample.
- Abnormal (positive) test results indicate
  - Microorganisms are present in the sample.

TEACH THE PATIENT

- Explain
  - Why the test is being performed.
  - What the patient will experience during the test.
  - That the patient will not feel any pain during the test.
  - That the patient should not use antibiotics or apply antiseptic to the throat before the test.

Sputum Culture 3

A sputum culture is ordered when the patient has a respiratory infection. A sample of sputum is collected in a sterile container and taken to the laboratory where the sample is placed in an environment conducive to the growth of microorganisms for 3 days. The sample is then examined to identify the presence and the type of microorganism. Once the microorganism is identified, a sensitivity test is usually performed to determine the medication that kills the microorganism.

Hint

Sputum is not saliva. Sputum is produced in the respiratory system.

WHAT IS BEING EXAMINED?

- Existence of microorganisms in the infected respiratory tract

HOW IS THE PROCEDURE PERFORMED?

- The sample is taken immediately after the patient awakens from sleep.
- The patient removes dentures.
- The patient rinses his/her mouth with water.

- The patient coughs deeply to produce the sputum.
- If the patient is unable to produce the sputum, the healthcare provider may loosen sputum in the lungs by using a mist inhaler. If this does not produce the sputum, then the healthcare provider collects the sputum using a bronchoscope that is inserted into the patient's airway.
- The sample(s) is then placed in a culture dish in an environment conducive to growing microorganism.
- After 3 days, tests are performed to identify the microorganism.
- Medication is applied to a portion of the culture to determine which medication kills the microorganism.

RATIONALE FOR THE TEST

- To assess the existence and type of microorganism in a patient with a respiratory infection
- To assess the medication used to treat the respiratory infection

NURSING IMPLICATIONS

- Determine if the patient
  - Has not used mouthwash prior to collecting the sample
  - Has not taken antibiotics prior to collecting the sample
  - Has eaten or drank 6 hours prior to collecting the sample, if the healthcare provider is using a bronchoscope to collect the sample

UNDERSTANDING THE RESULTS

- It takes a few seconds to collect the sample. Results are known within 3 days.
- Normal (negative) test results indicate
  - No microorganism is present in the sample.
- Abnormal (positive) test results indicate
  - Microorganisms are present in the sample.

TEACH THE PATIENT

- Explain
  - Why the test is being performed.
  - What the patient will experience during the test.

- That the patient will not feel any pain during the test.
- That the patient should not use antibiotics prior to the test.
- That the patient should not use mouthwash prior to collecting the sample.
- That the patient should cough deeply prior to taking the sample.
- That the patient should take the sample the first thing in the morning after awakening from sleeping.
- That the patient should make sure the sample is delivered to the laboratory immediately after it is collected.
- That sputum is different from saliva.

Sputum Cytology 4

Sputum cytology is a procedure that examines cells contained in a sputum sample to assess for asbestosis, pneumonia, respiratory infection, tuberculosis, and lung cancer.

Hint

Sputum is not saliva. Sputum is produced in the respiratory system.

WHAT IS BEING EXAMINED?

- Cells in the sputum

HOW IS THE PROCEDURE PERFORMED?

- The sample is taken immediately after the patient awakens from sleep.
- The patient removes dentures.
- The patient rinses his/her mouth with water.
- The patient coughs deeply to produce the sputum.
- If the patient is unable to produce the sputum, the healthcare provider may loosen sputum in the lungs by using a mist inhaler. If this doesn't produce the sputum, then the healthcare provider collects the sputum using a bronchoscope that is inserted into the patient's airway.
- The sample(s) is then viewed with a microscope to identify cells contained in the sputum.

RATIONALE FOR THE TEST

- To assess for lung cancer
- To assess for pneumonia
- To assess for asbestosis

NURSING IMPLICATIONS

- Determine if the patient
  - Has used mouthwash prior to collecting the sample
  - Has taken antibiotics prior to collecting the sample
  - Has eaten or drank 6 hours prior to collecting the sample, if the healthcare provider is using a bronchoscope to collect the sample

UNDERSTANDING THE RESULTS

- It takes a few seconds to collect the sample. Results are known within 3 days.
- Normal (negative) test results indicate
  - Normal cells in the sample
- Abnormal (positive) test results indicate
  - No respiratory cells in the sample

TEACH THE PATIENT

- Explain
  - Why the test is being performed.
  - What the patient will experience during the test.
  - That the patient will not feel any pain during the test.
  - That the patient should not use antibiotics prior to the test.
  - That the patient should not use mouthwash prior to collecting the sample.
  - That the patient should cough deeply prior to taking the sample.
  - That the patient should take the sample the first thing in the morning after awakening from sleep.

- That the patient should make sure the sample is delivered to the laboratory immediately after the sample is collected.
- That sputum is different from saliva.

Sinus X-ray 5

An X-ray of the sinus is ordered if the healthcare provider suspects that the patient has sinusitis or other conditions that might cause similar symptoms.

HINT

The healthcare provider may also order a CT scan of the head, which provides clearer images of the sinus than an X-ray.

WHAT IS BEING EXAMINED?

- Sinuses

HOW IS THE PROCEDURE PERFORMED?

- The patient signs a consent form.
- The patient sits in front of the X-ray machine.
- Several X-rays are taken, each at a different angle.

RATIONALE FOR THE TEST

- To identify blockage in the sinus
- To diagnose sinusitis

NURSING IMPLICATIONS

- Determine if the patient
    - Has signed a consent form
    - Can sit still

UNDERSTANDING THE RESULTS

- The procedure takes less than 30 minutes. Results are immediate.
- Normal test results indicate
  - Normal sinuses
- Abnormal test results indicate
  - A blockage is identified
  - The patient has sinusitis

TEACH THE PATIENT

- Explain
  - Why the test is being performed.
  - What the patient will experience during the test.
  - That the patient will not feel any pain during the test.

Tympanometry 6

The tympanometry is a test to measure the eardrum's response to pressure and sound and assess if there is fluid behind the eardrum. It is ordered if the patient is suspected of having otitis media, hearing loss, fluid behind the eardrum, or blocked eustachian tube.

HINT

This test is frequently performed in children.

WHAT IS BEING EXAMINED?

- The eardrum

HOW IS THE PROCEDURE PERFORMED?

- The tympanometer is placed into the ear canal.
- The tympanometer produces a tone that causes pressure change in the ear.
- The tympanometer measures the pressure change.

RATIONALE FOR THE TEST

- To identify fluid behind the eardrum
- To identify blockage of the eustachian tube
- To assess for otitis media
- To assess hearing loss

NURSING IMPLICATIONS

- Determine if the patient can lie still.

UNDERSTANDING THE RESULTS

- The procedure takes less than 5 minutes. Results are known immediately.
- Normal test results indicate
    - Normal pressure
- Abnormal test results indicate
    - Abnormal pressure

TEACH THE PATIENT

- Explain
    - Why the test is being performed.
    - What the patient will experience during the test.
    - That the patient will not feel any pain during the procedure.

Audiometric Testing 7

Audiometric testing determines the degree with which a patient can hear. Hearing loss is the patient's inability to hear or understand a range of sound frequencies due to either structural problems with the ear or neural problems preventing the brain from receiving impulses from the ear or properly interpreting those impulses. There are eight types of audiometric tests. These are

- Pure-tone audiometry: This test determines sound frequencies that can be heard by the patient.

- Whispered speech test: This test determines if the patient can hear low-volume sounds spoken behind him.
- Speech reception/word recognition: This test is used to identify sensorineural hearing loss.
- Tuning fork test: This test determines if the hearing problem is caused by sound being received by nerves in the ear or nerve problems.
- Auditory brain stem response (ABR): This test is used to detect sensorineural hearing problems.
- Otoacoustic emissions (OAE): This test identifies hearing problems in newborns.
- Vestibular test: This test identifies problems with the inner ear.
- Acoustic immittance: This test assess the middle ear's ability to conduct sound.

WHAT IS BEING EXAMINED?

- Hearing

HOW IS THE PROCEDURE PERFORMED?

- The patient signs a consent form.
- Pure-tone audiometry
  - The patient wears headphones that are connected to an audiometer.
  - The audiometer produces a series of tones.
  - The patient indicates which tones are heard.
- Whispered speech test
  - The patient places his/her finger in the ear canal of one ear.
  - The healthcare provider stands 2 ft behind the patient.
  - The healthcare provider whispers words.
  - The patient indicates if he/she can hear those words.
  - The test is repeated using the opposite ear.
- Speech reception/word recognition
  - The healthcare provider speaks familiar two-syllable words at varied degrees of loudness.

- The patient is asked to repeat those words.
- The results are measured as the patient's spondee threshold.

- Tuning fork test
  - The healthcare provider places a vibrating tuning fork either behind the ear or on the patient's head.
  - The patient indicates when he/she hears the tuning fork.
- Auditory brain stem response (ABR)
  - Electrodes are placed on the patient's ear lobe and scalp.
  - Earphones are placed in the patient's ears.
  - A clicking noise is sounded through the earphones.
  - The electrodes detect neural responses to the clicking noise and records the response on a graph.
- Otoacoustic emissions (OAE)
  - A microphone is placed in a newborn's ear.
  - A flexible probe is also placed in the baby's ear.
  - Sound is passed through the probe.
  - The microphone detects the response of inner ear to the sound.
- Vestibular test
  - The patient moves arms and legs, stands on one foot, walks heel-to-toe with his/her eyes open and then closed.
  - The patient's ability to keep his/her balance is assessed.
- Acoustic immittance
  - The ear canal is sealed closed using a soft-tip instrument.
  - Sound and air are directed toward the ear at different pressures.
  - A measurement is taken that registers how the middle ear relays sound.

RATIONALE FOR THE TEST

- To assess the patient's hearing
- To identify the underlying cause of hearing loss

NURSING IMPLICATIONS

- Determine if the patient
  - Has signed a consent form

- Can sit still
- Can respond to instructions
- Can indicate if sound is heard

UNDERSTANDING THE RESULTS

- The procedure takes less than 1 hour. Results are immediate.
- Normal test results indicate
  - Normal hearing
- Abnormal test results indicate
  - Structural or neurological problems cause reduced hearing ability.

TEACH THE PATIENT

- Explain
  - Why the procedure is being performed.
  - What the patient will experience during the procedure.
  - That the patient will not feel any pain during the procedure.

Summary

In this chapter you learned that the underlying cause of snoring can be remedied by performing an uvulopalatopharyngoplasty that removes excess tissue and radiofrequency palatoplasty that inhibits tissue vibration.

Sinusitis might be diagnosed by performing a sinus endoscopy, sinus X-ray, or a sinus aspiration, where a sample of sinus is sent to the laboratory for a culture and sensitivity test. The healthcare provider might also perform sinus surgery if the sinus is blocked.

The healthcare provider might order an electroencephalogram to identify abnormal electrical activities in the brain that may be a symptom of a serious neurological disorder.

Frequent throat infections are challenging to resolve; however, the healthcare provider might order a throat culture, sputum culture, or sputum cytology to identify the cause of the infection and the mediations that will kill the infecting microorganism. In addition, the healthcare provider may perform a procedure to prevent reinfection. The most common are a tonsillectomy tympanostomy tube, and a tympanocentesis.

Hearing loss is caused by a number of factors, including a buildup of cerumen in the ear canal, disorders of the eardrum, or a neurological problem. The healthcare provider canal perform tympanometry and audiometric tests to determine the cause of hearing loss.

Quiz

1. What would you expect the healthcare provider to order for a patient who experiences random seizures?
   a. An ambulatory EEG
   b. A helmet
   c. Restraints
   d. All of the above

2. What would your response be if a patient arriving for an EEG for sleep disorder tells you that he/she had a long nap before arriving for the test?
   a. Cancel the test, since the patient no longer has a sleeping disorder.
   b. Reschedule the test.
   c. Reschedule the test asking the patient not to sleep before the test.
   d. Wait 2 hours before conducting the test.

3. Why should a patient take a deep cough to produce a sputum sample?
   a. A deep cough loosens sputum.
   b. Sputum is produced in the respiratory system.
   c. A deep cough brings sputum into the oral cavity.
   d. All of the above.

4. What is the difference between sputum cytology and a sputum culture?
   a. A sputum cytology studies cells contained in the sputum, while a sputum culture identifies microorganism in the sputum.
   b. There is no difference.
   c. A sputum cytology identifies microorganism in the sputum, while a sputum culture studies cells contained in the sputum.
   d. None of the above.

5. The patient asks why it takes more than 1 day to get the results of a throat culture. What is your best response?
   a. The laboratory is backed up with other samples.
   b. Sufficient time must pass to allow the microorganism to grow.
   c. An outside laboratory is used.
   d. It takes time to determine what drug to use to kill the microorganism.

6. What is your response if a patient tells you that he/she spit into the sterile container when you asked for a sputum culture?
   a. Teach the patient how to properly produce a sputum sample.
   b. Sputum is not saliva.
   c. He/she contaminated the container.
   d. Take the sample to the laboratory quickly.

7. What is the best time to collect a sputum sample?
   a. Right before meals
   b. First thing in the morning
   c. Immediately after the patient awakens from sleep
   d. After the patient rinses his/her mouth with water

8. What is the purpose of tympanometry?
   a. To measure the eardrum's response to pressure and sound
   b. To identify the contents of fluid behind the eardrum
   c. To permit fluid behind the eardrum to drain
   d. To treat an ear infection

9. What test is performed to determine if the patient has sensorineural hearing loss?
   a. Pure-tone audiometry
   b. Speech reception/word recognition
   c. Whispered speech test
   d. Otoacoustic emissions test

10. What test is performed to identify hearing problems in a newborn?
    a. Pure-tone audiometry
    b. Speech reception/word recognition
    c. Whispered speech test
    d. Otoacoustic emissions test

Answers

1. a. An ambulatory EEG.
2. c. Reschedule the test asking the patient not to sleep before the test.
3. d. All of the above.
4. a. A sputum cytology studies cells contained in the sputum, while a sputum culture identifies microorganism in the sputum.
5. b. Sufficient time must pass to allow the microorganism to grow.
6. a. Teach the patient how to properly produce a sputum sample.
7. c. Immediately after the patient awakens from sleep.
8. a. To measure the eardrum's response to pressure and sound.
9. b. Speech reception/word recognition.
10. d. Otoacoustic emissions test.

CHAPTER 28

Vision Tests and Procedures

Light rays pass through the cornea, the pupil, and lenses which focus the ray of light on to the retina located at the back of the eye. When light rays are not properly focused on the retina, the patient is unable to see clearly.

Light rays focused in front of the retina cause myopia (nearsightedness), enabling patients to better see things near them than at a distance. Light rays focused behind the retina cause hyperopia (farsightedness), enabling patients to see things at a distance better than up close. When light rays are irregularly bent, images are blurred resulting in astigmatism.

In this chapter you will learn about tests that are used to diagnose problems with sight and disorders that can lead to loss of vision. You'll also learn about procedures that can be performed to treat vision disorders.

A standard vision test is used to assess the patient's ability to see close-up and at a distance. It also determines if the patient is experiencing peripheral vision difficulty or might have macular degeneration. In addition, the test determines the refractive error of the patient's eye, which determines the corrective lenses needed to restore the patient's eyesight.

Patients who have myopia might be able to have this condition fixed by having corneal ring implants or by reshaping with cornea using photorefractive keratectomy.

Glaucoma is caused by increased intraocular pressure. The healthcare provider assesses intraocular pressure by performing tonometry. If intraocular pressure is elevated, the healthcare provider may perform one of several procedures to relieve the pressure.

A trabeculectomy or a trabeculotomy can be performed to drain aqueous humor that might be backing up due to a blockage in the trabecular meshwork. Seton glaucoma surgery might be performed to insert a drainage tube in the eye to drain the aqueous humor.

Blood flow in the eye can be assessed by performing an eye angiogram, which detects a vitreous hemorrhage. A vitreous hemorrhage is treated by performing a vitrectomy.

A detached retina is a common problem. It occurs when the retina is detached from the back wall of the eye. When this occurs, the healthcare provider can repair the problem by performing a sclera buckling to relieve traction on the retina. A pneumatic retinopexy might be performed to push the retina back into the position by using a gas bubble.

These and other tests and procedures are covered in this chapter.

Learning Objectives

1. Vision Tests
2. Tonometry
3. Electronystagmogram (ENG)

Key Words

Amsler grid test
Applanation tonometry
Color vision test
Confrontation test
E-chart test
Electronic indentation tonometry
Fluorescein
Near test
Perimetry test
Pneumotonometry test
Refraction test
Schiotz tonometry
Snellen test
Tangent screen test
Tonometer
Trabecular meshwork
Vitreous hemorrhage

Vision Tests

The healthcare provider examines the patient's peripheral vision and ability to see near and far distances, along with his/her ability to distinguish colors by performing nine vision tests. These are

- Confrontation test: This test assesses the patient's peripheral vision by gazing at the healthcare provider's nose.
- Amsler grid test: This test assesses for macular degeneration.
- Perimetry test: This test assesses the patient's peripheral vision by flashing lights randomly in a perimeter.
- Tangent screen test: This test assesses the patient's peripheral vision by gazing at a concentric circle image.
- Snellen test: This test assesses the patient's ability to see distances.
- E-chart test: This test assesses the patients' ability to see distances when they are unable to read.
- Near test: This test assesses the patient's ability to see near distances.
- Color vision test: This test assesses the patient's ability to distinguish colors.
- Refraction test: This test measures the refractive error of the patient's eyes to determine lens that corrects his/her eyesight.

WHAT IS BEING EXAMINED?

- Eyesight

HOW IS THE PROCEDURE PERFORMED?

- Confrontation test
  - The healthcare provider sits 3 ft from where the patient is sitting.
  - The patient covers one eye.
  - The patient stares at the healthcare provider's nose with the other eye.
  - The healthcare provider moves his/her finger from outside the visual field to the center of the visual field.
  - The healthcare provider moves his/her finger from the center of the visual field to opposite outside of the visual field.
  - The patient signals when he/she first sees the healthcare provider's finger.
  - The test is repeated using the other eye.

- Amsler grid test
    - The patient covers one eye.
    - A 4-in square chart containing straight lines and a grid with a black dot at the center is held 14 in from the patient.
    - The patient focuses his/her eye on black dot on the chart.
    - The patient tells the healthcare provider if he/she can see the black dot, sees a dark spot other than the black dot, or sees the straight lines.
    - The patient should see the black dot and the straight lines.
- Perimetry test
    - The patient looks into the perimeter, focusing on a dot in the center of the perimeter screen.
    - Lights randomly flash at various locations on the perimeter screen.
    - The patient presses a button each time he/she sees a light on the screen.
    - The perimeter generates a printout, showing areas in the patient's peripheral vision where he/she did not see the light.
- Tangent screen test
    - The patient sits 6 ft away from the black screen that contains concentric circles and lines.
    - The patient covers one eye.
    - The patient focuses the other eye on a center of the circle.
    - A wand containing an object is moved from the outside of the circle to the inside of the circle.
    - The patient signals the healthcare provider when he/she sees the object.
    - The test is repeated using different-sized objects on the wand.
    - The test is repeated for the other eye.
- Snellen test
    - The patient is positioned 20 ft from the Snellen chart.
    - The patient covers one eye.
    - The patient reads the smallest row of letters on the chart.
    - Each row is designated with a visual acuity value.
    - A row designated as 20/20 visual acuity means a person with normal vision can read this line at a distance of 20 ft.
    - A row designated as 20/40 visual acuity means a person with normal vision can read this line at a distance of 40 ft.

- E-chart test
  - The patient covers one eye.
  - The patient reads the smallest E on the chart.
  - Each row is designated with a visual acuity value, the same as in the Snellen test.
- Near test
  - The patient is given a card that contains text.
  - The patient holds the card 14 in from his/her eyes.
  - The patient reads the text on the card.
  - The test is repeated with cards that have progressively smaller type size.
- Color vision test
  - The patient views pages that contain numbers, symbols, and paths that appear on a background of color dots.
  - The patient is asked to say out loud the number, symbol, or path.
  - The patient may have problems differentiating colors if he/she is unable to identify the number, symbol, or path correctly.
- Refraction test
  - The patient views the Snellen chart or E-chart and the cards used for the near test while looking through corrective lens.
  - The patient then notifies the healthcare provider when he/she can clearly read the chart and cards.
  - The healthcare provider then writes a prescription for those lenses.

RATIONALE FOR THE TEST

- To determine if the patient is color blind
- To correct the patient's vision with glasses or contact lenses
- To determine if the patient is losing his/her peripheral vision
- To determine if the patient has macular degeneration

NURSING IMPLICATIONS

- Determine if the patient
  - Has signed a consent form
  - Can sit still

- Can read
- Wears glasses or contact lenses
- Has an eye infection
- Can follow instructions
- Has his/her glasses or contact lenses with them

UNDERSTANDING THE RESULTS

- The procedure takes less than 1 hour. Results are immediate.
- Normal test results indicate
  - Normal eyesight
  - Normal peripheral vision
  - No macular degeneration
  - Can see all colors
- Abnormal test results indicate
  - Near sighted (can see near distances)
  - Far sighted (can see far distances)
  - Restricted peripheral vision
  - Possible macular degeneration
  - Color blind

TEACH THE PATIENT

- Explain
  - Why the procedure is being performed.
  - What the patient will experience during the procedure.
  - That the patient will not feel any pain during the procedure.
  - The patient should bring his/her glasses or contact lenses to the test.

Tonometry 2

Tonometry is a test for glaucoma that measures the intraocular pressure of the patient's eye by assessing the amount of pressure necessary to flatten the cornea. There are four methods used to perform tonometry. These are

- Pneumatonometry: This method uses a puff of air to measure intraocular pressure. No direct contact is made with the eye.
- Applanation tonometry: This method uses a tonometer to measure intraocular pressure.
- Electronic indentation tonometry: This method uses a tonometer that is connected to a computer to measure intraocular pressure.
- Schiotz tonometry: This method uses a plunger to measure intraocular pressure.

WHAT IS BEING EXAMINED?

- Intraocular pressure

HOW IS THE TEST PERFORMED?

- The patient removes contact lenses.
- Pneumatonometry
  - The patient sits.
  - The patient places his/her chin on a chin rest and forehead against a stabilizing bar of the slit lamp.
  - A light is shined into the eye.
  - A puff of air is blown into the eye.
  - The tonometer measures the change in the reflection of the light off the cornea when the puff of air strikes the eye.
  - Intraocular pressure is read from the tonometer.
- Applanation tonometry
  - The patient sits.
  - A local anesthetic is placed in the eye.
  - Fluorescein is administered to the eye to make the cornea visible.
  - The patient places his/her chin on a chin rest and forehead against a stabilizing bar of the slit lamp.
  - A tonometer is pressed against the cornea.
  - The tension dial on the tonometer records the intraocular pressure.
- Electronic indentation tonometry
  - The patient sits.
  - A local anesthetic is placed in the eye.

- The patient places his/her chin on a chin rest and forehead against a stabilizing bar of the slit lamp.
- A tonometer is pressed against the cornea.
- A click sounds when a reading is taken.
- Four readings are taken.
- Intraocular pressure is displayed on the computer screen.

- Schiotz tonometry
  - The patient lies on a table.
  - A local anesthetic is placed in the eye.
  - The patient stares at a fix point on the ceiling.
  - A tonometer is pressed against the cornea.
  - Intraocular pressure is displayed on the tonometer.

RATIONALE FOR THE PROCEDURE

- Assess for glaucoma

NURSING IMPLICATIONS

- Determine if the patient
  - Has not drunk alcohol 12 hours before the test
  - Has not drunk anything for 4 hours before the test
  - Can sit still
  - Has not smoked marijuana 24 hours before the test
  - Can respond to directions during the procedure
  - Can withstand a probe being directed at his/her eye
  - Has removed his/her contact lenses
  - Signed a consent form

UNDERSTANDING THE RESULTS

- The procedure takes less than 30 minutes. Results are immediate.
- Normal test results indicate
  - Normal intraocular pressure

- Abnormal test results indicate
  - High intraocular pressure
  - Low intraocular pressure

TEACH THE PATIENT

- Explain
  - Why procedure is being performed.
  - How the procedure is performed.
  - That the patient will not feel any pain during the procedure.
  - That the patient should not insert contact lenses for 2 hours following the test.
  - That the patient should not drink alcohol 12 hours before the test.
  - That the patient should not drink anything for 4 hours before the test.
  - That the patient should not smoke marijuana 24 hours before the test.
  - That the patient should not rub his/her eyes for 30 minutes following the test.

Electronystagmogram

Electronystagmogram (ENG) is a test that measures eye movement, both voluntary eye movement and nystagmus, to assess the patient's balance and the underlying cause of vertigo.

WHAT IS BEING EXAMINED?

- Balance

HOW IS THE TEST PERFORMED?

- The patient signs a consent form.
- Any obstruction in the ear canal is removed.
- Five electrodes are pasted on the patient's face.
- The room is darkened.
- During each of the following, electrical activities from the electrodes are recorded.

- The patient follows a moving light with his/her eyes while holding his/her head still.
- The patient closes his/her eyes and is asked to perform mental arithmetic.
- The patient opens his/her eyes and follows the back and forth movement of a pendulum.
- The patient is asked to look at objects that are moved in front of him/her.
- The patient is asked to move his/her head up and down and side to side.
- Cold water is placed in the patient's ear.
- Warm water is placed in the patient's ear.

RATIONALE FOR THE PROCEDURE

- To assess the underlying cause of loss of balance and vertigo
- To assess the inner ear

NURSING IMPLICATIONS

- Determine if the patient
  - Has not taken tranquilizers 5 days before the test
  - Has not drunk alcohol 5 days before the test
  - Has not ingested caffeine 5 days before the test
  - Has not eaten 4 hours before the test
  - Is not wearing facial makeup
  - Has not had a back or neck disorder

UNDERSTANDING THE RESULTS

- The procedure takes less than 2 hours. Results are known within a week.
- Normal test results indicate
  - Normal voluntary and involuntary eye movement
- Abnormal test results indicate
  - Abnormal involuntary eye movement

TEACH THE PATIENT

- Explain
  - Why procedure is being performed.
  - How the procedure is performed.
  - That the patient will not feel any pain during the procedure.
  - The patient should not take tranquilizers 5 days before the test.
  - The patient should not drink alcohol 5 days before the test.
  - The patient should not ingest caffeine 5 days before the test.
  - The patient should not eat 4 hours before the test.
  - The patient should not wear facial makeup during the test.

Summary

In this chapter you learned about tests that are used to diagnose problems with sight and disorders that can lead to loss of vision and about procedures that can be performed to treat vision disorders.

Patients undergo a standard vision test to assess their ability to see close-up, at a distance, and to assess their peripheral vision. During the test, the healthcare provider determines the patient's refractive error and then prescribes corrective lenses needed to restore the patient's eyesight. Some patients who have myopia choose to have this condition fixed by having corneal ring implants or by reshaping the cornea using photorefractive keratectomy.

Increased intraocular pressure causes glaucoma. The tonometry test measures intraocular pressure and, if elevated, the healthcare provider may perform a trabeculectomy or a trabeculotomy (children only) to drain aqueous humor that might be backing up due to a blockage in the trabecular meshwork. Seton glaucoma surgery might be performed to insert a drainage tube in the eye to drain the aqueous humor.

An eye angiogram assesses blood flow in the eye and can detect a vitreous hemorrhage. A vitreous hemorrhage is treated by performing a vitrectomy.

The retina can become detached from the back wall of the eye. It can be repaired by performing a sclera buckling to relieve traction on the retina. A pneumatic retinopexy might be performed to push the retina back into the position using a gas bubble.

Quiz

1. What is an E-chart used for?
    a. To assess the patient's ability to see distances when he/she is unable to read
    b. To assess the patient's ability to see distances
    c. To assess the patient's ability to see near him/her
    d. To assess the patient's ability to see near him/her when he/she is unable to read

2. What is a confrontation test?
    a. Examination of patient's peripheral vision by gazing at the healthcare provider's eyes
    b. Examination of patient's peripheral vision by gazing at the healthcare provider's nose
    c. Examination of patient's peripheral vision by gazing at the healthcare provider's hands
    d. None of the above

3. What tests the patient's peripheral vision by flashing lights randomly in a perimeter?
    a. Amsler grid test
    b. Perimetry test
    c. Tangent screen test
    d. Snellen test

4. What tests the patient's peripheral vision by gazing at a concentric circle image?
    a. Amsler grid test
    b. Perimetry test
    c. Tangent screen test
    d. Snellen test

5. How is the pneumotonometry test performed?
   a. It measures intraocular pressure directly on the eye.
   b. It uses a puff of air to measure intraocular pressure.
   c. It uses a probe to measure intraocular pressure.
   d. None of the above.

6. During the Schiotz tonometry test a tonometer is pressed against the cornea.
   a. True
   b. False

7. What test measures for macular degeneration?
   a. Amsler grid test
   b. Perimetry test
   c. Tangent screen test
   d. Snellen test

8. Why should a patient not take tranquilizers 5 days before the electronystagmogram (ENG) test?
   a. Tranquilizers might slow eye movement giving a false test result.
   b. Tranquilizers might increase eye movement giving a false test result.
   c. Tranquilizers might cause nausea and vomiting during the test.
   d. Tranquilizers might cause nausea and vomiting following the test.

9. The near test assesses the patient's ability to see long distances.
   a. True
   b. False

10. What is the purpose of a refraction test?
    a. To determine the corrective lens for prescribing to the patient
    b. To assess if the patient can see refracted light waves
    c. To assess if the patient can see refracted light waves at night
    d. All of the above

Answers

1. a. To assess patient's ability to see distances when he/she is unable to read.
2. b. Examination of patient's peripheral vision by gazing at the healthcare provider's nose.
3. b. Perimetry test.
4. c. Tangent screen test.
5. b. It uses a puff of air to measure intraocular pressure.
6. a. True.
7. a. Amsler grid test.
8. a. Tranquilizers might slow eye movement giving a false test result.
9. b. False.
10. a. To determine the corrective lens to prescribe to the patient.

Medical Tests and Procedures Demystified

Final Exam and Answers

1. What would you tell a patient who is about to undergo FBS test?
    a. The patient should fast for 8 hours before the test. All diabetes mellitus (DM) medications are withheld until the test is completed.
    b. The patient should eat a regular meal but refrain from ingesting sugar beverages.

c. The patient should eat a regular meal, including ingesting sugar beverages.

d. The patient should eat a bland meal and refrain from ingesting sugar beverages.

2. Before taking the prolactin test what must the patient do?
   a. Avoid the sun.
   b. Rest for 30 minutes.
   c. Exercise for 30 minutes.
   d. Avoid alcohol.
3. What negatively affects the immunoglobulins test?
   a. Recent blood transfusion.
   b. Radioactive scan 3 days before the test is administered.
   c. Recent vaccination.
   d. All of the above.
4. What should the patient do before the homocysteine test?
   a. Avoid eating and drinking except for water 12 hours before the test is administered.
   b. Walk 10 blocks before the test is administered.
   c. Avoid walking.
   d. Avoid driving.
5. What does the transcutaneous bilirubin meter measure?
   a. It is a handheld meter that helps determine if the newborn has jaundice.
   b. It measures the level of bilirubin by being placed on the skin.
   c. It measures the level of bilirubin in a newborn.
   d. All of the above.
6. What does the urodynamic test measure?
   a. Assesses bladder function
   b. Assesses the position of the urethra
   c. Assesses urethral pressure
   d. None of the above
7. Why would a patient who is scheduled for a CT scan stop taking Glucophage?
   a. Glucophage may react with contrast material.
   b. Glucophage reacts to X-rays.

c. Glucophage reacts to MRI.

d. There is no need to stop taking Glucophage if contrast material is administered before the CT scan.

8. What instructions would you give a patient who is scheduled for the renin assay test?

   a. Relax 2 hours before the first blood sample is taken.

   b. Ambulate for 2 hours after the first blood sample is taken.

   c. Sit upright when blood samples are taken.

   d. All of the above.

9. How does the dermabrasion improve the appearance of skin?

   a. It removes the outer layer of skin enabling new skin to grow.

   b. It removes the inner layer of skin enabling new skin to grow.

   c. It uses a chemical to remove the outer layer of skin enabling new skin to grow.

   d. All of the above.

10. What is done if a patient is anxious about being placed in a CT scanner?

    a. Administer a sedative per order.

    b. Cancel the test.

    c. Wait for a calmer moment to administer the test.

    d. Tell the patient to behave like an adult.

11. How do you assess the underlying cause of xerostomia?

    a. By using the thyroid gland scan

    b. By using the salivary gland scan

    c. By using the liver scan

    d. By using the bladder scan

12. What enzymes were found mostly in cardiac muscles?

    a. Troponin and CPK-MB

    b. CPK and creatinine phosphokinase

    c. CK and creatinine phosphokinase

    d. CKB and creatinine phosphokinase

13. What hormone signals the adrenal glands to release aldosterone?

    a. Renin

    b. Cortisol

c. Phenytoin
d. Dexamethasone

14. Why is the CEA test used to screen for cancer?
    a. The carcinoembryonic antigen is normally present during fetal development and is terminated at birth. It is present if there is a tumor.
    b. The carcinoembryonic antigen is excellent for diagnosing cancer.
    c. The carcinoembryonic antigen is used for early detection of cancer.
    d. None of the above.
15. What can cause a false-positive result from the bone mineral density test?
    a. The patient has arthritis.
    b. The patient has a T-score of 1.
    c. The patient has a Z-score of 1.
    d. The patient has an X-score of less than 1.
16. What should be done with the first urine during a 24-hour urine collection?
    a. It should be refrigerated.
    b. It should be stored in a gallon container.
    c. It should be discarded.
    d. All of the above.
17. Why is a perfusion CT ordered?
    a. To determine osteoporosis
    b. To assess the results of the KUB
    c. To determine blood supply to the brain
    d. None of the above
18. What is a first-pass scan?
    a. Captures images of blood going through the heart and lungs for the first time
    b. A type of cardiac blood pool scan
    c. A procedure that uses a radioactive tracer
    d. All of the above
19. What may the healthcare provider suspect if a patient lacks hexosaminidase A enzyme?
    a. May have multiple sclerosis
    b. May have Tay-Sachs disease

c. Is pregnant

d. Is infertile

20. Why stop taking Coumadin before a bone biopsy?

a. Coumadin increases coagulation time, thereby increasing the risk of bleeding during the procedure.

b. Coumadin decreases coagulation time, thereby increasing the risk of bleeding during the procedure.

c. Coumadin has a negative reaction to anesthetic administered during the procedure.

d. Coumadin has a negative reaction to the antibiotic that is administered following the procedure.

21. What does the total thyroxine test measure?

a. The total amount of thyroxine that is missing from the patient

b. The amount of thyroxine that is not bound to globulin

c. The amount of thyroxine that is attached to globulin and that is not bound to globulin

d. All of the above

22. What test helps diagnose precocious puberty?

a. FSH test

b. AFP

c. hCG

d. Inhibin A

23. Why would a patient be administered Retin-A following a chemical peel?

a. To reduce bleeding following the procedure

b. To prevent an infection

c. To encourage healing

d. None of the above

24. Why should a blood sample for prolactin be taken 3 hours after the patient awakens?

a. The test should always be taken immediately prior to lunch.

b. The patient requires a good night sleep before the test.

c. Prolactin levels are normally high when the patient first awakens.

d. None of the above.

25. What might you suspect if the patient's urine is cloudy?
    a. The patient is dehydrated.
    b. The patient has taken Pepto-Bismol.
    c. Bacterial infection.
    d. The patient is overhydrated.
26. What hormone does the renin cause to be released?
    a. Aldosterone
    b. Cortisol
    c. Phenytoin
    d. Dexamethasone
27. Why is the GHb test administered?
    a. To determine if the DM patient has maintained adequate blood glucose level for the previous 60 days
    b. To determine if the DM patient has maintained adequate blood glucose level for the previous 120 days
    c. To determine if the DM patient has maintained adequate blood glucose level for the previous 30 days
    d. None of the above
28. What is an E-chart used for?
    a. It assesses the patients' ability to see distances when they are unable to read.
    b. It assesses the patients' ability to see distances.
    c. It assesses the patients' ability to see near them.
    d. It assesses the patients' ability to see near them when they are unable to read.
29. What is the most accurate method of measuring bone mineral density?
    a. Dual-photo absorptiometry
    b. Quantitative computed tomography
    c. Dual-energy X-ray absorptiometry
    d. Ultrasound
30. What is a stress echocardiogram?
    a. A procedure that creates images of the heart
    b. A procedure that is performed after the patient has exercised

c. A procedure that is performed when the patient's heart is placed under stressed using a medication

d. All of the above

31. Why would a patient be asked to refrain from taking a sedative prior to a pulmonary function test?

a. A sedative may invalidate the test results because it might slow down respiration.

b. The patient will fall asleep during the test.

c. The patient may vomit during the test.

d. None of the above.

32. Aside from renal failure, what would a high BUN level indicate?

a. Shock

b. Urinary tract blockage

c. Respiratory tract bleeding

d. All of the above

33. What is the function of an MRA?

a. Assess the speed, direction, and flow of blood

b. Assess fluid content of the brain

c. Assess changes in the brain chemistry

d. All of the above

34. Why should you assess if the patient has taken antibiotics before testing for potassium?

a. Taking antibiotics indicates that the patient has a bacterial infection.

b. Some antibiotics contain potassium.

c. Patients who take antibiotics always have higher than normal levels of potassium.

d. Patients might vomit during the test.

35. What must be assessed before administering the BUN test?

a. Age of the patient

b. If the patient has taken diuretics

c. If the patient has ingested meat

d. All of the above

36. What allergy test requires that an allergen-containing pad be applied to the patient's skin?
    a. Skin patch test
    b. Skin prick test
    c. Intradermal test
    d. Q patch test
37. Why would a woman who is positive for BRCA1 receive annual MRI scans?
    a. The patient is at high risk for breast cancer and the MRI provides high detailed views of the patient's breasts.
    b. BRCA1 interferes with a normal mammogram.
    c. Only radio waves can penetrate the BRCA1.
    d. None of the above.
38. Why is not the Mantoux skin test used to diagnose tuberculosis?
    a. A positive result indicates that the patient has developed antibodies to *Mycobacterium tuberculosis* antigen possibly from a previous exposure to *M tuberculosis*.
    b. A negative result indicates that the patient has not developed antibodies to *M tuberculosis* antigen; however the immune system can take up to 10 weeks to develop the antibodies following the infection.
    c. Diagnosis is made using an X-ray.
    d. All of the above.
39. What test measures creatinine in 24-hour urine sample?
    a. Creatinine clearance test
    b. Creatinine level test
    c. BUN:creatinine level test
    d. None of the above
40. Why should a patient with kidney disease avoid being administered contrast material prior to an MRI?
    a. The patient might have difficulty excreting the contrast material.
    b. The contrast material contains metallic elements.
    c. Patients with kidney disease should never receive an MRI.
    d. None of the above.

41. Why should the patient avoid taking Coumadin prior to a cardiac procedure?
    a. The patient's bleeding time increases.
    b. There is a high risk of bleeding.
    c. The healthcare provider will need to take precautions to control bleeding that might occur during the procedure.
    d. All of the above.
42. What should the patient avoid before taking the creatinine clearance test?
    a. Strenuous exercise
    b. Eating meat
    c. Drinking coffee
    d. All of the above
43. What is the purpose of the monospot test?
    a. This test identifies heterophil antibodies that form between 2 and 9 weeks after the patient becomes infected.
    b. This test identifies heterophil antibodies that form between 1 and 9 weeks after the patient becomes infected.
    c. This test identifies heterophil antibodies that form between 2 and 4 weeks after the patient becomes infected.
    d. This test identifies heterophil antibodies that form between 2 and 9 weeks before the patient becomes infected.
44. What kind of hysterectomy removes the uterus, cervix, ovaries, and fallopian tubes?
    a. Total hysterectomy
    b. Radical hysterectomy
    c. Total hysterectomy with bilateral salpingo-oophorectomy
    d. Semi-hysterectomy
45. What is the purpose of administering tracer material in a lymph node biopsy?
    a. To highlight cancer cells.
    b. Tracer material helps map the route that cancer cells spread from the cancer site through the lymphatic system.
    c. To highlight noncancer cells.
    d. An incident report.

46. The spirometry test measures
    a. The amount of gasses that cross the alveoli per minute
    b. The patient's airway responses to allergens
    c. Volume and capacity of the lungs
    d. None of the above
47. What test is performed before drawing an arterial blood sample?
    a. Allen test
    b. Collection test
    c. Acid-alkaline test
    d. None of the above
48. What test is ordered to determine if the patient has a *Borrelia burgdorferi* bacteria infection?
    a. Indirect fluorescent antibody
    b. Enzyme-linked immunosorbent assay
    c. ELISA
    d. All of the above
49. Why would a healthcare provider order a transesophageal echocardiogram?
    a. To assess cardiac contraction under stress
    b. To generate an image of the heart that is not obstructed by bone
    c. To assess cardiac blood flow to the extremities
    d. To assess cardiac blood flow to the lungs
50. What microorganism causes a fishy odor in vaginosis?
    a. *Candida albicans*
    b. *Trichomonas vaginalis*
    c. Bacterial vaginosis
    d. None of the above
51. Normal results for a pulmonary function test
    a. Depend on the patient's age, height, sex, weight, and race
    b. Depend on laboratory standard
    c. Depend on if the patient wears dentures
    d. Depend on if the patient is a smoker

52. What is an excision biopsy?
    a. The entire skin lesion is removed.
    b. A piece of the skin lesion is removed.
    c. A circular sample of skin is removed.
    d. A few cells are removed from the top of the lesion.
53. Why is the direct Coombs test administered?
    a. On a patient to determine if an autoimmune response is occurring
    b. On a patient who received a blood transfusion to determine if there is a transfusion reaction
    c. On a newborn whose mother is Rh-negative to determine if the antibodies crossed the placenta into the newborn's blood
    d. All of the above
54. A patient with 20% carbon monoxide blood level
    a. Will show symptoms of carbon monoxide poisoning
    b. Will not show symptoms of carbon monoxide poisoning
    c. Will show slight symptoms of carbon monoxide poisoning
    d. None of the above
55. What is a vaginal vault prolapse?
    a. The upper segment of the vagina collapses and extends outside the vagina
    b. The pelvic wall collapses
    c. The pelvic ligaments are stretched
    d. The lower abdominal wall collapses
56. Why is a pillow placed under the patient's shoulders during a thyroid gland biopsy?
    a. This prevents the patient from moving during the procedure.
    b. This causes the thyroid gland to be pushed backward.
    c. This causes the thyroid gland to be pushed forward.
    d. This assures that the patient's airway is open during the procedure.
57. Diarrhea can result in
    a. Respiratory alkalosis
    b. Metabolic alkalosis
    c. Metabolic acidosis
    d. Respiratory acidosis

58. A perfusion scan involves
    a. A radioactive tracer being injected into a blood vessel
    b. Contrast material being injected into a blood vessel
    c. The patient inhaling a gas
    d. None of the above
59. What is a transurethral biopsy?
    a. A needle is inserted through the rectum to remove samples of the prostate tissue
    b. A needle is inserted through the urethra to remove samples of prostate tissue
    c. A needle is inserted in the skin between the anus and the scrotum to remove samples of the prostate tissue
    d. A needle is inserted between the third and fourth ribs to remove samples of lung tissue
60. What is used to hold the cervix in place during an endometrial biopsy?
    a. Braca clamp
    b. Spectrum clamp
    c. Tenaculum clamp
    d. Braca2 clamp
61. Alcoholism can result in
    a. Respiratory alkalosis
    b. Metabolic alkalosis
    c. Metabolic acidosis
    d. Respiratory acidosis
62. What test is performed if the healthcare provider suspects pulmonary emboli?
    a. Bronchoscopy
    b. Lung scan
    c. Pulmonary function test
    d. Thoracotomy
63. What is the purpose of the VDRL test?
    a. Identifies *Treponema pallidum* antibodies
    b. Identifies *T pallidum* antibodies

c. Identifies the anti-cardiolipin antibodies
d. Identifies the *T pallidum* bacterium

64. What test measure carboxyhemoglobin level in the blood?
    a. Carbon monoxide test
    b. Total carbon dioxide test
    c. Arterial blood gases
    d. None of the above
65. What is Holter monitoring?
    a. An ambulatory electrocardiogram
    b. Monitoring the patient's blood flow at rest
    c. Monitoring the patient's blood flow at sleep
    d. Monitoring the patient's blood flow after exercising
66. What is an orchiectomy?
    a. A procedure performed to assess for testicular cancer
    b. A procedure performed to remove tissue samples of the ovaries
    c. A procedure performed to remove tissue samples of the kidneys
    d. A procedure performed to remove tissue samples of the lung
67. A mastopexy is a
    a. Breast augmentation
    b. Breast reduction
    c. Breast lift
    d. None of the above
68. What is the purpose of a cardiac blood pool scan?
    a. Measures the ejection fraction of the heart
    b. Measure the amount of blood that pools in the heart
    c. Measures the amount of blood that pools on the extremities
    d. Measures the amount of blood that pools in that coronary arteries
69. A breast ultrasound is commonly ordered
    a. In place of a mammogram
    b. If there is a suspicious result on a mammogram
    c. Always in combination with a mammogram
    d. Only when performing breast augmentation

70. What is the purpose of the urea breath test?
    a. Determines the presence of *Helicobacter pylori* in stool.
    b. Determines the presence of *H pylori* in the blood.
    c. Determines the presence of *H pylori* in the stomach.
    d. Estimates the alcohol content of blood.
71. Why should a breast-feeding mother avoid undergoing a PET scan?
    a. The mother's breasts will be tender 24 hours following the PET scan.
    b. The tracer is likely to pass to the baby in breast milk.
    c. The PET scan places the mother under extreme stress that decreases the volume of breast milk.
    d. None of the above.
72. What does the serum osmolality test measure?
    a. The volume of blood in the body
    b. Antidiuretic hormone
    c. The number of particles of substances that are dissolved in the serum
    d. The dose of a vasopressin
73. What is the difference between sputum cytology and a sputum culture?
    a. A sputum cytology studies cells contained in the sputum and a sputum culture identifies microorganism in the sputum.
    b. There is no difference.
    c. A sputum cytology identifies microorganism in the sputum and a sputum culture studies cells contained in the sputum.
    d. None of the above.
74. What is the purpose of the color duplex Doppler test in men?
    a. To assess the velocity and direction of blood flowing through the penis
    b. To assess size of an erection
    c. To assess if the patient had an erection during sleep
    d. To assess if the patient has urine in his bladder
75. The patient who is 8 months pregnant and is scheduled to receive a transabdominal ultrasound
    a. Does not have to have a full bladder for the test.
    b. Must have a full bladder for the test.

c. Cannot have their bladder filled using a urinary catheter.

d. Must have their bladder filled using a urinary catheter.

76. What might an increase in bilirubin levels from an amniocentesis taken after the 20th week indicate?

a. The fetus is healthy.

b. Fetal blood cells are attacking the mother's antibodies.

c. Fetal blood cells are being attacked by the mother's antibodies.

d. The fetus' lungs have matured.

77. Why is the reticulocyte count test ordered?

a. To screen for anemia

b. To screen for risk of bleeding

c. To assess the vitamin B_{12} level

d. To assess the vitamin B_6 level

78. What can produce a false-positive result in the bone mineral density test?

a. The patient has arthritis.

b. The patient has a T-score of 1.

c. The patient has a Z-score of 1.

d. The patient has an X-score of less than 1.

79. What is the purpose of tympanometry?

a. To measure the eardrum's response to pressure and sound

b. To identify the contents of fluid behind the eardrum

c. To permit fluid behind the eardrum to drain

d. To treat an ear infection

80. A PET scan helps the healthcare provider diagnose

a. Transient ischemic attack

b. Multiple sclerosis

c. Cancer

d. All of the above

81. What is the purpose of using pilocarpine during a sweat test?

a. Pilocarpine helps to draw sweat from the newborn.

b. Pilocarpine prevents the newborn from sweating.

c. Pilocarpine is not used in the sweat test.

d. Pilocarpine protects the skin from electrodes that are placed on the skin during the test.

82. What happens if the patient is unable to drink a large volume of fluid before a transabdominal ultrasound?

a. The test is cancelled.

b. An MRI is ordered.

c. Sterile water is inserted into the bladder using a urinary catheter.

d. A CT scan is ordered.

83. What is a Z-score?

a. A score that compares the patient's conductive velocity with that of people of his/her own age, sex, and race

b. A score that compares the patient's bone mineral density with that of a healthy 30 years old

c. A score that compares the patient's bone mineral density with that of people of his/her own age, sex, and race

d. None of the above

84. Why is the fructose level in semen measured?

a. Fructose prevents sperm motility

b. Fructose provides energy for sperm

c. Fructose reduces sperm count

d. Fructose increase semen volume

85. Why would the uric acid blood test be ordered?

a. To assess treatment for hypouricemia

b. To screen for uric acid kidney stones

c. To assess for inflammation

d. To assess for infection

86. What test is performed to identify hearing problems in a newborn?

a. Pure-tone audiometry

b. Speech reception/Word recognition

c. Whispered speech test

d. Otoacoustic emissions test

87. What is liquefaction time?
    a. The time necessary for semen to liquefy
    b. The time necessary for semen to dehydrate
    c. The time necessary for sperm to liquefy
    d. The time necessary for sperm to dehydrate
88. Why would the prothrombin time test be ordered?
    a. Risk for bleeding
    b. Therapeutic level of Coumadin
    c. Vitamin K deficiency
    d. All of the above
89. Why is oxytocin administered during a contraction stress test?
    a. To lower blood press of the fetus
    b. To stop contractions
    c. To induce contractions
    d. To keep the fetus from moving
90. During the biophysical profile test performed early in the trimester, the mother should
    a. Have an empty bladder before the test
    b. Have a full bladder before the test
    c. Have a bowel movement prior the test
    d. None of the above
91. A fetal ultrasound produces a
    a. Sonogram
    b. An X-ray of the fetus
    c. An X-ray of the mother
    d. None of the above
92. What test is performed to determine if the patient has sensorineural hearing loss?
    a. Pure-tone audiometry
    b. Speech reception/Word recognition
    c. Whispered speech test
    d. Otoacoustic emissions test

93. Why is transillumination performed in men?
    a. Highlights growths found during a testicle examination
    b. Visualizes the prostate gland through the urethra
    c. To scan the prostate gland
    d. To scan the bladder
94. Why is not a cranial ultrasound performed after 18 months of age?
    a. The fontanelles are closed.
    b. The fontanelles remain open.
    c. The test is too painful for the child to undergo.
    d. Complications from a premature birth would have already manifested.
95. When is a contraction stress test not performed?
    a. The mother had a cesarean section.
    b. The mother has placenta previa.
    c. The mother has placenta abruptio.
    d. All of the above.
96. Why should a patient take a deep cough to produce a sputum sample?
    a. A deep cough loosens sputum.
    b. Sputum is produced in the respiratory system.
    c. A deep cough brings sputum into the oral cavity.
    d. All of the above.
97. What should the nurse do if the patient drank a large cup of coffee the morning of the PET scan?
    a. Tell the patient this is unacceptable and continue with the PET scan.
    b. Notify the healthcare provider.
    c. Ask the patient to drink 5 cups of water prior to the PET scan.
    d. Take a sample of the patient's blood for testing.
98. What is the purpose of a chemistry screen?
    a. Assess the CMAC only.
    b. Assess blood carrying capability of blood only.
    c. Assess blood serum only.
    d. Assess the patient's overall health.

99. How does the snap gauge work?
    a. It measures the size of an erection.
    b. The snap gauge consisting of a film is placed around the penis. The film snaps when the patient has erection.
    c. It measures the length of time of an erection.
    d. All of the above.
100. An elevated C-reactive protein level prior to surgery indicates
    a. Higher risk of infection following surgery
    b. Lower risk of infection following surgery
    c. Surgery is no longer necessary
    d. All the above

Answers

1. a. The patient fasts for 8 hours. All DM medications are withheld until the test is completed.
2. b. Rest for 30 minutes.
3. d. All of the above.
4. a. Avoid eating and drinking except for water 12 hours before the test is administered.
5. d. All of the above.
6. a. Assesses bladder function.
7. a. Glucophage may react with contrast material.
8. d. All of the above.
9. a. It removes the outer layer of skin enabling new skin to grow.
10. a. Administer a sedative per order.
11. b. By using the salivary gland scan.
12. a. Troponin and CPK-MB.
13. a. Renin.
14. a. The carcinoembryonic antigen is normally present during fetal development and is terminated at birth. It is present if there is a tumor.

15. a. The patient has arthritis.
16. c. It should be discarded.
17. c. To determine blood supply to the brain.
18. d. All of the above.
19. b. May have Tay-Sachs disease.
20. b. Coumadin decreases coagulation time, thereby increasing the risk of bleeding during the procedure.
21. c. The amount of thyroxine that is attached to globulin and that is not bound to globulin.
22. a. FSH test.
23. c. To encourage healing.
24. c. Prolactin levels are normally high when the patient first awakens.
25. c. Bacterial infection.
26. a. Aldosterone.
27. b. To determine if the DM patient has been maintaining adequate blood glucose level for the previous 120 days.
28. a. To assess patients' ability to see distances when they are unable to read.
29. c. Dual-energy X-ray absorptiometry.
30. d. All of the above.
31. a. A sedative may invalidate the test results because it might slow down respiration.
32. d. All of the above.
33. a. To assess the speed, direction, and flow of blood.
34. b. Some antibiotics contain potassium.
35. d. All of the above.
36. a. Skin patch test.
37. a. The patient is at high risk for breast cancer and the MRI provides high, detailed views of the patient's breasts.
38. d. All of the above.
39. a. Creatinine clearance test.
40. a. The patient might have difficulty excreting the contrast material.
41. d. All of the above.
42. d. All of the above.

43. a. This test identifies heterophil antibodies that form between 2 and 9 weeks after the patient becomes infected.
44. c. Total hysterectomy with bilateral salpingo-oophorectomy.
45. b. Tracer material helps map the route through which cancer cells spread from the cancer site through the lymphatic system.
46. c. Volume and capacity of the lungs.
47. a. Allen test.
48. d. All of the above.
49. b. To generate an image of the heart that is not obstructed by bone.
50. c. Bacterial vaginosis.
51. a. Depend on the patient's age, height, sex, weight, and race.
52. a. The entire skin lesion is removed.
53. d. All of the above.
54. b. Will not show symptoms of carbon monoxide poisoning.
55. a. The upper segment of the vagina collapses and extends outside the vagina.
56. c. This causes the thyroid gland to be pushed forward.
57. b. Metabolic alkalosis.
58. a. A radioactive tracer being injected into a blood vessel.
59. b. A needle is inserted through the urethra to remove samples of prostate tissue.
60. c. Tenaculum clamp.
61. c. Metabolic acidosis.
62. b. Lung scan.
63. c. Identifies the anti-cardiolipin antibodies.
64. a. Carbon monoxide test.
65. a. An ambulatory electrocardiogram.
66. a. A procedure performed to assess for testicular cancer.
67. c. Breast lift.
68. a. Measures the ejection fraction of the heart.
69. b. If there is a suspicious result on a mammogram.
70. c. Determines the presence of *H pylori* in the stomach.
71. b. The tracer is likely to pass to the baby in breast milk.

72. c. The number of particles of substances that are dissolved in the serum.
73. a. A sputum cytology studies cells contained in the sputum and a sputum culture identifies microorganism in the sputum.
74. a. To assess the velocity and direction of blood flowing through the penis.
75. a. Does not have to have a full bladder for the test.
76. c. Fetal blood cells are being attacked by the mother's antibodies.
77. a. To screen for anemia.
78. a. The patient has arthritis.
79. a. To measure the eardrum's response to pressure and sound.
80. d. All of the above.
81. a. Pilocarpine helps to draw sweat from the newborn.
82. c. Sterile water is inserted into the bladder using a urinary catheter.
83. c. A score that compares the patient's bone mineral density with that of people his/her own age, sex, and race.
84. b. Fructose provides energy to sperm.
85. b. To screen for uric acid kidney stones.
86. d. Otoacoustic emissions test.
87. a. The time necessary for semen to liquefy.
88. d. All of the above.
89. c. To induce contractions.
90. b. Have a full bladder before the test.
91. a. Sonogram.
92. b. Speech reception/Word recognition.
93. a. Highlights growths found during a testicle examination.
94. a. The fontanelles are closed.
95. d. All of the above.
96. d. All of the above.
97. b. Notify the healthcare provider.
98. d. Assess the patient's overall health.
99. b. The snap gauge consisting of a film is placed around the penis. The film snaps when the patient has erection.
100. a. Higher risk of infection following surgery.

INDEX

A

B

C

F

G

H

L

M

N

O

Q

R

S

T

U